WeChat Mini Program Shop In Action

微信小程序商城开发实战

唐 磊 编著

北京航空航天大学出版社
BEIHANG UNIVERSITY PRESS

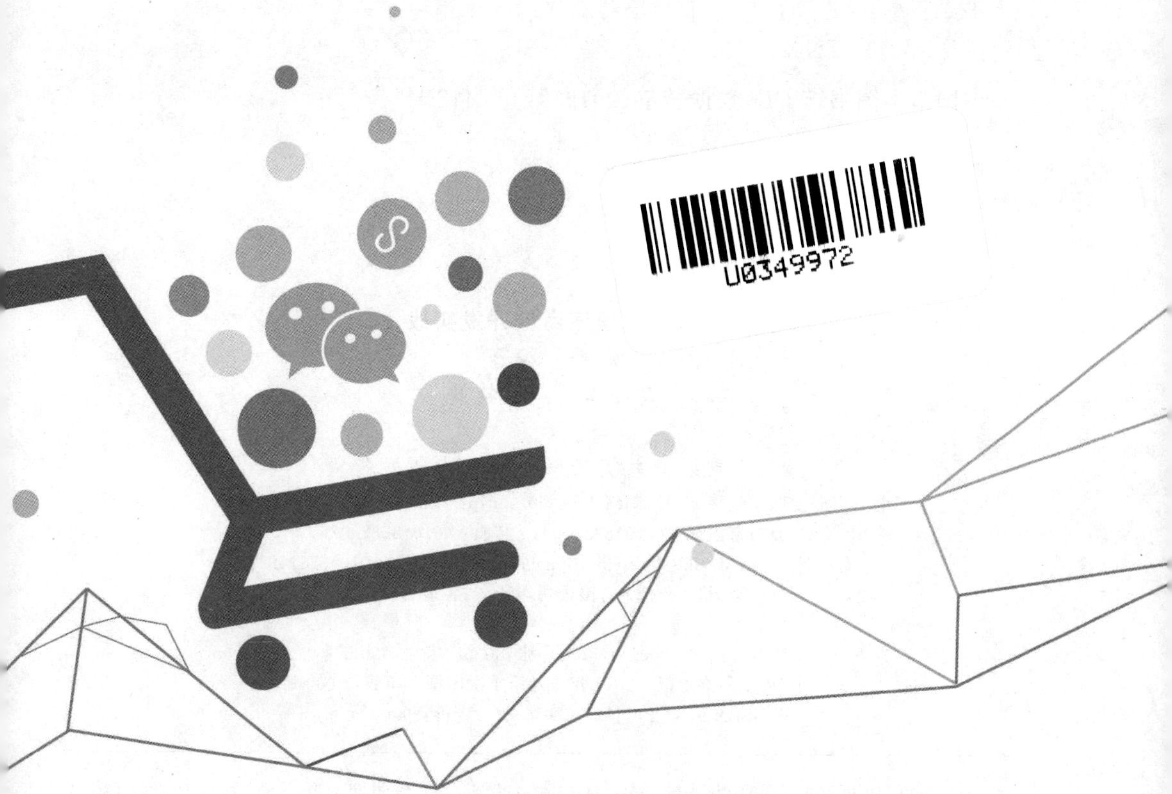

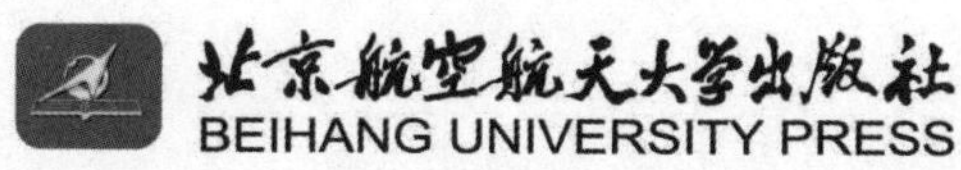

内容简介

本书分三篇，系统地介绍了小程序开发基础、核心框架和商城项目实战。第一篇为基础篇，包括小程序入门和小程序框架等基础知识；第二篇为高级篇，包括小程序框架组件、小程序框架 API 和小程序服务端开发思路等相关知识；第三篇为实战篇，包括小程序商城需求分析和数据库设计、小程序商城前端程序开发和小程序商城后端程序开发等商城项目实战开发内容。本书由浅入深、循序渐进地讲解技术知识，借助丰富的图表示例以及详实的代码说明，带领读者从 0 到 1 全面认知微信小程序实战项目开发，读者只需扎实理解和具体实践，即可快速开发出微信小程序商城这个最具商业价值的应用。

本书适合对微信小程序开发感兴趣的读者自学，同时可供小程序开发人员、前端开发者、培训机构和企业内训使用。

图书在版编目(CIP)数据

微信小程序商城开发实战/唐磊编著. -- 北京 : 北京航空航天大学出版社，2019.4

ISBN 978-7-5124-2973-4

Ⅰ. ①微… Ⅱ. ①唐… Ⅲ. ①移动终端—应用程序—程序设计 Ⅳ. ①TN929.53

中国版本图书馆 CIP 数据核字(2019)第 052412 号

微信小程序商城开发实战

唐磊 编著

责任编辑 剧艳婕

*

北京航空航天大学出版社出版发行

北京市海淀区学院路 37 号(邮编 100191) http://www.buaapress.com.cn

发行部电话：(010)82317024 传真：(010)82328026

读者信箱：emsbook@buaacm.com.cn 邮购电话：(010)82316936

艺堂印刷（天津）有限公司印装 各地书店经销

*

开本：710×1 000 1/16 印张：30.5 字数：650 千字

2019 年 4 月第 1 版 2019 年 4 月第 1 次印刷 印数：3 000 册

ISBN 978-7-5124-2973-4 定价：89.00 元

前　言

在当今移动互联网大潮中，微信应用凭借其庞大的用户基数和极强的用户黏性牢牢地占据着移动 App 应用的头把交椅之位，据相关统计，微信月活跃用户已达 10 亿之多，且占据着中国用户 30%以上的应用时间，微信已经是当之无愧的超级 App。大树底下好乘凉，含着金钥匙诞生的微信小程序，背靠微信这棵大树，从微信官方首次透露之日起，就受到了广大用户和开发者的重点关注。

小程序自 2017 年 1 月 9 日正式上线以来，经过将近两年的发展，市面上已经涌现出了诸多优秀的小程序应用，如：跳一跳、拼多多、摩拜单车和肯德基+等。而在这些小程序应用中，最具商业价值的则是电商类的小程序应用，借助微信的社交引流再加上小程序“无需安装、用完即走”的技术赋能，社交电商小程序开辟出了移动互联网电商市场的新局面。正是基于电商小程序的潜在商业价值和庞大的市场需求，我选择微信小程序商城这个实战项目案例为各位读者详细讲解如何打造一款小而美的电商小程序应用以飨读者。

本书的编写思路力求符合程序员自学的最佳路径，先进行小程序技术知识框架基础以及核心的由浅入深、由点及面的全面梳理和认知，再结合具体的小程序商城实战项目案例来综合实践前面所学的技术知识点。我没有选择太过简单的小程序项目案例，而是选择了小程序商城这个经典实用和综合性强，且极具商业价值的实战项目来进行讲解，这样能更好地让自学者对整个小程序项目有一个更高维度的认知和理解，同时该实战项目又可以很方便地用作实际的互联网创业或商业项目来落地应用，可谓一举两得，兼具学习用途和商业用途。

全书共分三大篇，第一篇为小程序开发基础篇，主要讲解了小程序概念、市场分析、入门 DEMO 体验和框架基础等，读者需要重点掌握的是小程序框架基础知识部分；第二篇为小程序开发高级篇，主要讲解了小程序框架组件、API 和服务端开发思路，建议读者快速阅读框架组件和 API 部分的知识，形成大致印象即可，知道如何查阅这些技术要点，而对于小程序服务端开发则重点在于理解小程序前后端如何通过接口进行数据交互的开发模式；第三篇为小程序开发实战篇，我按照项目需求分析、项目数据库设计、小程序前端程序开发和后端程序开发详细讲解了微信小程序商城应用的完整实战开发过程，并罗列了具体代码示例，建议读者首先对项目需求和数据库设计形成初步理解之后，再对小程序前端程序开发部分进行具体的学习，小程序前端这块是该项目的重中之重，也是对前面两篇技术知识点的综合运用和实践。至于小程序后端程序开发部分，如果读者是专注做前端开发的，建议可以大致了解下后端开发的技术知识，方便以后进行前后端协作开发时的快速沟通和理解开发需求。如果读者是全栈开发者或是项目经理，建议重点阅读后端开发这一部分的内容，学习和参考均可。

此外，我还有一个多年学习技术的经验——“技术学习九字真诀”，也希望在此分享给各位：学习技术，重点掌握该技术的三点，其一为“是什么”，知道该技术的基本概念，最好结合一些具体的生活化的实例来理解抽象的技术知识；其二为“为什么”，理解该技术的核心原理，是否对技术有深度的认知重点在于是否对该技术的核心原理理解透彻；其三为“怎么用”，掌握并实际运用该技术是学习技术的最终目的和最佳实践方式，当你知道了该技术怎么用、用在什么业务场景时，想必你对技术的掌握就已经达到了一定的高度。

本书读者对象包括：前端开发工程师、软件项目经理、微信应用开发者、全栈开发爱好者和计算机软件相关专业的学生。

在写作本书的过程中，得到了众多好友和同行的鼓励，在此由衷感谢；同时也感谢北京航空航天大学出版社的编辑的全力支持；最后感谢家人的理解和支持，让我能全身心地投入到这本书籍的创造中。谨以此书，献给自己的宝贝女儿唐悠然，希望她未来能健康快乐、悠然自得。

请到北京航空航天大学出版社官网的“下载专区（随书资料）”下载本书源码。

唐　磊

2019 年 2 月

目录

小程序开发基础篇

第1章 初识小程序 …… 2

1.1 小程序概述 …… 2
1.2 什么是小程序 …… 2
1.3 小程序的优劣势 …… 4
1.4 小程序市场前景 …… 4
1.5 小程序商业价值分析 …… 5

第2章 小程序入门 …… 10

2.1 开发前的准备 …… 10
2.1.1 注册微信公众平台成为开发者 …… 11
2.1.2 开发工具下载与安装 …… 17
2.2 体验第一个小程序 …… 20
2.2.1 开发模式选择 …… 20
2.2.2 创建 Hello World 小程序 …… 21
2.2.3 如何获取小程序 AppID …… 23
2.3 微信开发者工具使用介绍 …… 24
2.3.1 开发工具界面图解 …… 24
2.3.2 小程序调试工具 …… 27
2.3.3 如何预览小程序 …… 31
2.3.4 上传和发布小程序 …… 32

第3章 小程序框架基础 …… 35

3.1 理解小程序开发框架 …… 35
3.1.1 小程序框架原理 …… 35

3.1.2 小程序目录结构和文件构成 …… 37
3.1.3 小程序配置 …… 38
3.1.4 小程序运行和加载机制 …… 41
3.1.5 小程序生命周期和线程架构 …… 42
3.2 WXML:小程序版 HTML …… 43
3.2.1 标签与属性 …… 43
3.2.2 数据绑定 …… 45
3.2.3 条件渲染 …… 50
3.2.4 列表渲染 …… 50
3.2.5 模板、引用和事件 …… 53
3.2.6 WXML 与 HTML 的区别 …… 58
3.3 WXSS:小程序版 CSS …… 58
3.3.1 选择器与优先级 …… 59
3.3.2 盒子模型 …… 60
3.3.3 内联样式 …… 61
3.3.4 尺寸单位与样式导入 …… 62
3.3.5 WXSS 与 CSS 的区别 …… 63
3.4 WXS:小程序版 JavaScript …… 63
3.4.1 小程序 JavaScript …… 63
3.4.2 同步和异步 …… 64
3.4.3 模块化 …… 65

小程序开发高级篇

第 4 章 小程序框架组件 …… 69

4.1 视图容器组件 …… 69
4.1.1 视图容器 …… 69
4.1.2 可滚动视图区域 …… 71
4.1.3 滑块视图容器 …… 74
4.1.4 可移动视图容器 …… 77
4.1.5 覆盖在原生组件上的文本视图 …… 80
4.1.6 覆盖在原生组件上的图片视图 …… 80
4.2 基础内容组件 …… 83
4.2.1 图　标 …… 83
4.2.2 文　本 …… 85
4.2.3 富文本 …… 87
4.2.4 进度条 …… 88
4.3 表单组件 …… 89
4.3.1 按　钮 …… 89

4.3.2 多项选择器 …… 93
4.3.3 表 单 …… 94
4.3.4 输入框 …… 96
4.3.5 标 签 …… 100
4.3.6 滚动选择器 …… 104
4.3.7 单项选择器 …… 116
4.3.8 滑动选择器 …… 118
4.3.9 开关选择器 …… 120
4.3.10 多行输入框 …… 121
4.4 导航组件 …… 123
4.4.1 页面导航 …… 123
4.4.2 功能页导航 …… 125
4.5 媒体组件 …… 127
4.5.1 音 频 …… 127
4.5.2 图 片 …… 129
4.5.3 视 频 …… 135
4.5.4 相 机 …… 139
4.5.5 实时音视频播放 …… 140
4.5.6 实时音视频录制 …… 143
4.6 地图组件 …… 145
4.7 画布组件 …… 148
4.8 开放能力组件 …… 151
4.8.1 开放数据 …… 151
4.8.2 网页容器 …… 151
4.8.3 广 告 …… 152
4.8.4 公众号关注 …… 153

第 5 章 小程序框架 API …… 154

5.1 网络 API …… 155
5.1.1 发起请求 …… 155
5.1.2 上传下载 …… 156
5.1.3 长连接 WebSocket …… 158
5.2 媒体 API …… 160
5.2.1 音视频 …… 160
5.2.2 图 片 …… 166
5.2.3 录 音 …… 170
5.3 文件 API …… 172
5.4 数据缓存 API …… 176
5.5 位置 API …… 180

5.6 设备 API ………………………………………………………… 182
5.6.1 加速计、蓝牙、罗盘 ………………………………………… 182
5.6.2 电量、性能、屏幕 ………………………………………… 184
5.6.3 联系人、电话、扫码 ………………………………………… 186
5.7 界面 API ………………………………………………………… 187
5.7.1 动 画 ………………………………………………………… 187
5.7.2 交 互 ………………………………………………………… 190
5.7.3 导航栏 ………………………………………………………… 193
5.7.4 置顶、滚动、下拉刷新 ………………………………………… 194
5.8 开放接口 API ………………………………………………………… 196
5.8.1 授权、登录、用户信息 ………………………………………… 196
5.8.2 微信支付 ………………………………………………………… 201
5.8.3 模板消息、统一服务消息 ………………………………………… 203
5.9 其他 API ………………………………………………………… 207
5.9.1 基 础 ………………………………………………………… 207
5.9.2 转 发 ………………………………………………………… 207
5.9.3 系统信息 ………………………………………………………… 209

第 6 章 小程序服务端开发思路 ……………………………………… 211

6.1 服务端开发语言和框架选择建议 ………………………………… 211
6.1.1 开发语言和框架的选择 ………………………………………… 211
6.1.2 新选择——小程序·云开发 ………………………………………… 213
6.2 数据库设计思路 ………………………………………………………… 214
6.2.1 小程序项目数据库选型 ………………………………………… 214
6.2.2 数据库设计方法和建议 ………………………………………… 216
6.3 服务端接口开发思路 ………………………………………………… 217
6.3.1 RESTful API 介绍 ………………………………………… 217
6.3.2 后端 API 开发技巧和建议 ………………………………………… 218
6.4 服务端后台管理系统开发思路 ………………………………………… 219
6.4.1 后台管理系统页面设计建议 ………………………………………… 219
6.4.2 后台管理系统功能开发注意要点 ………………………………… 220

小程序开发实战篇

第 7 章 小程序商城需求分析和数据库设计 ………………………… 223

7.1 项目需求分析 ………………………………………………………… 223
7.1.1 项目背景概述 ………………………………………………………… 223
7.1.2 业务需求分析 ………………………………………………………… 223
7.1.3 产品需求分析 ………………………………………………………… 225

7.1.4 产品结构图 …… 228
7.2 项目数据库设计 …… 229
7.2.1 数据库设计概述 …… 229
7.2.2 数据库表结构设计 …… 230

第8章 小程序商城前端程序开发 …… 237

8.1 小程序商城前端页面开发概述 …… 237
8.2 项目全局基础模块代码清单 …… 237
8.2.1 全局入口 app.js …… 237
8.2.2 全局配置 app.json …… 241
8.2.3 全局样式 app.wxss …… 243
8.2.4 项目配置 project.config.json …… 244
8.3 首页代码清单 …… 245
8.3.1 wxml 模板代码 …… 246
8.3.2 wxss 样式代码 …… 247
8.3.3 js 逻辑代码 …… 257
8.3.4 json 配置代码 …… 261
8.4 商品分类页代码清单 …… 261
8.4.1 wxml 代码 …… 262
8.4.2 wxss 样式代码 …… 263
8.4.3 js 逻辑代码 …… 267
8.4.4 json 配置代码 …… 269
8.5 商品详情页代码清单 …… 270
8.5.1 wxml 模板代码 …… 270
8.5.2 wxss 样式代码 …… 272
8.5.3 js 逻辑代码 …… 290
8.5.4 json 配置代码 …… 300
8.6 购物车页代码清单 …… 300
8.6.1 wxml 模板代码 …… 300
8.6.2 wxss 样式代码 …… 302
8.6.3 js 逻辑代码 …… 311
8.6.4 json 配置代码 …… 317
8.7 订单确认页代码清单 …… 317
8.7.1 wxml 模板代码 …… 318
8.7.2 wxss 样式代码 …… 320
8.7.3 js 逻辑代码 …… 328
8.7.4 json 配置代码 …… 333
8.8 订单详情页代码清单 …… 333
8.8.1 wxml 模板代码 …… 333

8.8.2 wxss 样式代码 …… 335
8.8.3 js 逻辑代码 …… 341
8.8.4 json 配置代码 …… 343
8.9 地址管理页代码清单 …… 343
8.9.1 wxml 模板代码 …… 343
8.9.2 wxss 样式代码 …… 344
8.9.3 js 逻辑代码 …… 346
8.9.4 json 配置代码 …… 350
8.10 保存地址页代码清单 …… 350
8.10.1 wxml 模板代码 …… 351
8.10.2 wxss 样式代码 …… 352
8.10.3 js 逻辑代码 …… 353
8.10.4 json 配置代码 …… 358
8.11 优惠券页代码清单 …… 358
8.11.1 wxml 模板代码 …… 359
8.11.2 wxss 样式代码 …… 359
8.11.3 js 逻辑代码 …… 363
8.11.4 json 配置代码 …… 365
8.12 个人中心页代码清单 …… 365
8.12.1 wxml 模板代码 …… 366
8.12.2 wxss 样式代码 …… 367
8.12.3 js 逻辑代码 …… 371
8.12.4 json 配置代码 …… 374
8.13 我的订单列表页代码清单 …… 374
8.13.1 wxml 模板代码 …… 375
8.13.2 wxss 样式代码 …… 376
8.13.3 js 逻辑代码 …… 379
8.13.4 json 配置代码 …… 387
8.14 搜索页代码清单 …… 387
8.14.1 wxml 模板代码 …… 388
8.14.2 wxss 样式代码 …… 389
8.14.3 js 逻辑代码 …… 391
8.14.4 json 配置代码 …… 394

第 9 章 小程序商城后端程序开发 …… 395

9.1 小程序商城后端接口开发概述 …… 395
9.2 小程序后端接口开发 …… 395
9.2.1 接口公共父类代码清单 …… 395
9.2.2 登录接口代码清单 …… 399

9.2.3 首页接口代码清单 …… 405
9.2.4 商品分类接口代码清单 …… 407
9.2.5 商品接口代码清单 …… 409
9.2.6 购物车接口代码清单 …… 417
9.2.7 订单接口代码清单 …… 425
9.2.8 下单结算接口代码清单 …… 434
9.2.9 微信支付接口代码清单 …… 449
9.2.10 地址接口代码清单 …… 454
9.2.11 优惠券接口代码清单 …… 460
9.2.12 用户接口代码清单 …… 463
9.3 项目小结 …… 474
参考文献 …… 475

小程序开发基础篇

第1章

初识小程序

说起小程序，可能需要追溯到2016年1月11日，微信之父张小龙时隔多年之后公开亮相并解读了微信的四大价值观，同时首次提及了微信研发团队内部正在研究一种用于替代服务号且能提供更好服务的新形态，即微信小程序。

2016年9月21日，微信小程序正式开启内测。在微信生态下，"触手可及、用完即走"的微信小程序立即引起了业界的广泛关注。随后，腾讯云也为小程序在云端服务器方面提供了技术支持。

2017年1月9日0点，备受关注的第一批微信小程序正式低调上线，移动互联网用户开始体验到各种各样的小程序所带来的"触手可及式"的应用服务。

1.1 小程序概述

小程序可以简单地理解为：一种无需下载安装的"即开型"应用，打开即可使用，不用等待太久，随时随地，触手可及。

1.2 什么是小程序

微信小程序，简称小程序，英文名 Mini Program，是一种不需要下载安装即可使用的应用，它实现了应用"触手可及"的梦想，用户扫一扫或搜一下即可打开应用，也体现了"用完即走"的理念。小程序其实解决了以往 APP 应用的一些老大难问题：安装卸载麻烦、跨平台不兼容、应用太多占用过多系统资源和应用打开率过低等。小程序解决了这些 APP 应用的用户痛点，给用户呈现出了一个可以随时可用、用完即走、无需关心安装和卸载，并且支持跨平台快速开发的"即开型"轻量级应用。微信小程序官网首页如图 1-1 所示。

微信小程序目前已经全面开放申请，主体类型可以是企业、政府、媒体、其他组织或个人。小程序、订阅号、服务号、企业微信（原企业号）是并行的体系。微信公众平台账号的分类如图 1-2 所示。

微信公众平台 | 小程序

微信小程序

小程序是一种新的开放能力，开发者可以快速地开发一个小程序。小程序可以在微信内被便捷地获取和传播，同时具有出色的使用体验。

开放注册范围

个人

企业

政府

媒体

其他组织

图 1-1　微信小程序官网首页

服务号

给企业和组织提供更强大的业务服务与用户管理能力，帮助企业快速实现全新的公众号服务平台。

订阅号

为媒体和个人提供一种新的信息传播方式，构建与读者之间更好的沟通与管理模式。

小程序

一种新的开放能力，可以在微信内被便捷地获取和传播，同时具有出色的使用体验。

企业微信　原企业号

企业的专业办公管理工具。与微信一致的沟通体验，提供丰富免费的办公应用，并与微信消息、小程序、微信支付等互通，助力企业高效办公和管理。

图 1-2　微信公众平台账号分类

1.3 小程序的优劣势

小程序的优势比较显而易见，如：

① 无需下载安装，无需注册，用完即走，不会过多地占用手机内存；

② 小程序可以跨平台（苹果或安卓系统）运行，其开发门槛和成本比 APP 开发要低；

③ 打开速度比 H5（HTML5）要快，接近于原生 APP 的体验；

④ 安卓手机可以直接添加小程序至手机桌面，打开方式和 APP 一样方便。

当然小程序也不是完美无缺的，它也有一些自身的劣势：

① 小程序目前无法直接分享至朋友圈，只能分享给微信好友或微信群；

② 小程序的二维码不支持长按识别，只能通过微信扫一扫才能打开；

③ 小程序没有 PUSH 功能，不能给用户推送消息和个人相关的通知消息；

④ 小程序没有用户体系，用户不需要注册，因此用户用完即走，很难形成用户体系。

1.4 小程序市场前景

据腾讯官方报告显示，目前国内微信活跃用户数已突破 10 亿，且微信用户数量一直处于平稳增长的状态，而微信小程序累计用户数量的增长态势则猛于微信。2017 年 1 月 9 日，小程序正式上线，当月小程序用户数量仅有 200 万。但是由于初期小程序接口存在诸多限制，各类开发者表现得并不积极。从 2017 年 3 月开始，小程序不断地快速更新迭代，在一年多的时间里，小程序推出了上百项全新功能，开放入口几十种，其中包括用户分享、公众号等多个重量级入口，使得小程序用户也呈现出不断攀升的态势。

另据移动微电商有赞创始人白鸦在 2018 年的一次小程序峰会上透露和预测：

① 2018 年 1 月 15 日，“2018 微信公开课 PRO”上曾公布，每天有 1.7 亿用户使用小程序。现在，这个数字已经变成了 2.5 亿。

② 2018 年底，小程序用户规模会与 2013 年底中国网民总数持平，达到至少 6.18 亿用户。

③ 未来最火的 2 个关键词会是微信互联网和小程序互联网。今年（2018 年），可能大部分平台都会做小程序，比如今日头条小程序、淘宝小程序、微博小程序、百度小程序等。

同时，我们通过微信月活跃用户数变化图（如图 1－3 所示），也可以清晰地看出

图 1－3　微信月活跃用户数变化

微信所具备的庞大用户数规模以及高活跃度，为小程序的发展奠定了坚实的用户基础。所以就小程序的市场前景而言，是被广泛看好的，这一点毋庸置疑。

微信小程序的出现很完美地解决了原生 APP 应用与 WEB APP 应用的短板，以“无需安装、用完即走”的全新体验，快速解决了用户的实际需求痛点，势必会成为移动互联网下半场的核心领导者。

1.5　小程序商业价值分析

小程序最适合的应用场景应该是“低频刚需”类应用。目前市面上的 APP 其实使用率都不是特别高，有的只是偶尔被使用一下，但是用户又不想删除，担心某一天会用到，而小程序正好切准了这个痛点。一个不到 2MB 大小的小程序，可以做到接近于原生 APP 的优良体验，因此那些使用频率低且只是偶尔有刚性需求的 APP 就会被小程序顺理成章地取代。

根据 QuestMobile 发布的 2018 年微信小程序用户规模调研数据，此处摘取了其中 TOP15 的排名榜单，如表 1－1 所列。

表 1-1　2018 年微信小程序用户规模 TOP15

排名	小程序名称	一级行业	累计活跃用户数/万
1	跳一跳	手机游戏	38914
2	拼多多	移动购物	23332
3	成语猜猜看	手机游戏	14703
4	头脑王者	手机游戏	13341
5	摩拜单车	旅游出行	13229
6	农行微服务	金融理财	12209
7	欢乐斗地主	手机游戏	11284
8	腾讯视频	移动视频	10943
9	收款小账本	实用工具	9770
10	王卡申请助手	电话通讯	9748
11	京东购物	移动购物	8620
12	成语消消看	手机游戏	8303
13	挑战答题王	手机游戏	8092
14	肯德基＋	生活服务	7271
15	流量充值优惠推荐	电话通讯	6983

从上面的榜单可以看出，小程序在手机游戏、移动购物、旅游出行、实用工具和生活服务类等“低频刚需”应用场景中具备庞大的用户规模，且非常易于凸显出小程序的潜在商业价值。下面我们简略地看一下这几大类中的头部小程序应用。

“跳一跳”是微信的一个小程序游戏（如图 1-4 所示），依靠操作小人蓄力跳跃进行游玩。按照小人跳跃盒子的数量，以及特殊盒子加分项计算得分。有单人模式和

图 1-4　跳一跳

双人模式两种。从“跳一跳”小程序的走红可以看出，益智休闲类的轻量级小游戏很适合用小程序来做，用户可以快速轻松地体验到小游戏的乐趣。

“拼多多”是一家专注于 C2B 拼团的第三方社交电商平台（如图 1－5 所示）。用户通过发起和朋友、家人、邻居等的拼团，可以以更低的价格购买优质商品。其中，通过沟通分享形成的社交理念，形成了“拼多多”独特的新社交电商思维。“拼多多”的迅速崛起，很大程度上依赖于微信社交的强大势能，通过分享给微信好友或微信群，轻松达到了为“拼多多”电商平台引流的效果。

图 1－5　拼多多

“摩拜单车”为用户提供了一种互联网短途出行的解决方案，借助无桩借还车模式的智能硬件，可以实现让人们通过智能手机就能快速租用和归还一辆单车，用可负担的价格来完成一次几公里的市内短途骑行。“摩拜单车”自从 2017 年 3 月底全面接入微信以来，其月活跃用户量环比增速超过 200%，每日新增注册用户超过 50%来自微信小程序，由此可见小程序的便捷性给摩拜用户带来了很高的用户增量，如图 1－6 所示。

“收款小账本”小程序是一个智慧收银系统（如图 1－7 所示），“收款小账本”小程序专门针对使用面对面收款码的老板和店员打造，通过“收款小账本”小程序可帮助用户更好地使用微信二维码收款功能，让老板和店员方便收款、轻松对账。“收款小账本”这类实用工具型小程序，旨在为用户提升效率，节省时间，市面上的查询类小程序也属于此类。

图 1-6　摩拜单车

图 1-7　收款小账本

“肯德基+”小程序是一款手机自助点餐类微信小程序(如图 1-8 所示),借助该小程序,用户无需排队即可在线点餐,随时随地,更轻更方便。“肯德基+”小程序可以自动定位距离用户最近的餐厅位置,方便用户在线点餐,在“肯德基+”小程序中还能随时查询订单详情,了解订单最新状态。“肯德基+”这类生活服务类小程序,完美地诠释了“用完即走”的小程序的初心,因为点餐是一个低频刚需的操作,只有在需要的时候才会用,所以用小程序来做自助点餐是再合适不过的选择。

图 1-8　肯德基+

看过了这些经典的爆款小程序,读者应该对小程序的应用场景和商业价值有了一定的认知。如果说微信公众号改变了中国的媒体业,那么小程序将有可能改变中国的商业。在一个人人都用微信的时代,社交是刚需,微信连接着人与人之间最具价值的一环。在移动互联网的下半场,小程序借助微信庞大的用户规模,将传统中心化流量平台模式下的流量稀释,再重新聚合成一个个独立但又紧密联系的小型中心。想必未来小程序会在微信生态中形成独有的“分散式中心化”模式。而适应该模式发展的小程序开发者,将拥有更多构建高转化率场景的机会。届时,小程序会成为新零售、移动电商、智慧餐饮等行业的流量高地,推动各大应用平台争夺流量战场,进而进行新一轮的洗牌。而本书选择“小程序商城”这一实战开发案例,也是基于去中心化的小程序电商将成为未来的一大风口这一大趋势的判断,在本书的后半部分将为读者呈现该实战案例的详细分析和代码解读。

第 2 章

小程序入门

我们在学习软件开发的时候都写过"Hello World"程序，这是一个入门级的DEMO。与此相似，在准备开始深入学习小程序开发之前，我们也需要来尝试开发和体验一下小程序版的"Hello World"。

2.1 开发前的准备

首先，我们需要知道一些小程序开发的常用网址和工具，列举如下：

(1) 微信公众平台

https://mp.weixin.qq.com

(2) 微信小程序主页以及注册入口

https://mp.weixin.qq.com/cgi-bin/wx

(3) 小程序开发文档

https://developers.weixin.qq.com/miniprogram/dev/index.html

(4) 小程序开发工具下载

https://developers.weixin.qq.com/miniprogram/dev/devtools/download.html

(5) 微信小程序设计指南

https://developers.weixin.qq.com/miniprogram/design/index.html

(6) 小程序体验 DEMO

https://developers.weixin.qq.com/miniprogram/dev/demo.html

(7) 小游戏开发文档

https://developers.weixin.qq.com/minigame/dev/index.html

其中微信小程序主页以及注册入口最为重要，该页提供了小程序开发文档、工具、设计指南、体验 DEMO 和注册的重要入口链接，建议保存该网址，方便查阅。微信小程序主页中的小程序接入流程和注册入口如图 2-1 所示。

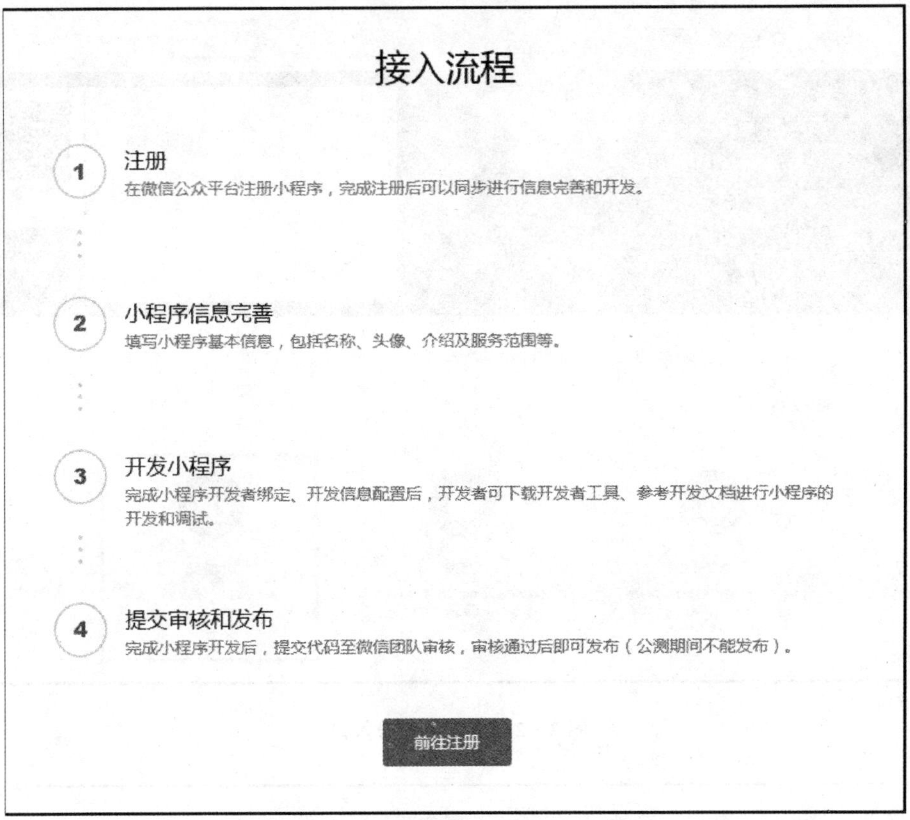

接入流程

1 注册
在微信公众平台注册小程序，完成注册后可以同步进行信息完善和开发。

2 小程序信息完善
填写小程序基本信息，包括名称、头像、介绍及服务范围等。

3 开发小程序
完成小程序开发者绑定、开发信息配置后，开发者可下载开发者工具、参考开发文档进行小程序的开发和调试。

4 提交审核和发布
完成小程序开发后，提交代码至微信团队审核，审核通过后即可发布（公测期间不能发布）。

前往注册

图 2-1　小程序接入流程和注册入口

2.1.1　注册微信公众平台成为开发者

要进行小程序开发，需要先在微信公众平台注册，成为开发者。微信公众平台（网址：https://mp.weixin.qq.com）主页上有两个小程序注册入口，如图 2-2 所示。

点击“立即注册”，弹出如图 2-3 所示界面，选择“小程序”。

进入到如图 2-4 所示的小程序注册页面开始注册，该页面网址为：https://mp.weixin.qq.com/wxopen/waregister? action=step1&token=&lang=zh_CN。这里需要注意的是：作为小程序登录账号，请填写未被微信公众平台注册、未被微信开放平台注册、未被个人微信号绑定的邮箱。

填写完注册信息后，点击“注册”按钮，来到如图 2-5 所示的页面，此页面提示用户进入邮箱激活刚注册的公众平台账号。

进入注册邮箱中，可以看见微信系统发送了一份提示激活微信小程序的邮件，点击邮件内容中的激活链接即可完成激活，如图 2-6 所示。

图 2-2　小程序注册入口

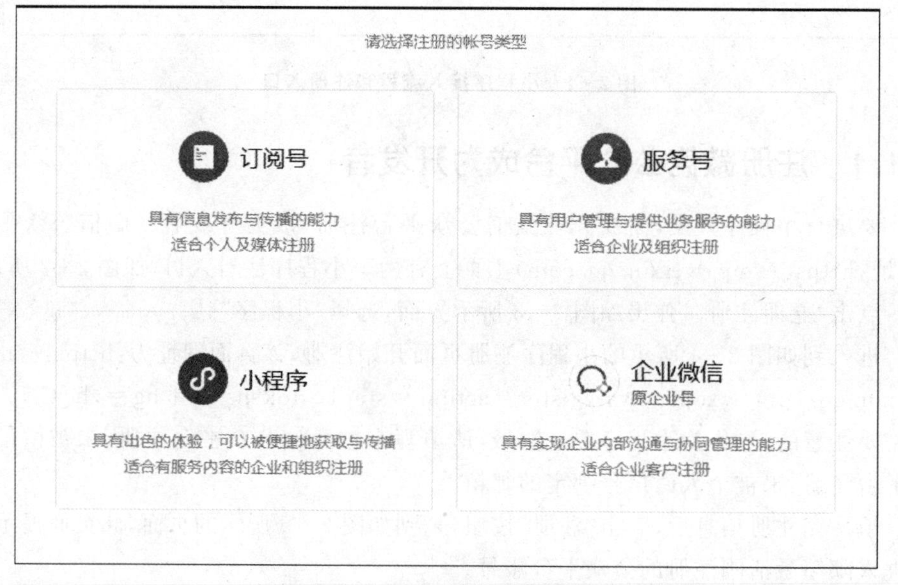

图 2-3　微信公众平台账号类型

然后再回到微信公众平台中，进行信息登记，如图 2-7 所示。

小程序注册

①帐号信息 …… ②邮箱激活 …… ③信息登记

每个邮箱仅能申请一个小程序

已有微信小程序？立即登录

邮箱

作为登录帐号，请填写未被微信公众平台注册，未被微信开放平台注册，未被个人微信号绑定的邮箱

密码

字母、数字或者英文符号，最短8位，区分大小写

确认密码

请再次输入密码

验证码 DPAM 换一张

你已阅读并同意《微信公众平台服务协议》及《微信小程序平台服务条款》

注册

图 2-4　小程序注册

图 2-5　邮箱激活

主体类型分为：个人、企业、政府、媒体和其他组织。需要注意的是，个人类型暂不支持微信认证、微信支付以及高级接口能力。如果是做商城类涉及实际支付的小程序，那么此处建议用户注册企业主体，这样一步到位，避免后续的麻烦。当然如果是为了学习研究小程序，没有企业资料，可以先暂时注册个人类型的小程序，先快速上手开发一些 DEMO 来体验一下小程序的功能特性也是可以的。接着我们需要在如图 2-8 所示的页面中填写主体信息。

"请激活你的微信小程序"

发件人：weixinteam <weixinteam@tencent.com>

时　间：2018年

收件人：@qq.com>

请认准此标志，拥有此标志的才是发自腾讯公司的系统邮件。

微信公众平台·小程序

你好，

感谢你注册微信小程序。

你的登录邮箱为：@qq.com。请点击以下链接激活账号，

如果以上链接无法点击，请将上面的地址复制到你的浏览器地址栏进入微信公众平台。

Kero Zheng

图 2-6　请激活您的微信小程序

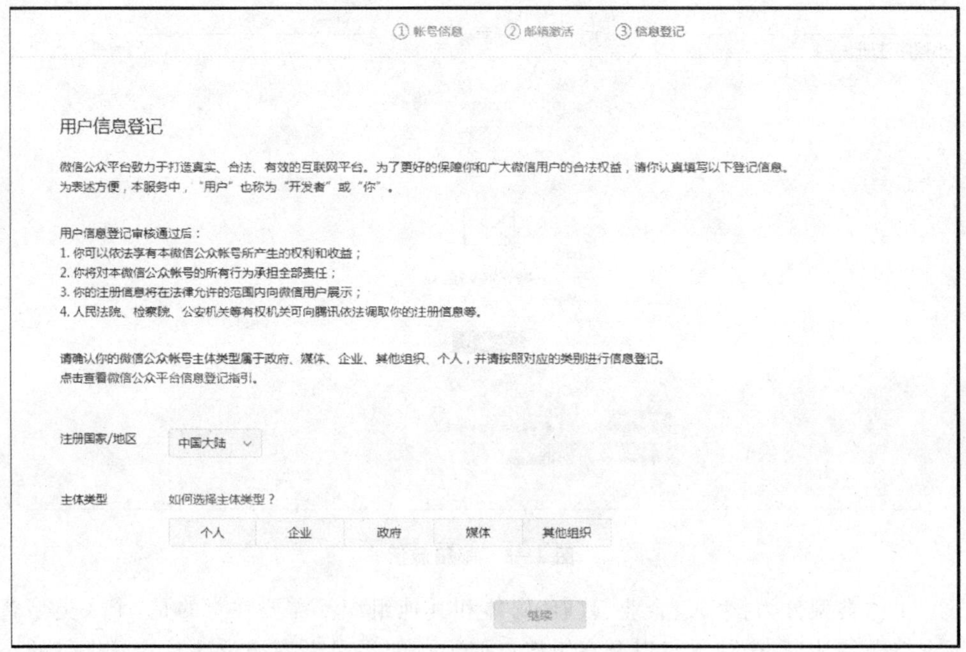

图 2-7　信息登记主体类型的选择

然后需要用管理员微信扫二维码验证，成功后点击如图 2-9 所示的"继续"按钮。

个人	企业	政府	媒体	其他组织

个人类型包括：由自然人注册和运营的公众帐号。

帐号能力：个人类型暂不支持微信认证、微信支付及高级接口能力。

主体信息登记

身份证姓名

信息审核成功后身份证姓名不可修改；如果名字包含分隔号“·”，请勿省略。

身份证号码

请输入您的身份证号码。一个身份证号码只能注册5个小程序。

管理员手机号码　29秒后可重发

请输入您的手机号码，一个手机号码只能注册5个小程序。

短信验证码　无法接收验证码？

验证码应为6位数字

请输入手机短信收到的6位验证码

管理员身份验证

请用管理员本人微信扫描二维码。本验证方式不扣除任何费用。注册后，扫码的微信号将成为该帐号的管理员微信号。

继续

图 2-8　填写主体信息(个人主体)

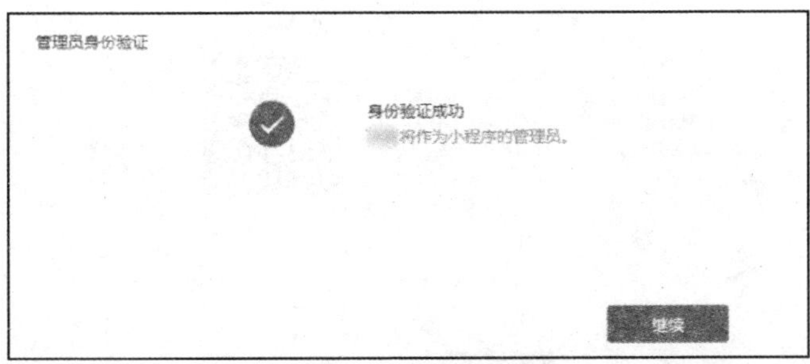

图 2-9　管理员身份验证

如果是企业主体，则需要填写如图 2－10 所示的信息。目前企业类型账号可选择两种主体验证方式：一是需要用公司的对公账户向腾讯公司打款来验证主体身份。打款信息在提交主体信息后可以查到；二是通过微信认证验证主体身份，需支付 300 元的认证费。认证通过前，小程序部分功能暂无法使用。

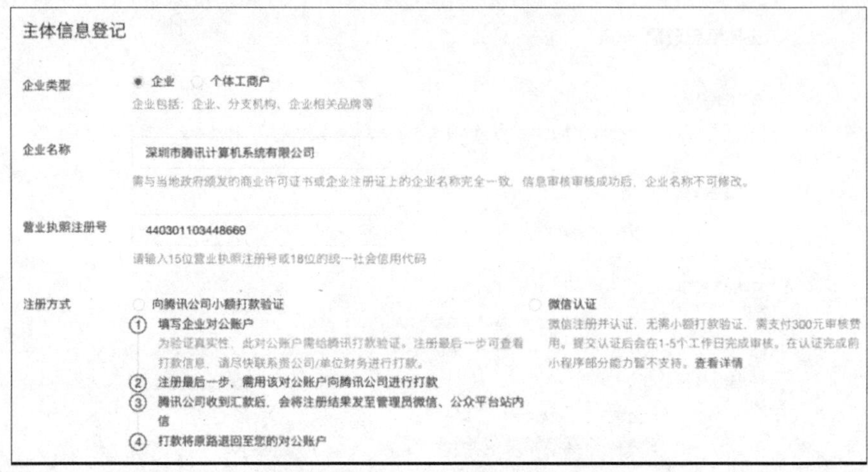

图 2－10　填写主体信息(企业主体)

接着完成管理员信息登记，此处会提示：确认主体信息提交后不可修改，如图 2－11 所示。

图 2－11　确认主体信息

点击“确定”之后，会有信息提交成功的提示，如图 2－12 所示。

图 2－12　信息提交成功

2.1.2　开发工具下载与安装

微信研发团队给小程序开发者提供了专用的开发工具。在小程序主页中找到“开发支持”，点击其中的“开发者工具”，如图 2－13 所示。进入到如图 2－14 所示的开发者工具页面。这个页面是小程序开发工具的概览页面，左侧菜单项分别对应该开发工具的详细分类及介绍。

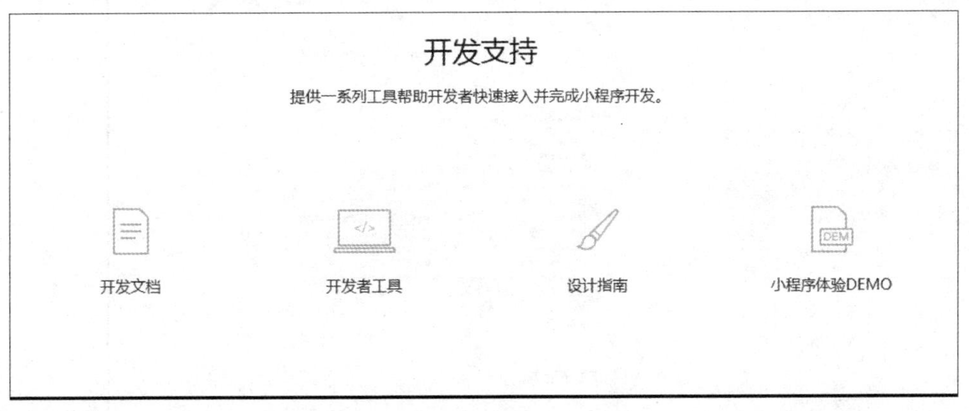

图 2－13　小程序开发支持

点击图 2－14 中的“微信开发者工具”文字链接，即可跳转到该开发工具的最新

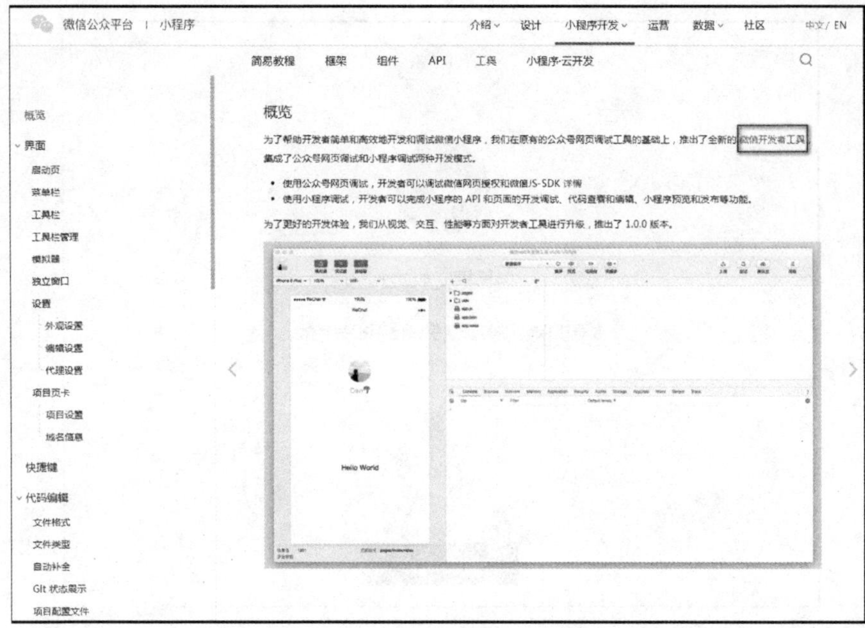

图 2-14 小程序开发工具概览

版下载地址页，如图 2-15 所示。

图 2-15 小程序开发工具下载地址页

该开发工具的下载地址为：https://developers.weixin.qq.com/miniprogram/

dev/devtools/download.html。微信一共提供了 3 个版本：Windows 64 位、Windows 32 位和 Mac OS 供开发者选择，可以根据自己电脑的操作系统选择相应的下载包。然后双击下载安装包，会出现安装向导，如图 2－16 所示。

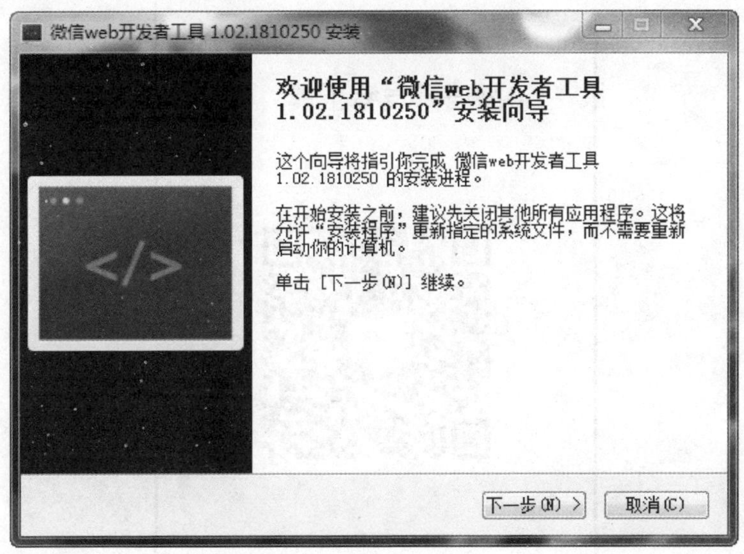

图 2－16 安装向导

点击“下一步”按钮，完成小程序开发工具的安装，如图 2－17 所示。

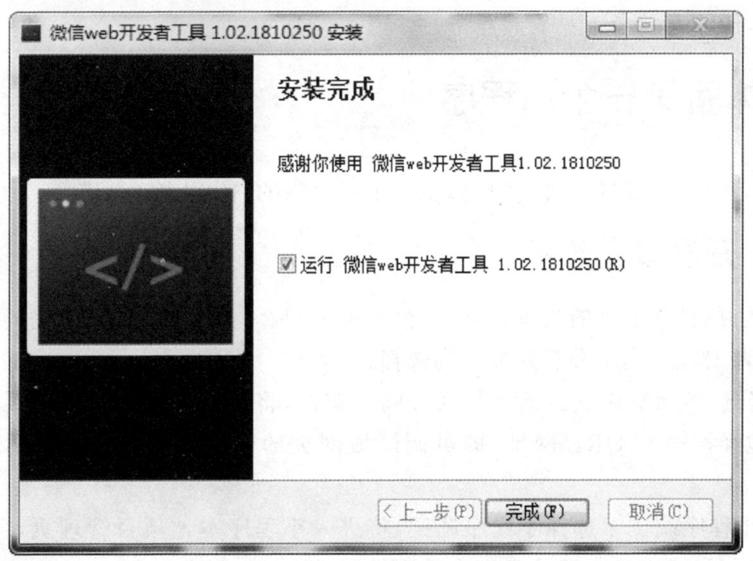

图 2－17 安装完成

点击“完成”按钮，会出现微信开发者工具的登录页，如图 2－18 所示。在登录页,可以使用微信扫码登录开发者工具,开发者工具将使用这个微信账号的信息进行小程序的开发和调试。

图 2-18 登录页

至此,微信开发者工具安装完毕。

2.2 体验第一个小程序

安装完微信开发者工具,立马体验下小程序版的"Hello World"!

2.2.1 开发模式选择

用微信扫码登录页的二维码之后,会出现如图 2-19 所示的模式选择页面。

开发者工具提供两种开发模式的选择:

(1) 公众号网页调试。选择公众号网页调试,将直接进入公众号网页项目调试界面,在地址栏输入 URL 网址,即可调试该网页的微信授权以及微信"JS-SDK"功能。

(2) 小程序调试。选择小程序调试,将进入小程序本地项目管理页,可以新建、删除本地项目,或者选择进入已存在的本地项目。

很显然,我们这里做的是小程序开发,因此选择小程序项目即可。

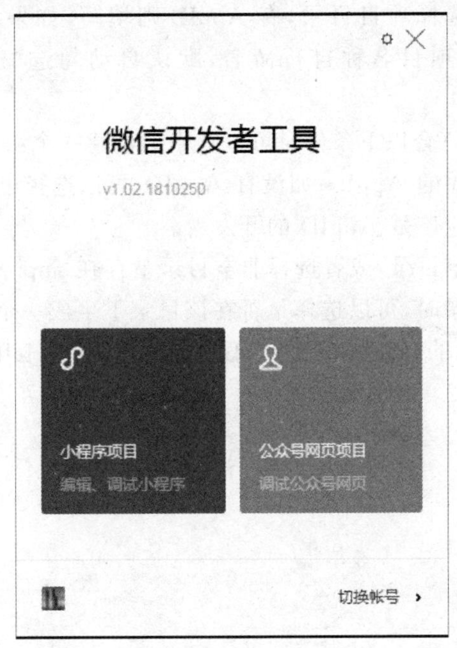

图 2-19　模式选择

2.2.2　创建 Hello World 小程序

选择了小程序调试模式之后，新建一个"Hello World"小程序项目，如图 2-20

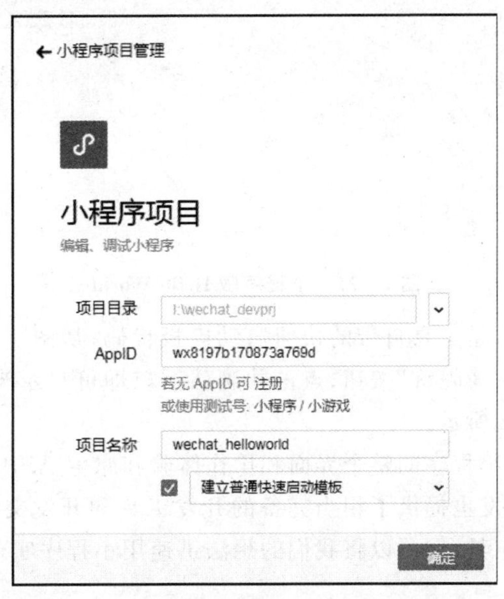

图 2-20　新建小程序项目

所示。这里首先需要选择项目目录，有 AppID 则填写，如果没有可以点击图中的“小程序”获得测试号，项目名称自行命名，默认自动勾选“建立普通快速启动模板”。

需要说明的是，当符合以下条件时可以在本地创建一个小程序项目：

① 需要一个小程序的 AppID；如没有 AppID，可以选择申请使用测试号。

② 登录的微信号应该是 AppID 的开发者。

③ 需要选择一个空目录，或者选择非空目录下存在 app. json 或者 project. config. json。当选择空目录时，可以选择是否在该目录下生成一个简单的项目。

点击图 2 - 20 所示的“确认”按钮，等程序加载完毕，“Hello World”版的小程序就呈现在眼前了，如图 2 - 21 所示。

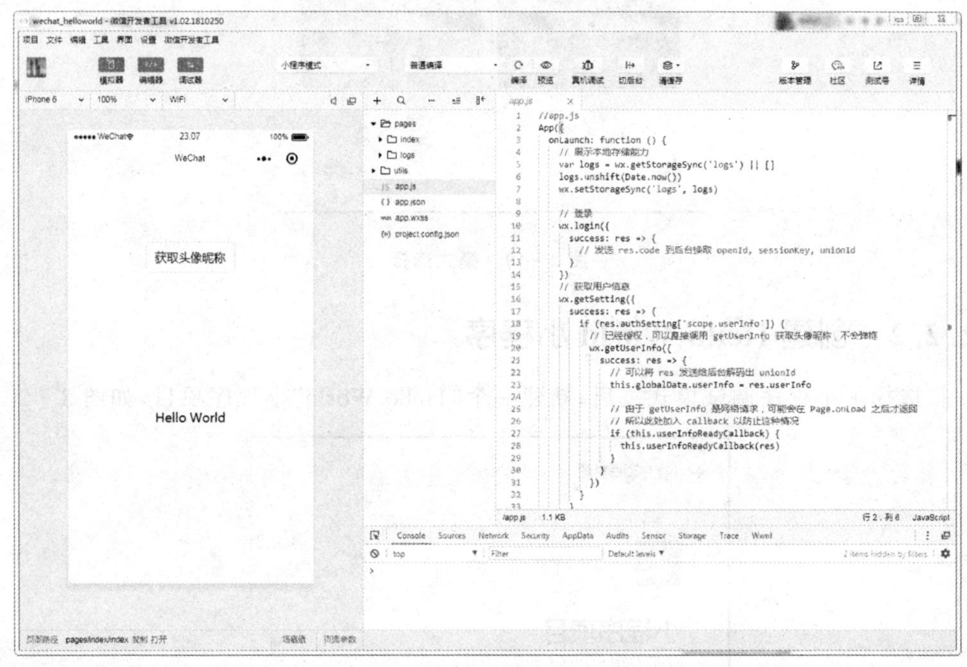

图 2 - 21 小程序版 Hello World

这个程序很简单，是工具自带的快速启动模板代码，从图 2 - 21 中可以看到，有一个很简单的“获取头像昵称”按钮，点击并确认授权即可显示当前开发者的微信头像和昵称，如图 2 - 22 所示。

是不是很简单？小程序的整个界面和操作体验和原生 APP 相比而言非常接近，而且微信为小程序开发也提供了相当完备的开发工具和开发文档，我们只需要遵循其开发规则，依赖其工具，就可以将我们的想法迅速用小程序实现。

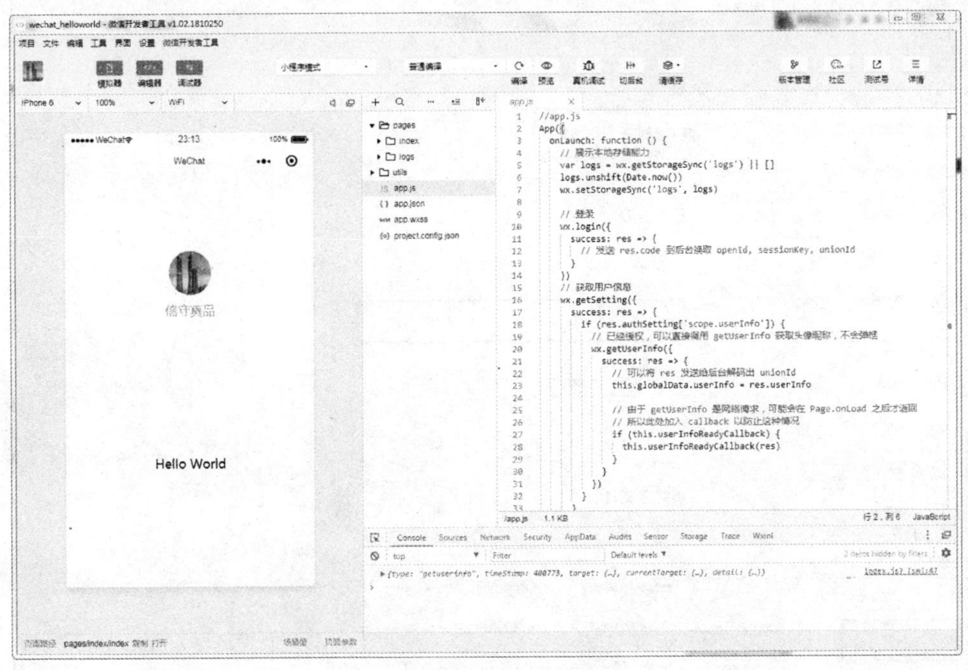

图 2-22　获取头像昵称

2.2.3　如何获取小程序 AppID

刚刚在创建小程序项目的时候，有一处需要填写 AppID，我们填写的是小程序测试号 AppID 来做 DEMO 演示，如果我们以后需要运用在实际上线的项目中，那必须填写注册小程序后微信公众平台生成的唯一 AppID。登录微信公众平台（https://mp.weixin.qq.com）之后，点击左侧“设置”再点击右侧“开发设置”即可查看到 AppID，如图 2-23 所示。

需要注意的是，小程序的 AppID 不可使用微信订阅号或服务号的 AppID，它们互相不可通用。测试用的 AppID 仅仅只能做 DEMO 测试开发体验使用，功能受限且无法上传发布小程序应用。

图 2-23　查看小程序 AppID

2.3　微信开发者工具使用介绍

"工欲善其事必先利其器"，在进行小程序实际开发之前，需要先来简单了解下，官方提供的微信开发者工具的基本功能。

2.3.1　开发工具界面图解

下面先对微信开发者工具进行简单的分区介绍。如图 2-24 所示，开发者工具主界面，从上到下、从左到右分别为：菜单栏、工具栏、模拟器、编辑器和调试器 五大部分。

(1) 菜单栏和 Windows 的文件管理器菜单基本一样，包含"项目""文件""编辑""工具""界面""设置"和"微信开发者工具"这几个菜单项。其中"微信开发者工具"菜单中可以快速切换登录用户，可以选择调试开发者工具或编辑器。

(2) 工具栏提供了常用功能的快捷操作按钮。点击用户头像可以打开个人中心，在这里可以便捷地切换用户和查看开发者工具收到的消息。工具栏中间，可以选择普通编译，也可以新建并选择自定义条件进行编译和预览；通过切换后台按钮，可以模拟小程序进入后台的情况；工具栏上还提供了清除缓存的快速入口；工具栏右侧

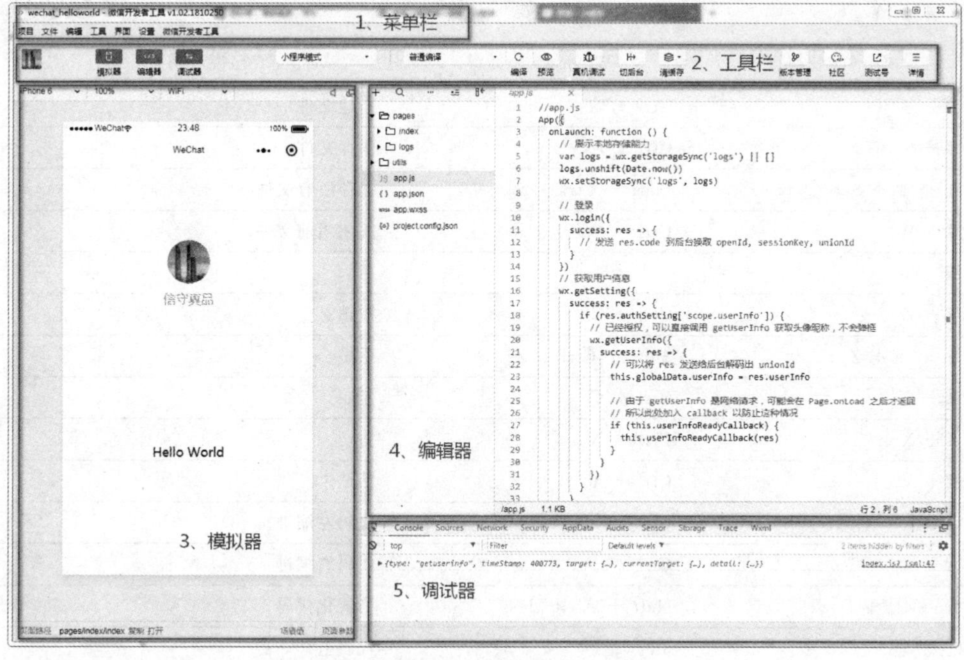

图 2-24　开发工具分区介绍

是开发辅助功能的区域，在这里可以上传代码、申请测试、上传腾讯云、查看项目信息等。

(3) 模拟器可以模拟小程序在微信客户端的表现。小程序的代码通过编译后可以在模拟器上直接运行。开发者可以选择不同的设备，也可以添加自定义设备来调试小程序在不同尺寸机型上的适配问题。在模拟器底部的状态栏，可以直观地看到当前运行小程序的场景值、页面路径及页面参数。

(4) 编辑器左侧为代码文件夹目录树，可点击展开和收起。右侧为代码文本编辑器。

(5) 调试器和网页浏览器的调试器非常类似，熟悉网页调试的开发者很容易上手。

微信开发者工具常用快捷键如表 2-1 所列。

表 2-1　微信开发者工具常用快捷键

Mac OS 快捷键	Windows 快捷键	说　明
⌘ + Q		退出开发者工具
⇧ + ⌘ + N	Shift + Ctrl + N	新建项目
⇧ + ⌘ + W	Shift + Ctrl + W	关闭当前项目
文　件		

续表 2 - 1

Mac OS 快捷键	Windows 快捷键	说　明
⌘+ N	Ctrl + N	新建文件
⌘+ S	Ctrl + S	保存文件
⇧ + ⌘+ S	Shift + Ctrl + S	保存所有文件
⌘+ W	Ctrl + W	关闭当前文件
编　辑		
⌘+ Z	Ctrl + Z	撤销
⇧ + ⌘+ Z	Shift + Ctrl + Z	重做
⌘+ C	Ctrl + C	复制
⌘+ X	Ctrl + X	剪切
⌘+ V	Ctrl + V	粘贴
⌘+ [	Ctrl + [	代码左缩进
⌘+]	Ctrl +]	代码右缩进
⇧ + □ + F	shift + Alt + F	格式化代码
□ + □	alt + □	代码上移一行
□ + □	alt + □	代码上移一行
⇧ + □ + □	shift + Alt + □	复制并向上粘贴
⇧ + □ + □	shift + Alt + □	复制并向下粘贴
⌘+ P	Ctrl + P	跳到文件
⌘+ E	Ctrl + E	跳到最近文件
⌘+ □ + □	Ctrl + Alt + □	打开当前文件编辑器左边的文件
⌘+ □ + ➡	Ctrl + Alt + ➡	打开当前文件编辑器右边的文件
⌘+ F	Ctrl + F	文件内搜索
⇧ + ⌘+ F	shift + Ctrl + F	项目内搜索
⇧ + ⌘+ R	shift + Ctrl + R	焦点在编辑器内，表示替换
工　具		
⌘+ B	Ctrl + B	编译项目
⌘+ R	Ctrl + R	焦点在编辑器外，编译项目
⇧ + ⌘+ P	shift + Ctrl + P	预览代码
⇧ + ⌘+ U	shift + Ctrl + U	上传代码
界　面		
⌘+ ,	Ctrl + ,	打开设置窗口

2.3.2 小程序调试工具

小程序调试工具分为 7 大功能模块：Wxml、Console、Sources、Network、Appdata、Storage 和 Sensor，下面将一一展开介绍。

Wxml 模块用于帮助开发者开发 wxml 转化后的界面。在这里可以看到真实的页面结构以及结构对应的 wxss 属性，同时可以通过修改对应的 wxss 属性，在模拟器中实时看到修改的情况（仅为实时预览，无法保存到文件）。通过调试模块左上角的选择器，还可以快速定位页面中组件对应的 wxml 代码，如图 2－25 所示。

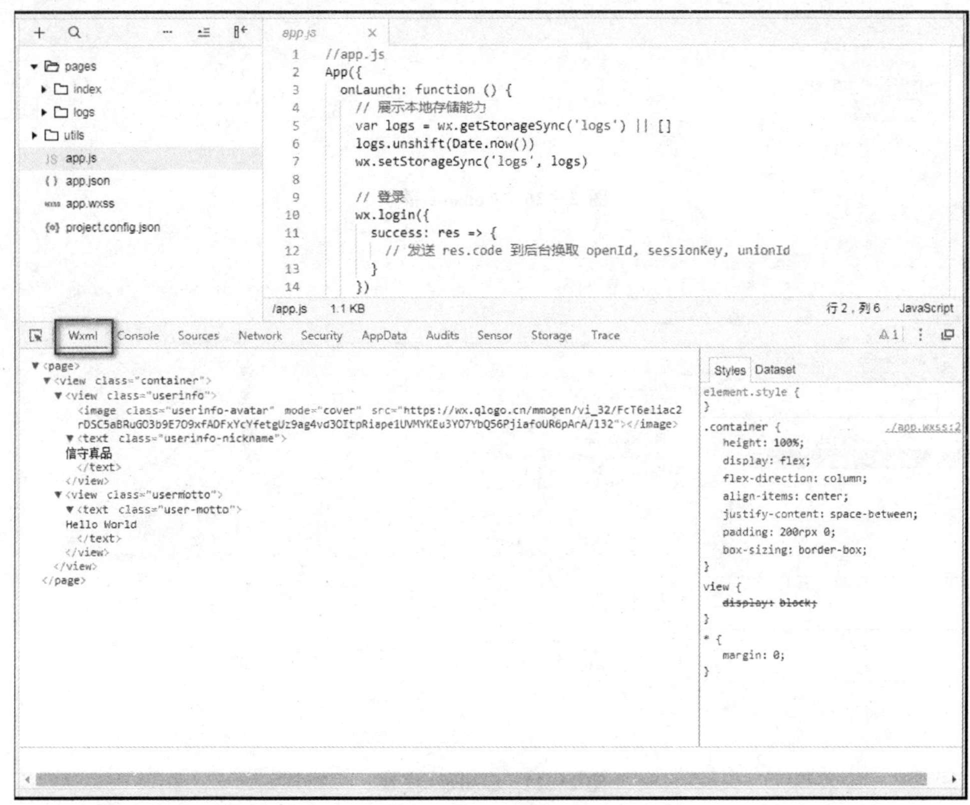

图 2－25 Wxml 模块

Console 模块有两大功能：开发者可以在此输入和调试代码；小程序的错误提示输出会显示在此处，如图 2－26 所示。

Sources 模块用于显示当前项目的脚本文件，同浏览器开发不同，微信小程序框架会对脚本文件进行编译，所以在 Sources 模块中，开发者看到的文件是经过处理之后的脚本文件，开发者的代码都会被包裹在 define 函数中；并且对于 Page 代码，在尾部会有 require 的主动调用，如图 2－27 所示。

Network 模块用于观察和显示 request 和 socket 的请求情况。注意，uploadFile

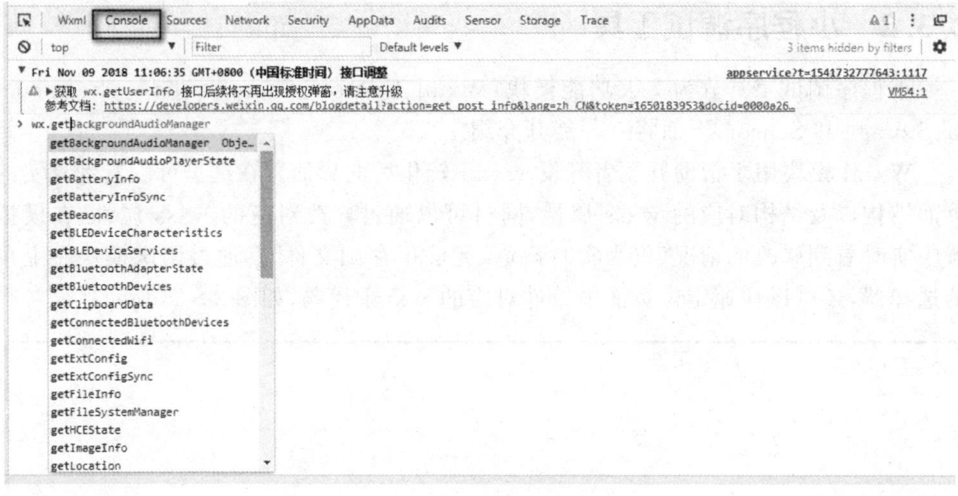

图 2-26　Console 模块

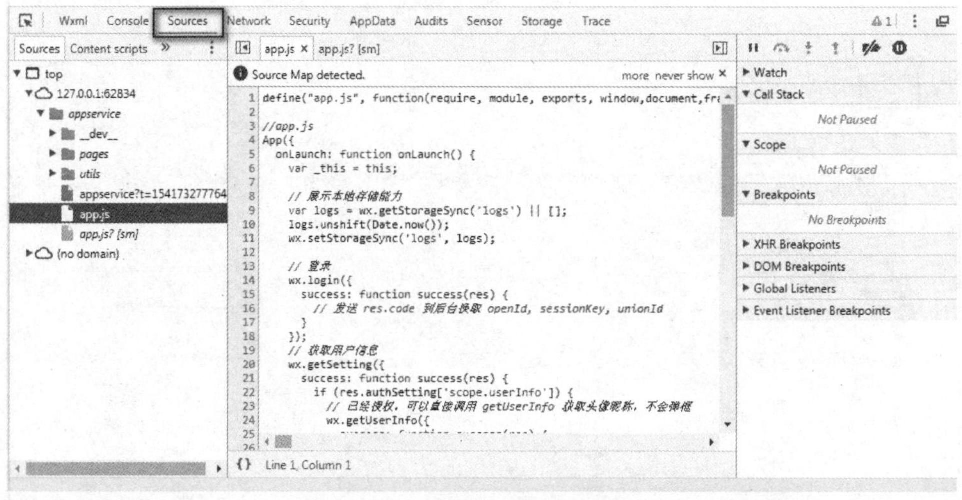

图 2-27　Sources 模块

和 downloadFile 暂时不支持在 Network 模块中查看，如图 2-28 所示。

AppData 模块用于显示当前项目运行时小程序 AppData 的具体数据，实时地反映项目数据情况，可以在此处编辑数据，并及时地反馈到界面上，如图 2-29 所示。

Storage 模块用于显示当前项目使用 wx. setStorage 或者 wx. setStorageSync 后的数据存储情况。非常方便的是，用户可以直接在 Storage panel 上对数据进行删除（按 Delete 键）、新增、修改操作，如图 2-30 所示。

Sensor 模块有两大功能：模拟地理位置和模拟移动设备表现，用于调试重力感应 API，如图 2-31 所示。

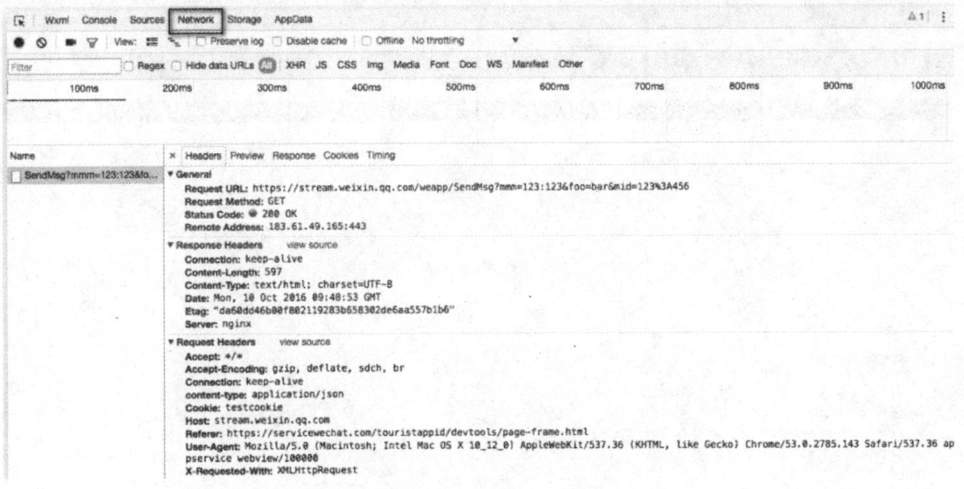

图 2－28 Network 模块

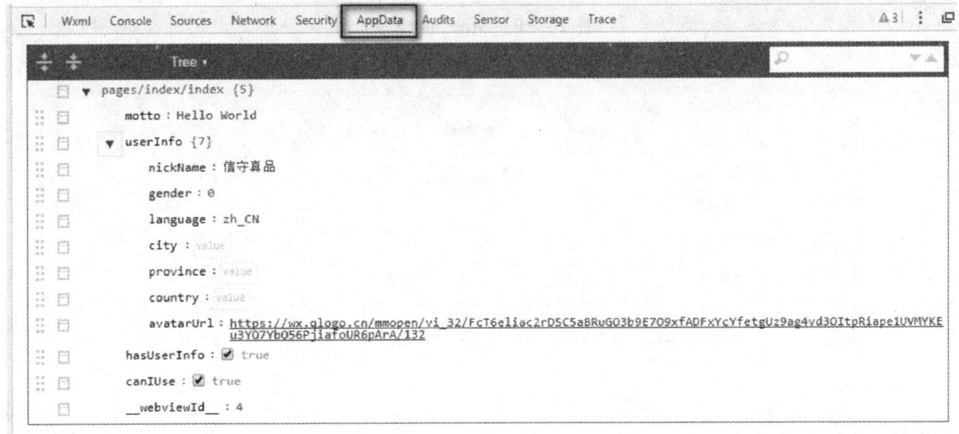

图 2－29 AppData 模块

最后，如果需要远程真机调试小程序，那么可以点击工具栏中的“真机调试”按钮，真机远程调试功能可以直接利用开发者工具，通过网络连接，对手机上运行的小程序进行调试，帮助开发者更好地定位和查找在手机上出现的问题。此时，工具会将本地代码进行处理打包并上传，就绪后，使用手机客户端扫描二维码即可弹出调试窗口，开始远程调试，如图 2－32 所示。

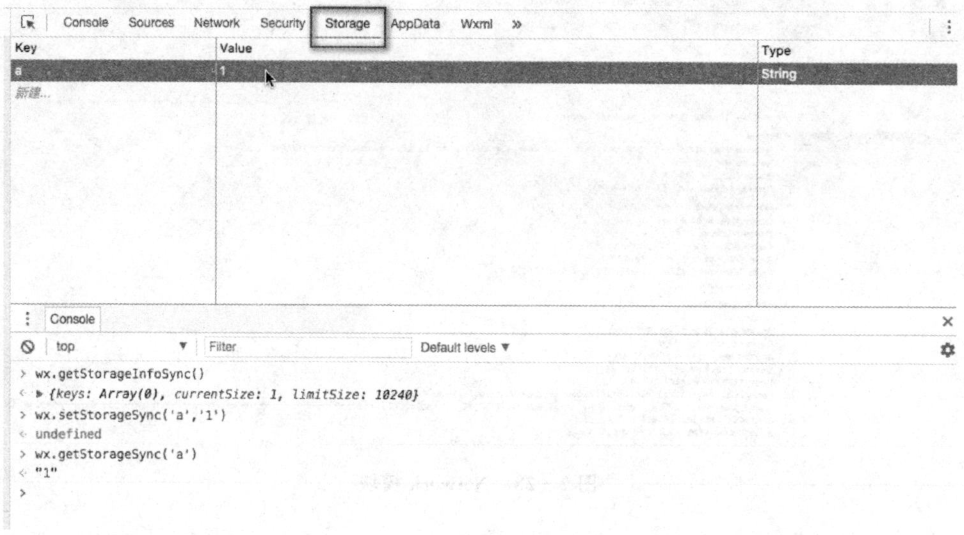

图 2-30 Storage 模块

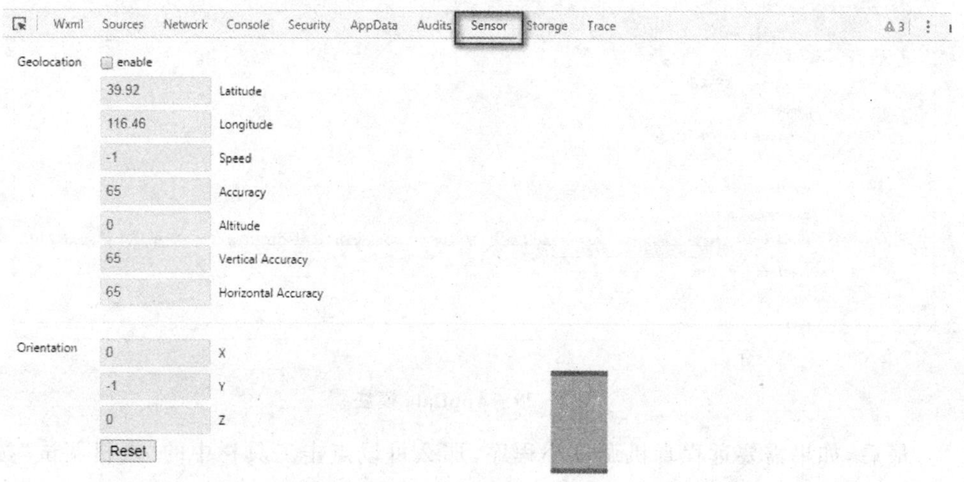

图 2-31 Sensor 模块

图 2－32　真机调试

2.3.3　如何预览小程序

小程序开发好之后需要预览小程序在真机上的效果，可以点击工具栏中的预览按钮，最新版默认优先采用自动预览模式。自动预览可以实现编写小程序时的快速预览，免去了每次查看小程序效果时都要扫码或者使用小程序开发助手的麻烦，如图 2－33 所示。

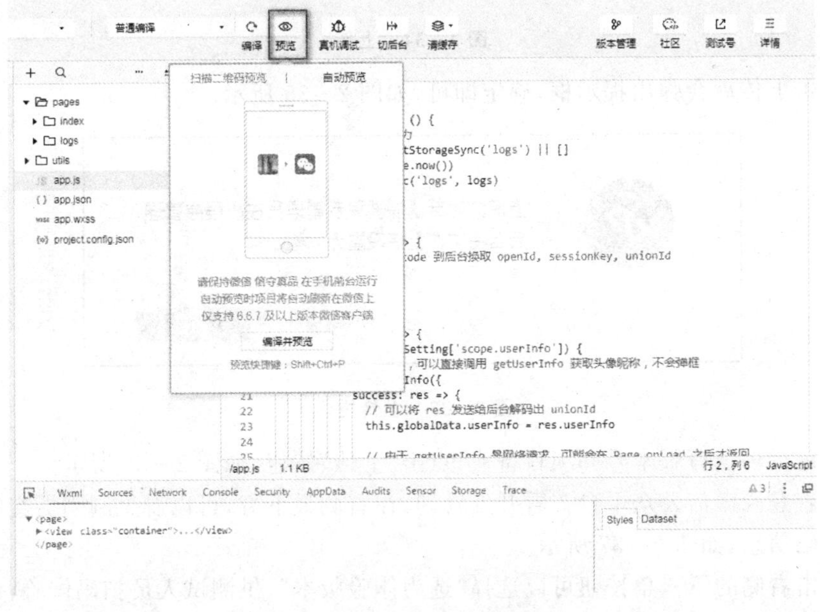

图 2－33　自动预览

注意，自动预览功能仅限于登录开发者工具的同账号微信使用，且需要配合6.6.7及以上的微信客户端版本。如需换回普通预览模式，只需要点击"扫描二维码预览"标签即可。

2.3.4 上传和发布小程序

在菜单栏"工具"中选择"上传"即可将小程序代码上传至微信服务器，如图2-34所示。

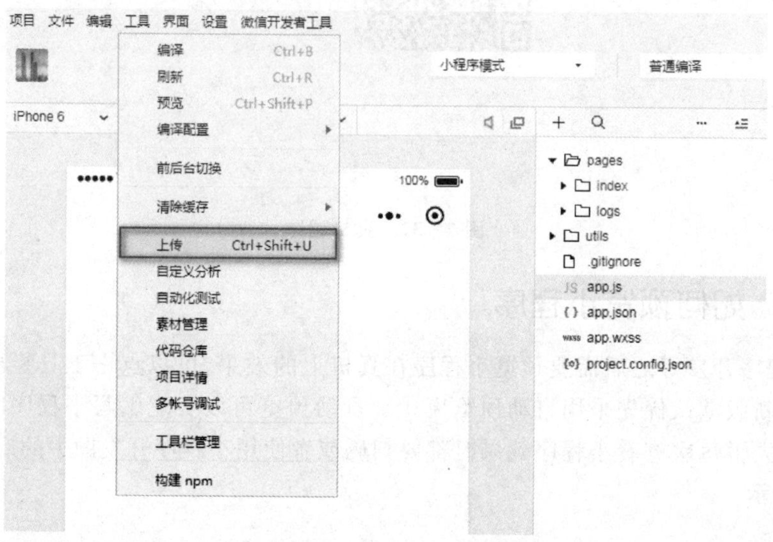

图2-34 上传

点击上传后会弹出提示框，确定即可，如图2-35所示。

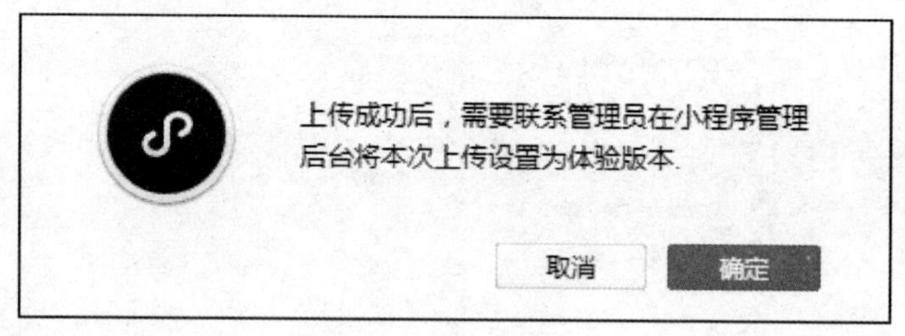

图2-35 上传提示

确定后再填写版本号和项目备注，点击"上传"按钮，如图2-36所示。

接着登录微信公众平台查看开发管理，在右侧最下方，可以看见刚刚提交的开发版本代码信息，如图2-37所示。

点击右侧的下三角按钮可以选择"选为体验版本"，供测试人员扫码体验，也可以删除，如图2-38所示。点击图中的"提交审核"按钮即可提交该版本的代码给微信

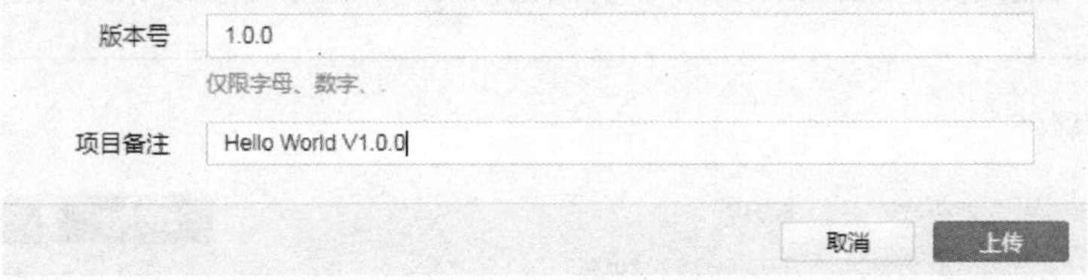

图 2－36　填写版本号和项目备注

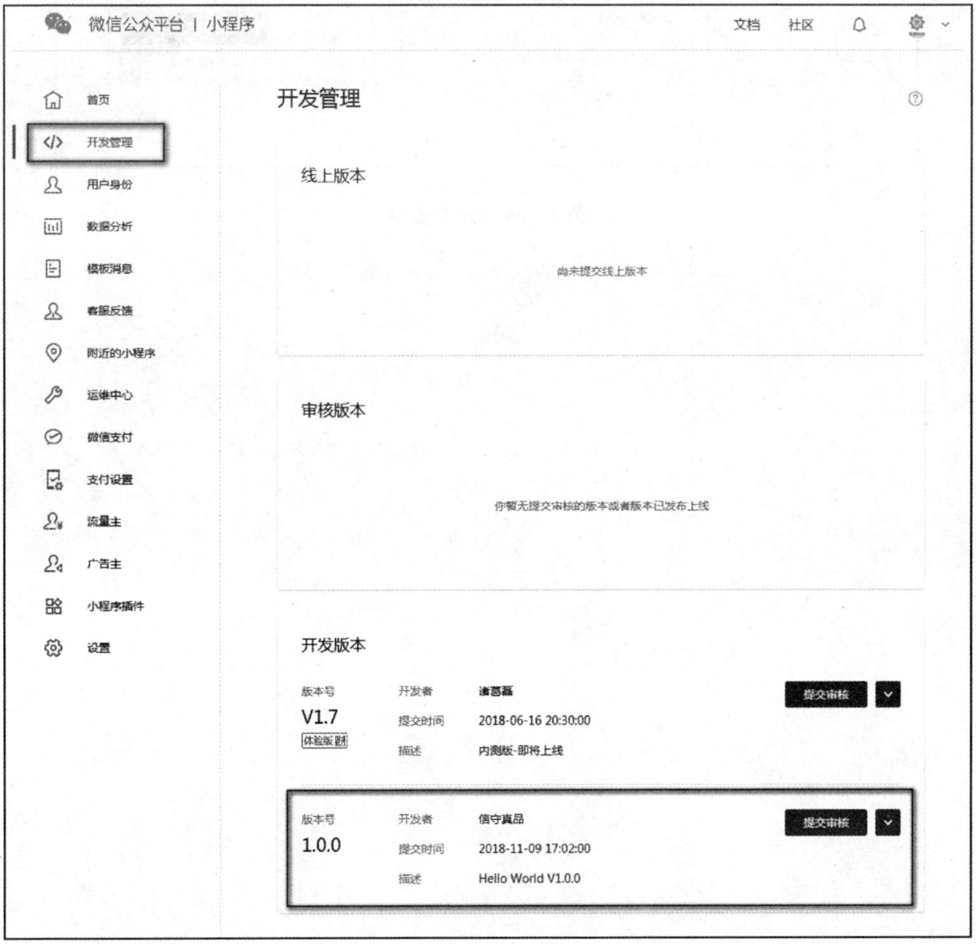

微信公众平台 | 小程序
文档　社区

首页
开发管理
用户身份
数据分析
模板消息
客服反馈
附近的小程序
运维中心
微信支付
支付设置
流量主
广告主
小程序插件
设置

开发管理

线上版本

尚未提交线上版本

审核版本

你暂无提交审核的版本或者版本已发布上线

开发版本

版本号　V1.7　体验版
开发者　谢蕾蕾
提交时间　2018-06-16 20:30:00
描述　内测版-即将上线
提交审核

版本号　1.0.0
开发者　信守真品
提交时间　2018-11-09 17:02:00
描述　Hello World V1.0.0
提交审核

图 2－37　开发管理

审核团队进行上线审核使用。

代码审核通过之后需要开发者手动点击发布，小程序才会发布到线上提供服务。

图 2－38　开发版本

第3章

小程序框架基础

本章主要讲解小程序框架的基础内容：框架原理、WXML 模板、WXSS 样式和小程序 JavaScript 脚本等核心内容。通过这些内容的学习和理解，读者可以很方便地掌握关于小程序框架基础的核心知识要点。

3.1 理解小程序开发框架

小程序的开发和之前常见的 H5WEB 开发有很大的相似性，因此，熟悉 H5 WEB 开发的开发者对小程序开发应该更容易上手。H5 前端技术这几年突飞猛进的发展使得前端开发框架异常繁荣，这其中涌现出了很多基于 MVVM 模式打造的前端框架，如：Vue、Angular 等；而小程序框架同样也是基于 MVVM 模式打造的，微信团队为小程序提供的框架命名为“MINA 应用框架”。

3.1.1 小程序框架原理

微信小程序 MINA 框架通过封装微信客户端提供的文件系统、网络通信、任务管理、数据安全等基础功能，对上层提供一整套 JavaScript API，让开发者能够非常方便地使用微信客户端提供的各种基础功能与能力，快速构建一个应用。通过图 3-1 所示的框架图，可以看到整个 MINA 框架包含两大部分：View 视图层和 App Service 逻辑层。

在 View 视图层中，wxml 是 MINA 提供的一套类似 html 标签的语言以及一系列基础组件。开发者使用 wxml 文件来搭建页面的基础视图结构，使用 wxss 文件来控制页面的展现样式。App Service 应用逻辑层是 MINA 的服务中心，由微信客户端启用异步线程单独加载运行。页面渲染所需的数据、页面交互处理逻辑都在 App Service 中实现。MINA 框架中的 App Service 使用 Java Script 来编写交互逻辑、网络请求和数据处理，但不能使用 Java Script 中的 DOM 操作。小程序中的各个页面可以通过 App Service 实现数据管理、网络通信、应用生命周期管理和页面路由。

MINA 框架的核心是一个响应的数据绑定系统。框架可以让数据与视图非常

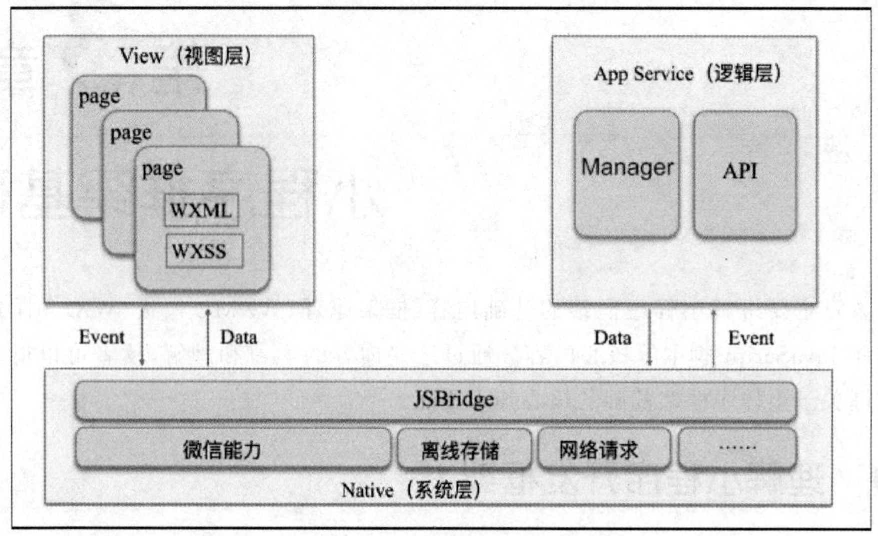

图 3-1 小程序 MINA 框架图

简单地保持同步。当修改数据的时候，只需要在逻辑层修改数据，视图层就会做相应的更新。该框架体现了 MVVM 的设计思想，MVVM 模式示意如图 3-2 所示。

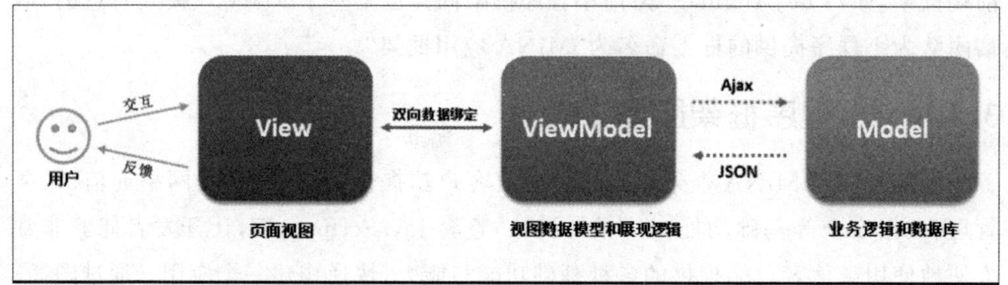

图 3-2 MVVM 模式

MVVM 源自于经典的 Model-View-Controller(MVC)模式。MVVM 的出现促进了 GUI 前端开发与后端业务逻辑的分离，极大地提高了前端开发效率。MVVM 的核心是 ViewModel 层，它就像是一个中转站(value converter)，负责转换 Model 中的数据对象来让数据变得更容易管理和使用，该层向上与视图层进行双向数据绑定，向下与 Model 层通过接口请求进行数据交互，起承上启下的作用。

3.1.2 小程序目录结构和文件构成

小程序包含一个描述整体程序的 app 和多个描述各自页面的 page，这是小程序最重要的组成部分，其他还包括一个公用工具类库 utils 和一个项目 IDE 配置：project.config.json，小程序目录结构如图 3-3 所示。

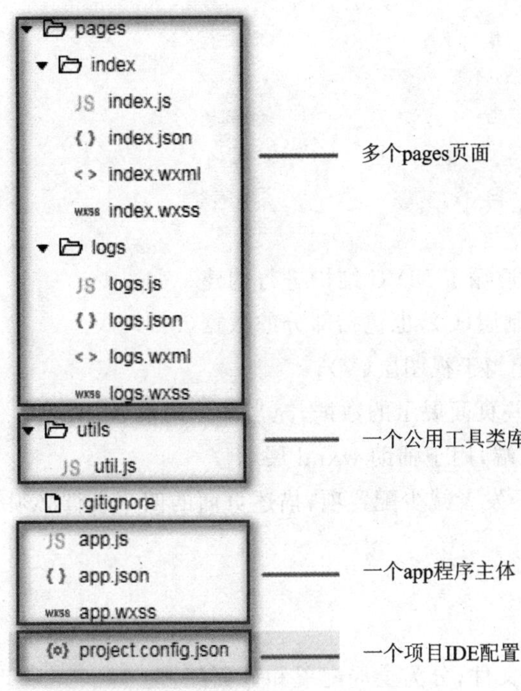

图 3-3 小程序目录结构

一个小程序主体部分由三个文件组成，且必须放在项目的根目录。小程序主体文件如表 3-1 所列。

表 3-1 小程序主体文件

文件	必需	作用
app.js	是	小程序逻辑
app.json	是	小程序公共配置
app.wxss	否	小程序公共样式表

一个小程序页面由四个文件组成，如表 3-2 所列。

表 3-2 小程序页面文件

文件类型	必需	作用
js	是	页面逻辑
wxml	是	页面结构
json	否	页面配置
wxss	否	页面样式表

小程序页面设计基本上也遵循了 MVC 结构进行构建。如：

- js：页面逻辑，相当于控制层(C)，也包括部分的数据(M)；
- wxml：页面结构展示，相当于视图层(V)；
- json：页面配置，配置一些页面展示的数据，充当部分的模型(M)；
- wxss：页面样式表，纯前端，用于辅助 wxml 展示。

需要注意的是，为了方便开发者减少配置项，描述页面的四个文件必须具有相同的路径与文件名。

3.1.3 小程序配置

小程序有两种重要的配置文件，分为全局配置和页面配置。

1. app.json 应用的全局配置文件

小程序根目录下的 app.json 文件用来对小程序进行全局配置，决定页面文件的路径、决定窗口表现、设置网络超时时间和设置多 tab 等。

以下是一个包含了部分常用配置选项的 app.json 配置示例代码 ：

```
{
    "pages": [
    "pages/index/index",
    "pages/logs/index"
  ],
    "window": {
    "navigationBarTitleText": "Demo"
  },
    "tabBar": {
    "list": [{
```

```
    "pagePath": "pages/index/index",
    "text": "首页"
      }, {
    "pagePath": "pages/logs/logs",
    "text": "日志"
      }]
  },
    "networkTimeout": {
    "request": 10000,
    "downloadFile": 10000
  },
    "debug": true,
    "navigateToMiniProgramAppIdList": [
    "wxe5f52902cf4de896"
  ]
}
```

从配置代码示例来看，配置遵循 Key - Value 的 JSON 格式，清晰明了，参看 app.json 配置项列表的描述即可知晓配置的大概意义，如表 3 - 3 所列。

表 3 - 3　app.json 配置项列表

属性	类型	必填	描述	支持版本
pages	String Array	是	页面路径列表	
window	Object	否	全局的默认窗口表现	
tabBar	Object	否	底部 tab 栏的表现	
networkTimeout	Object	否	网络超时时间	
debug	Boolean	否	是否开启 debug 模式，默认关闭	
functionalPages	Boolean	否	是否启用插件功能页，默认关闭	2.1.0
subpackages	Object Array	否	分包结构配置	1.7.3
workers	String	否	Worker 代码放置的目录	1.9.90
requiredBackgroundModes	String Array	否	需要在后台使用的能力，如「音乐播放」	
plugins	Object	否	使用到的插件	1.9.6
preloadRule	Object	否	分包预下载规则	2.3.0

有关更具体的配置解释描述定义，请参看微信小程序官方文档中框架分类的配置一栏，笔者在此不再过多赘述。

2. page.json 页面的全局配置文件

每一个小程序页面也可以使用.json 文件来对本页面的窗口表现进行配置。页面的配置只能设置 app.json 中部分 window 配置项的内容，页面中的配置项会覆盖 app.json 的 window 中相同的配置项。页面配置文件示例代码如下：

```
{
  "navigationBarBackgroundColor": "#ffffff",
  "navigationBarTextStyle": "black",
  "navigationBarTitleText": "微信接口功能演示",
  "backgroundColor": "#eeeeee",
  "backgroundTextStyle": "light"
}
```

参看如表 3-4 所列的页面配置项列表即可知晓页面配置的具体情况。

表 3-4 页面配置项列表

属性	类型	默认值	描述
navigationBarBackgroundColor	HexColor	#000000	导航栏背景颜色，如 `#000000`
navigationBarTextStyle	String	white	导航栏标题颜色，仅支持 `black` / `white`
navigationBarTitleText	String		导航栏标题文字内容
backgroundColor	HexColor	#ffffff	窗口的背景色
backgroundTextStyle	String	dark	下拉 loading 的样式，仅支持 `dark` / `light`
enablePullDownRefresh	Boolean	false	是否全局开启下拉刷新。 详见 Page.onPullDownRefresh
onReachBottomDistance	Number	50	页面上拉触底事件触发时距页面底部距离，单位为px。 详见 Page.onReachBottom
disableScroll	Boolean	false	设置为 `true` 则页面整体不能上下滚动；只在页面配置中有效，无法在 `app.json` 中设置该项

3.1.4 小程序运行和加载机制

1. 小程序运行机制

小程序启动分为冷启动和热启动，区别在于：

- 冷启动用户第一次打开或打开后被微信主动销毁后，再次启动，小程序需重新加载启动；
- 热启动：已打开过该小程序，且在一定时间内再次打开，小程序从后台切换到前台。

小程序被微信销毁的情况有两种：

- 超时销毁，现在一般是 5 min；
- 如果短时间连续超过 2 次的系统警告也会销毁，这个短时间目前是 5 s。

2. 小程序加载机制

小程序通过网络 CDN 更新 Package 代码包，同时通过 Ajax 请求与 Web 服务器交互以获取 JSON 数据，如图 3.4 所示。

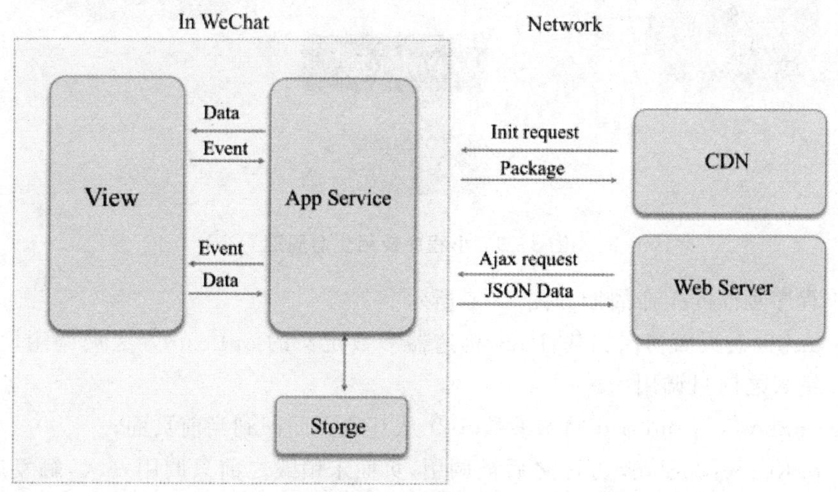

图 3-4 小程序加载机制

微信小程序采用的是类似离线包加载的方案，当用户第一次打开小程序时会先下载好所有的代码，然后再加载页面；当用户再次进入小程序时，会直接使用已下载的代码，省去了代码下载的过程，打开速度更快。小程序在版本 1.7.3 及以上基础库时开始支持分包加载，不支持的版本默认使用整包的方式。分包加载可以理解为：打开小程序，默认先加载主包代码；进入分包页面时，再加载对应分包代码。

3.1.5 小程序生命周期和线程架构

1. 小程序生命周期

小程序生命周期分为:应用生命周期和页面生命周期。

小程序应用的生命周期如图 3-5 所示。

➢ onLaunch:启动时初始化;

➢ onShow:从后台唤醒,切换到前台;

➢ onHide:切换到后台;

➢ onError:脚本错误或 API 调用失败;

➢ globalData:应用全局数据。

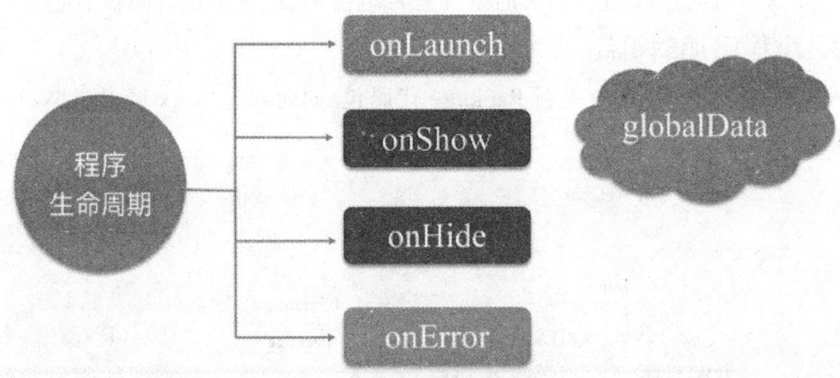

图 3-5 小程序应用生命周期

小程序页面的生命周期如图 3-6 所示。

➢ onLoad:页面初次加载,Page 构造器参数定义的 onLoad 方法被调用,页面未销毁之前只调用一次;

➢ onShow:① onLoad 回调参数 ;② 从其他页面回到当前页面;

➢ onReady:onShow 方法之后被调用,页面未销毁之前只调用一次,触发后逻辑层和视图层可交互;

➢ onHide:在当前页打开新的页面;

➢ onUnload:关闭当前页;

➢ data:当前页面数据。

2. 小程序线程架构

小程序两大线程:AppService Thread 和 View Thread,协同负责完成小程序生命周期的调用,如图 3-7 所示。其中 View 线程负责解析渲染页面(wxml 和 wxss),而 AppServer 线程负责运行 js。由于 js 不跑在 web-view 里,不能直接操纵 DOM 和 BOM,这就是小程序没有 window 全局变量的原因。

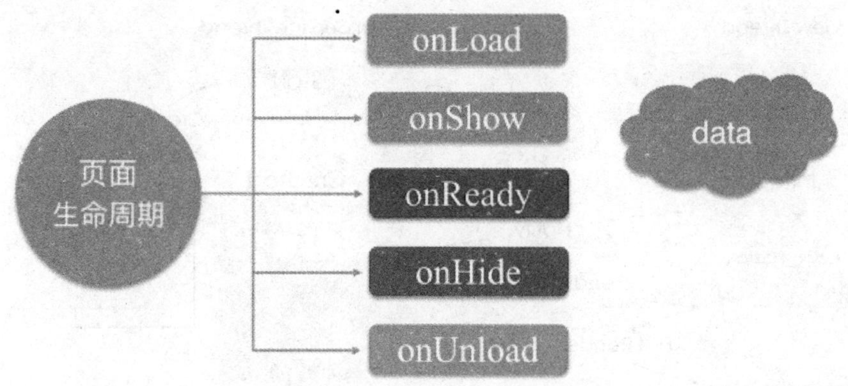

图 3-6 小程序页面生命周期

理解小程序的线程架构之后，基本上可以归纳出一个小程序开发的大致步骤：

(1) 创建小程序实例(定义、配置及页面执行关联)。即编写 3 个 App 前缀的文件，它们共同定义了整个小程序的主体逻辑、生命周期及页面构成、样式等。小程序实例将由 AppServer 线程运行。

(2) 创建页面(页面结构与事务处理逻辑)。在小程序中，一个完整的页面(page)是由.js、.json、.wxml、.wxss 这四个文件组成的。小程序页面由 View 线程执行。

3.2 WXML：小程序版 HTML

WXML(WeiXin Markup Language)是框架设计的一套标签语言，结合基础组件和事件系统，可以构建出页面的结构。

3.2.1 标签与属性

WXML 标签的语法规则如下：

- 所有元素都必须闭合标签：

  <text> Hello World </text>

- 所有元素都必须正确嵌套：

  <view>(开始标签 1)<text>(开始标签 2)Hello </text>(结束标签 2)</view>(结束标签 1)

- 属性值必须使用引号包围：

  <text id = "myText"(参数值必须使用引号包围) > myText </text>

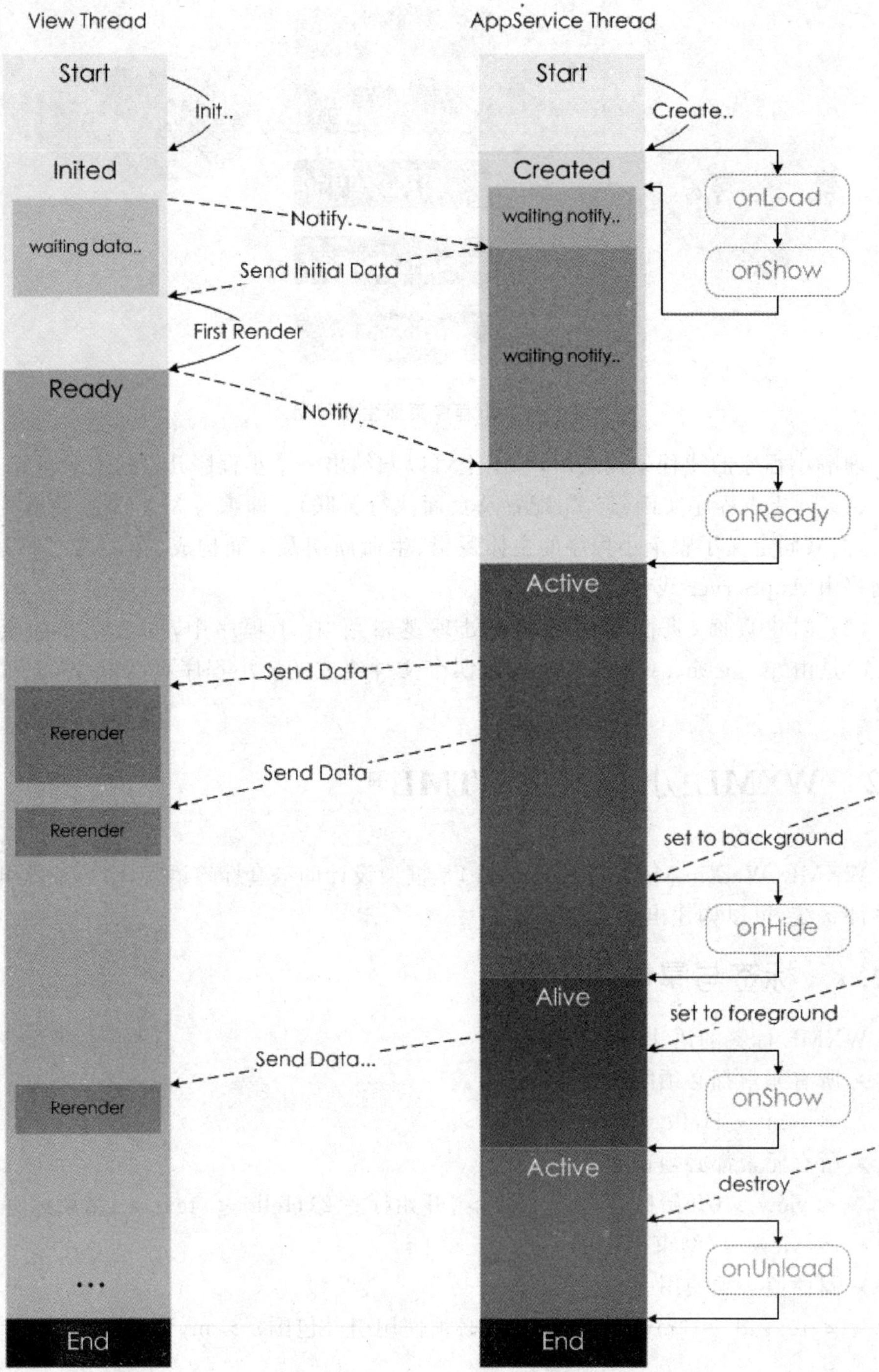

View Thread
AppService Thread
Start
Init..
Inited
waiting data..
Start
Create..
Created
waiting notify..
onLoad
onShow
Notify
Send Initial Data
First Render
waiting notify..
Ready
Notify
onReady
Active
Send Data
Rerender
Send Data
Rerender
set to background
onHide
Alive
set to foreground
Send Data...
Rerender
onShow
Active
destroy
onUnload
...
End
End

图 3-7　小程序线程架构

➢ 标签必须使用小写，如：

< text(小写) > Hello World < /text >

➢ WXML 中连续多个空格会被合并为 1 个空格，如：

< text > Hello(此处 7 个空格)World < /text >

< text > Hello(合并为 1 个空格) World < /text >

小程序框架为开发者提供了一系列基础组件，供开发者进行快速开发。组件是视图层的基本组成单元。一个组件通常包括“开始标签”和“结束标签”，属性用来修饰这个组件，内容在两个标签之内。因为一些属性几乎被所有的组件共同使用，故这些属性被抽离出来形成了公共属性，如表 3-5 所列。

表 3-5 公共属性

属性名	类型	描述	注解
id	String	组件的唯一标识	保持整个页面唯一
class	String	组件的样式类	在对应的 WXSS 中定义的样式类
style	String	组件的内联样式	可以动态设置的内联样式
hidden	Boolean	组件是否显示	所有组件默认显示
data-*	Any	自定义属性	组件上触发事件时，会发送给事件处理函数
bind* / catch*	EventHandler	组件的事件	详见事件

每个组件也有各自定义的属性，可以对该组件的功能或样式进行修饰，请参考官方文档查阅各个组件的定义，此处不再赘述。

3.2.2 数据绑定

数据绑定功能使得程序在运行过程中，具备动态改变渲染界面的能力，从而达到了更好的用户体验效果。在 WEB 开发中，需要借助 JavaScript 并通过 DOM 接口来实现界面的动态更新，而在小程序中，则是使用 WXML 语言提供的数据绑定功能来实现的。

1. 简单绑定

WXML 中的动态数据均来自对应 Page 的 data。数据绑定使用 Mustache 语法(双大括号)将变量包起来，具体的数据绑定代码示例如下。

内 容：

```
<!-- wxml -->
<view> {{ message }} </view>
//page.js
Page({
  data：{
    message：'Hello WXML! '
  }
})
```

组件属性(需要在双引号之内)：

```
<!-- wxml -->
<view id="item-{{id}}"> </view>
//page.js
Page({
  data：{
    id：0
  }
})
```

控制属性(需要在双引号之内)：

```
<!-- wxml -->
<view wx:if="{{condition}}"> </view>
//page.js
Page({
  data：{
    condition：true
  }
})
```

关键字(需要在双引号之内)：

true：boolean 类型的 true，代表真值；

false：boolean 类型的 false，代表假值。

```
<checkbox checked="{{false}}"> </checkbox>
```

注意：不要直接写 checked="false"，其计算结果是一个字符串，转成 boolean 类型后则代表真值。

2. 运　算

可以在“{{}}”内进行简单的运算，支持下面几种方式。

三元运算：

```
<!-- wxml -->
```

```
< view hidden = "{{flag ? true : false}}" > Hidden < /view >
```

算数运算：

```
< !-- wxml -->
< view > {{a + b}} + {{c}} + d < /view >
//page.js
Page({
  data: {
    a: 1,
    b: 2,
    c: 3
  }
})
```

view 中的内容为“3 ＋ 3 ＋ d”。

逻辑判断：

```
< !-- wxml -->
< view wx:if = "{{length > 5}}" >  < /view >
```

字符串运算：

```
< !-- wxml -->
< view >{{"hello" + name}}< /view >
//page.js
Page({
  data:{
    name: 'MINA'
  }
})
```

数据路径运算：

```
< !-- wxml -->
< view >{{object.key}} {{array[0]}}< /view >
//page.js
Page({
  data: {
    object: {
      key: 'Hello '
    },
    array: ['MINA']
  }
})
```

3. 组 合

可以在 Mustache 内直接进行组合，构成新的对象或者数组，如下所示。

数组：

```
<!-- wxml -->
<view wx:for="{{[zero, 1, 2, 3, 4]}}"> {{item}} </view>
//page.js
Page({
  data: {
    zero: 0
  }
})
```

最终组合成数组“[0, 1, 2, 3, 4]”。

对 象：

```
<!-- wxml -->
<template is="objectCombine" data="{{for: a, bar: b}}"> </template>
//page.js
Page({
  data: {
    a: 1,
    b: 2
  }
})
```

最终组合成的对象是“{for: 1, bar: 2}”，也可以用扩展运算符将一个对象展开，如下所示：

```
<!-- wxml -->
<template is="objectCombine" data="{{...obj1, ...obj2, e: 5}}"> </template>
//page.js
Page({
  data: {
    obj1: {
      a: 1,
      b: 2
    },
    obj2: {
      c: 3,
      d: 4
    }
```

```
    }
})
```

最终组合成的对象是“{a：1，b：2，c：3，d：4，e：5}”。如果对象的 key 和 value 相同，也可以间接地表达，如下所示：

```
<!-- wxml -->
<template is="objectCombine" data="{{foo, bar}}"></template>
//page.js
Page({
  data: {
    foo: 'my-foo',
    bar: 'my-bar'
  }
})
```

最终组合成的对象是“{foo：'my－foo'，bar:'my－bar'}”。

注意：上述方式可以随意组合，但是如存在变量名相同的情况，后边的变量会覆盖前面，如下所示：

```
<!-- wxml -->
<template is="objectCombine" data="{{...obj1, ...obj2, a, c: 6}}"></template>
//page.js
Page({
  data: {
    obj1: {
      a: 1,
      b: 2
    },
    obj2: {
      b: 3,
      c: 4
    },
    a: 5
  }
})
```

最终组合成的对象是“{a：5，b：3，c：6}”。

注意：花括号和引号之间如果有空格，将最终被解析成字符串。

```
<view wx:for="{{[1,2,3]}} ">
  {{item}}</view>
```

等同于：

```
<view wx:for="{{[1,2,3] + ''}}">
  {{item}}</view>
```

3.2.3 条件渲染

条件渲染相当于在小程序页面上输出条件语句。

(1) wx:if

在框架中,使用 wx:if="{{condition}}" 来判断是否需要渲染该代码块:

```
<view wx:if="{{condition}}"> True </view>
```

也可以用 wx:elif 和 wx:else 来添加一个 else 块:

```
<view wx:if="{{length > 5}}"> 1 </view>
<view wx:elif="{{length > 2}}"> 2 </view>
<view wx:else> 3 </view>
```

(2) block wx:if

因为 wx:if 是一个控制属性,需要将它添加到一个标签上。如果要一次性判断多个组件标签,可以使用一个 <block/> 标签将多个组件包装起来,并在上边使用 wx:if 控制属性。

```
<block wx:if="{{true}}">
  <view> view1 </view>
  <view> view2 </view>
</block>
```

注意:"<block/>"并不是一个组件,它仅仅是一个包装元素,不会在页面中做任何渲染,只接受控制属性。

(3) wx:if 与 hidden 的比较

因为 wx:if 之中的模板也可能包含数据绑定,所以当 wx:if 的条件值切换时,框架有一个局部渲染的过程,因为它会确保条件块在切换时销毁或重新渲染。同时 wx:if 也是惰性的,如果初始渲染条件为 false,框架什么也不做,在条件第一次变成真的时候才开始局部渲染。相比之下,hidden 就简单多了,组件始终会被渲染,只是简单的控制显示与隐藏。一般来说,wx:if 有更高的切换消耗,而 hidden 有更高的初始渲染消耗。因此,如果在需要频繁切换的情景下,用 hidden 更好,如果在运行时条件不大可能改变则 wx:if 较好。

3.2.4 列表渲染

列表渲染是将一个数组内的所有数据依次展示在界面上,相当于在小程序页面上输出循环语句。循环迭代在程序中会经常用到,常用场景有:文章列表和商品列表等，建议读者重点掌握该知识点。

（1）wx:for

在组件上使用 wx:for 控制属性绑定一个数组，即可使用数组中各项的数据重复渲染该组件。数组的当前项的下标变量名默认为“index”，数组当前项的变量名默认为“item”，代码示例如下：

```
<!-- wxml -->
<view wx:for="{{array}}">
  {{index}}: {{item.message}}
</view>
//page.js
Page({
  data: {
    array: [{
      message: 'foo',
    }, {
      message: 'bar'
    }]
  }
})
```

使用“wx:for-item”可以指定数组当前元素的变量名，使用“wx:for-index”可以指定数组当前下标的变量名：

```
<view wx:for="{{array}}" wx:for-index="idx" wx:for-item="itemName">
  {{idx}}: {{itemName.message}}
</view>
```

wx:for 也可以嵌套，下面是一个九九乘法表：

```
<view wx:for="{{[1, 2, 3, 4, 5, 6, 7, 8, 9]}}" wx:for-item="i">
  <view wx:for="{{[1, 2, 3, 4, 5, 6, 7, 8, 9]}}" wx:for-item="j">
    <view wx:if="{{i <= j}}">
      {{i}} * {{j}} = {{i * j}}
    </view>
  </view>
</view>
```

（2）block wx:for

类似 block wx:if，也可以将 wx:for 用在 <block/> 标签上，以便渲染一个包含多节点的结构块。例如：

```
<block wx:for="{{[1, 2, 3]}}">
  <view> {{index}}: </view>
```

```
<view> {{item}} </view>
</block>
```

(3) wx:key

如果列表中项目的位置会动态改变或者有新的项目添加到列表中，并且希望列表中的项目保持自己的特征和状态（如 <input/> 中的输入内容，<switch/> 的选中状态），需要使用 wx:key 来指定列表中项目的唯一标识符。

wx:key 的值以两种形式提供：

➢ 字符串，代表在 for 循环的 array 中 item 的某个 property，该 property 的值需是列表中唯一的字符串或数字，且不能动态改变。

➢ 保留关键字“*this”代表在 for 循环中的 item 本身，这种表示需要 item 本身是一个唯一的字符串或者数字，如：当数据改变触发渲染层重新渲染的时候，会校正带有 key 的组件，框架会确保他们被重新排序，而不是重新创建，以确保使组件保持自身的状态，并且提高列表渲染时的效率。

需要注意的是，如不提供 wx:key，则会报一个 warning，如果明确知道该列表是静态的，或者不必关注其顺序，可以选择忽略。示例代码如下：

```
<!-- wxml -->
<switch wx:for="{{objectArray}}" wx:key="unique" style="display: block;"> {{item.id}} </switch>
<button bindtap="switch"> Switch </button> <button bindtap="addToFront"> Add to the front </button>
<switch wx:for="{{numberArray}}" wx:key="*this" style="display: block;"> {{item}} </switch>
<button bindtap="addNumberToFront"> Add to the front </button>

//page.js
Page({
  data: {
    objectArray: [
      {id: 5, unique: 'unique_5'},
      {id: 4, unique: 'unique_4'},
      {id: 3, unique: 'unique_3'},
      {id: 2, unique: 'unique_2'},
      {id: 1, unique: 'unique_1'},
      {id: 0, unique: 'unique_0'},
    ],
    numberArray: [1, 2, 3, 4]
  },
  switch: function(e) {
```

```
    const length = this.data.objectArray.length
    for (let i = 0; i < length; ++i) {
      const x = Math.floor(Math.random() * length)
      const y = Math.floor(Math.random() * length)
      const temp = this.data.objectArray[x]
      this.data.objectArray[x] = this.data.objectArray[y]
      this.data.objectArray[y] = temp
    }
    this.setData({
      objectArray: this.data.objectArray
    })
  },
  addToFront: function(e) {
    const length = this.data.objectArray.length
    this.data.objectArray = [{id: length, unique: 'unique_' + length}].concat(this.
data.objectArray)
    this.setData({
      objectArray: this.data.objectArray
    })
  },
  addNumberToFront: function(e){
    this.data.numberArray = [ this.data.numberArray.length + 1 ].concat(this.data.
numberArray)
    this.setData({
      numberArray: this.data.numberArray
    })
  }
})
```

注意：当 wx:for 的值为字符串时，会将字符串解析成字符串数组；花括号和引号之间如果有空格，将最终被解析成为字符串。

3.2.5 模板、引用和事件

1. 模 板

WXML 提供了模板(template)，可以在模板中定义代码片段，然后在不同的地方调用。定义模板，使用 name 属性，作为模板的名字。然后在 < template/ > 内定义代码片段，如：

```
<!--
  index: int
  msg: string
```

```
  time: string
-->
<template name="msgItem">
  <view>
    <text> {{index}}: {{msg}} </text>
    <text> Time: {{time}} </text>
  </view>
</template>
```

使用模板，使用 is 属性，声明需要使用的模板，然后将模板所需要的 data 传入，如：

```
<!-- wxml -->
<template is="msgItem" data="{{...item}}"/>
//page.js
Page({
  data: {
    item: {
      index: 0,
      msg: 'this is a template',
      time: '2016-09-15'
    }
  }
})
```

is 属性可以使用 Mustache 语法来动态地决定具体需要渲染哪个模板：

```
<template name="odd">
<view> odd </view>
</template>
<template name="even">
<view> even </view>
</template>

<block wx:for="{{[1, 2, 3, 4, 5]}}">
   <template is="{{item % 2 == 0 ? 'even' : 'odd'}}"/>
</block>
```

模板拥有自己的作用域，只能使用 data 传入的数据以及模板定义文件中定义的 <wxs /> 模块。

2. 引　用

WXML 提供两种文件引用方式：import 和 include。

import 可以在该文件中使用目标文件定义的 template。如，在 item.wxml 中定义了一个叫 item 的 template：

```
<!-- item.wxml -->
<template name="item">
<text>{{text}}</text>
</template>
```

在 index.wxml 中引用了 item.wxml，就可以使用 item 模板：

```
<import src="item.wxml"/>
<template is="item" data="{{text:'forbar'}}"/>
```

import 有作用域的概念，即只会 import 目标文件中定义的 template，而不会 import 目标文件 import 的 template。如：C import B，B import A，在 C 中可以使用 B 定义的 template，在 B 中可以使用 A 定义的 template，但是 C 不能使用 A 定义的 template。

```
<!-- A.wxml -->
<template name="A">
  <text> A template </text>
</template>

<!-- B.wxml -->
<import src="a.wxml"/>
<template name="B">
  <text> B template </text>
</template>

<!-- C.wxml -->
<import src="b.wxml"/>
<template is="A"/>    <!-- Error! Can not use tempalte when not import A. -->
<template is="B"/>
```

include 可以将目标文件中除了 <template/> <wxs/> 外的整个代码引入，相当于拷贝到 include 位置，如：

```
<!-- index.wxml -->
<include src="header.wxml"/>
<view> body </view>
<include src="footer.wxml"/>

<!-- header.wxml -->
<view> header </view>

<!-- footer.wxml -->
```

```
< view > footer < /view >
```

3. 事　件

事件是视图层到逻辑层的通讯方式，它可以将用户的行为反馈到逻辑层进行处理。事件可以绑定在组件上，当达到触发事件，就会执行逻辑层中对应的事件处理函数。事件对象可以携带额外信息，如 id、dataset 和 touches。

事件的使用方式。在组件中绑定一个事件处理函数，如 bindtap，当用户点击该组件时会在该页面对应的 Page 中找到相应的事件处理函数。如：

```
< view id = "tapTest" data - hi = "WeChat" bindtap = "tapName" > Click me! < /view >
```

在相应的 Page 定义中写上相应的事件处理函数，参数是 event，如：

```
Page({
  tapName: function(event) {
    console.log(event)
  }
})
```

可以看到 log 出来的信息大致如下：

```
{
"type":"tap",
"timeStamp":895,
"target": {
"id": "tapTest",
"dataset":  {
"hi":"WeChat"
    }
  },
"currentTarget":  {
"id": "tapTest",
"dataset": {
"hi":"WeChat"
    }
  },
"detail": {
"x":53,
"y":14
  },
"touches":[{
"identifier":0,
"pageX":53,
"pageY":14,
```

```
"clientX":53,
"clientY":14
  }],
"changedTouches":[{
"identifier":0,
"pageX":53,
"pageY":14,
"clientX":53,
"clientY":14
  }]
}
```

事件分为冒泡事件和非冒泡事件：

- 冒泡事件：当一个组件上的事件被触发后，该事件会向父节点传递；
- 非冒泡事件，当一个组件上的事件被触发后，该事件不会向父节点传递。

WXML 的冒泡事件列表，如表 3-6 所列。

表 3-6　WXML 冒泡事件列表

类型	触发条件	最低版本
touchstart	手指触摸动作开始	
touchmove	手指触摸后移动	
touchcancel	手指触摸动作被打断，如来电提醒，弹窗	
touchend	手指触摸动作结束	
tap	手指触摸后马上离开	
longpress	手指触摸后，超过350ms再离开，如果指定了事件回调函数并触发了这个事件，tap事件将不被触发	1.5.0
longtap	手指触摸后，超过350ms再离开（推荐使用longpress事件代替）	
transitionend	会在 WXSS transition 或 wx.createAnimation 动画结束后触发	
animationstart	会在一个 WXSS animation 动画开始时触发	
animationiteration	会在一个 WXSS animation 一次迭代结束时触发	
animationend	会在一个 WXSS animation 动画完成时触发	
touchforcechange	在支持 3D Touch 的 iPhone 设备，重按时会触发	1.9.90

除上表之外，其他组件自定义事件如无特殊声明都是非冒泡事件，如 < form/ > 的 submit 事件，< input/ > 的 input 事件，< scroll-view/ > 的 scroll 事件(详见各个组件)。

关于事件绑定和冒泡。事件绑定的写法同组件的属性一样，以 key、value 的形式展现。

key 以 bind 或 catch 开头，然后跟上事件的类型，如 bindtap、catchtouchstart。自基础库版本 1.5.0 起，在非原生组件中，bind 和 catch 后可以紧跟一个冒号，其含义不变，如 bind:tap、catch:touchstart。value 是一个字符串，需要在对应的 Page 中定义同名函数，不然当触发事件的时候会报错。bind 事件绑定不会阻止冒泡事件向上冒泡，catch 事件绑定可以阻止冒泡事件向上冒泡。

如在下面这个例子中，点击 inner view 会先后调用 handleTap3 和 handleTap2(因为 tap 事件会冒泡到 middle view，而 middle view 阻止了 tap 事件冒泡，不再向父节点传递)，点击 middle view 会触发 handleTap2，点击 outer view 会触发 handleTap1。

```
< view id = "outer" bindtap = "handleTap1" >
  outer view
 < view id = "middle" catchtap = "handleTap2" >
    middle view
 < view id = "inner" bindtap = "handleTap3" >
      inner view
 < /view >
 < /view >
 < /view >
```

3.2.6 WXML 与 HTML 的区别

WXML 与 HTML 的区别主要有以下几点：

- WXML 无 DOM 树，小程序运行在 JS Core 内，没有 DOM 树和 window 对象，没有办法使用相关 API；
- 组件封装不同，WXML 对组件进行了重新封装，为后续性能优化提供了支持，也能避免开发者写出低质低效的代码；
- 开发工具有限制，WXML 只能在微信开发者工具中进行预览，而 HTML 可以在浏览器内预览。

3.3 WXSS：小程序版 CSS

WXSS(WeiXin Style Sheets)是一套样式语言，用于描述 WXML 的组件样式。

WXSS 用来决定 WXML 的组件应该怎么显示。为了适应广大的前端开发者，WXSS 具有 CSS 大部分特性。同时为了更适合开发微信小程序，WXSS 对 CSS 进行了扩充以及修改。

3.3.1 选择器与优先级

借助选择器，我们可以设置不同类型、不同 ID、不同组件的样式，小程序目前支持的选择器有以下几种，如表 3－7 所列。需要注意的是，WXSS 不支持级联选择器。

表 3－7 WXSS 选择器

选择器	样例	样例描述
.class	.intro	选择所有拥有 class="intro" 的组件
#id	#firstname	选择拥有 id="firstname" 的组件
element	view	选择所有 view 组件
element, element	view, checkbox	选择所有文档的 view 组件和所有的 checkbox 组件
::after	view::after	在 view 组件后边插入内容
::before	view::before	在 view 组件前边插入内容

选择器的用法如下：

- 标签内样式：动态样式，如特定场景下的属性设置；
- ID 选择器：针对某个特定的组件进行属性的设置；
- 类选择器：针对某一类组件进行属性的设定；
- 标签选择器：针对全局的某一种组件进行属性的设定。

选择器的优先级决定了哪个选择器被优先选择，WXSS 选择器的优先级如图 3－8 所示。

优先级优先规则为：权重越高，优先级越高；相同权重以后面出现的选择器为最后的规则。权重计算举例如下：

```
view.content
```

权重为 11(view 权重为 1，.content 权重为 10)；

```
view#content
```

权重为 101(view 权重为 1，#content 权重为 100)。

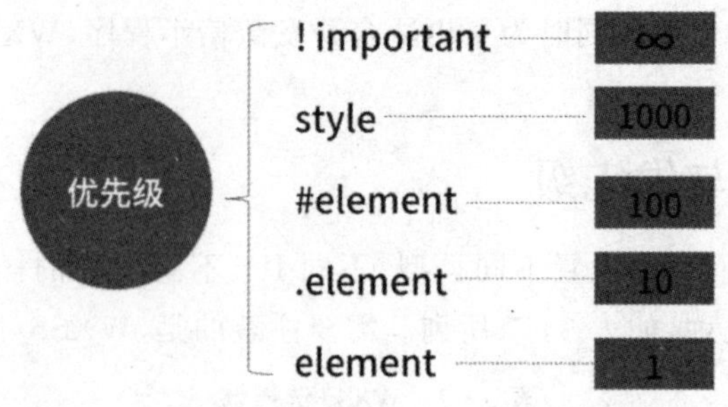

图 3-8　WXSS 选择器优先级

3.3.2　盒子模型

小程序的盒子模型采用的是 CSS3 规范所提出的弹性盒布局模型。Flexbox 布局方式相比传统盒模型而言，可以快速实现诸如垂直居中、弹性宽度和高度等场景，以及更多排版需求，比如：竖排元素或元素左侧/右侧竖向对齐。

在进行 Flexbox 布局时，核心的属性包括容器和项目，其中容器为 Flexbox 中外部的容器，项目为 flexbox 中内部的项目，如图 3-9 所示。

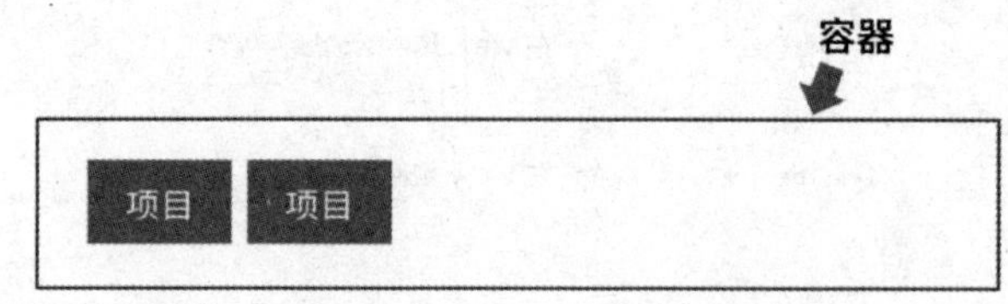

图 3-9　Flexbox 核心属性

通过设置 flex 容器的属性，可以设置容器内各项目的样式，Flex 容器属性如下：

- flex-direction：项目元素排列方向；
- flex-wrap：项目元素排列方式；
- justify-content：项目在主轴上的对齐方式；
- align-items：项目在交叉轴上的对齐方式；
- align-content：多行项目的排列方式。

通过设置 flex 项目属性，我们可以控制各项目自己的属性，Flex 项目属性如下：

- order：项目的排列顺序；
- flex-grow：项目的放大比例；
- flex-shrink：项目的缩小比例；
- flex-basis：项目在主轴上的空间；
- align-self：项目的对齐方式。

3.3.3 内联样式

框架组件上支持使用 style、class 属性来控制组件的样式。

- class：用于指定样式规则，其属性值是样式规则中类选择器名(样式类名)的集合，样式类名不需要带上“.”，样式类名之间用空格分隔。示例如下：

```
< view class = "normal_view" / >
```

- style：接收动态的样式，在运行时会进行解析，请尽量避免将静态的样式写进 style 中(静态的样式统一写到 class 中)，以免影响渲染速度。代码示例如下：

```
< ! -- wxml -- >
< view wx:for = "{{data}}" class = "block" style = "{{item.style}}" > Block  {{index}}
< /view >

//pages/style/style.wxss
.block{
    background - color: #eee;
    margin: 10rpx 0rpx;
}

// pages/style/style.js
Page({

  /**
   * 页面的初始数据
   */
  data: {
      data:[
          {
"title":"Hi, WXSS",
"style":"color:#ff00ff"
          },
          {
"title": "Hello, WXSS",
"style": "color:#00ff00"
          }
      ]
  },

})
```

页面效果图，如图 3 - 10 所示。

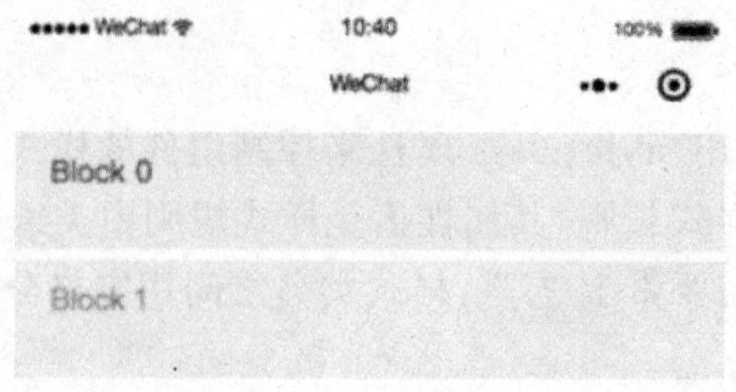

图 3-10 内联样式

3.3.4 尺寸单位与样式导入

小程序引入 rpx(responsive pixel)作为 WXSS 的尺寸单位，rpx 可以根据屏幕宽度进行自适应，规定屏幕宽为 750rpx。如在 iPhone 6 上，屏幕宽度为 375px，共有 750 个物理像素，则 750rpx = 375px = 750 物理像素，1rpx = 0.5px = 1 物理像素，尺寸换算表如表 3-8 所列。

表 3-8 尺寸换算

设备	rpx换算px (屏幕宽度/750)	px换算rpx (750/屏幕宽度)
iPhone5	1rpx = 0.42px	1px = 2.34rpx
iPhone6	1rpx = 0.5px	1px = 2rpx
iPhone6 Plus	1rpx = 0.552px	1px = 1.81rpx

建议在开发微信小程序时，设计师用 iPhone 6 作为视觉稿的标准。需要注意的是，在较小的屏幕上不可避免地会有一些毛刺，请在开发时尽量避免这种情况。小程序 WXSS 使用"@import"语句可以导入外联样式表，"@import"后跟需要导入的外联样式表的相对路径，用";"表示语句结束。示例代码如下：

```
/** common.wxss **/
.small-p {
  padding:5px;
}

/** app.wxss **/
@import "common.wxss";
.middle-p {
  padding:15px;
}
```

3.3.5 WXSS 与 CSS 的区别

WXSS 与 CSS 的区别主要有以下几点：

- 引入了新的尺寸单位 rpx；
- 与选择器相比 CSS 有缩减；
- 有全局样式和局部样式的明确区分；
- 可自动优化。

3.4 WXS:小程序版 JavaScript

WXS(WeiXin Script)是小程序的一套脚本语言，结合 WXML 可以构建出页面的结构。

3.4.1 小程序 JavaScript

EMCAScript 是一种标准化的程序设计语言，而 javascript 是它的一种实现，小程序 javascript 是由 EMCAScript 和小程序框架、小程序 API 共同实现的。在开发小程序时，需要根据实际情况选择使用 WXS 还是 javascript，WXS 是专门用于 WXML 页面的，如果用户有在页面中使用 JavaScript 脚本的需求时可以使用，但是 WXS 是不能被其他 javascript 文件引用的。

需要注意的是：

- wxs 不依赖于运行时的基础库版本，可以在所有版本的小程序中运行；
- wxs 与 javascript 是不同的语言，有自己的语法，并不和 javascript 一致；
- wxs 的运行环境和其他 javascript 代码是隔离的，wxs 中不能调用其他 javascript 文件中定义的函数，也不能调用小程序提供的 API；
- wxs 函数不能作为组件的事件回调；
- 由于运行环境的差异，在 iOS 设备上，小程序内的 wxs 会比 javascript 代码快 2～20 倍；在 Android 设备上二者运行效率无差异。

以下是一些使用 WXS 的简单示例。

页面渲染：

```
<!-- wxml -->
<wxs module="m1">
var msg = "hello world";

module.exports.message = msg;
</wxs>

<view> {{m1.message}} </view>
```

页面输出：hello world。

数据处理：

```
// page.js
Page({
  data: {
    array: [1, 2, 3, 4, 5, 1, 2, 3, 4]
  }
})
<!-- wxml -->
<!-- 下面的 getMax 函数，接受一个数组，且返回数组中最大的元素的值 -->
<wxs module = "m1">
var getMax = function(array) {
  var max = undefined;
  for (var i = 0; i < array.length; ++i) {
    max = max === undefined ?
      array[i] :
      (max >= array[i] ? max : array[i]);
  }
  return max;
}

module.exports.getMax = getMax;
</wxs>

<!-- 调用 wxs 里面的 getMax 函数，参数为 page.js 里面的 array -->
<view> {{m1.getMax(array)}} </view>
```

页面输出：5。

小程序 javascript 目前可以运行在三大平台：

➢ iOS 平台、包括 iOS9/iOS10/iOS11；

➢ Android 平台；

➢ 小程序 IDE。

需要注意的是，iOS9 和 iOS10 对 ES6 不完全兼容。此处可以借助 IDE 提供的 ES6 转 ES5 功能实现转码，以确保在所有的运行环境下都能得到很好的执行，如图 3-11 所示。

3.4.2 同步和异步

下面简单介绍下 javascript 中的同步调用和异步调用，方便在以后的开发中更好的理解小程序代码。

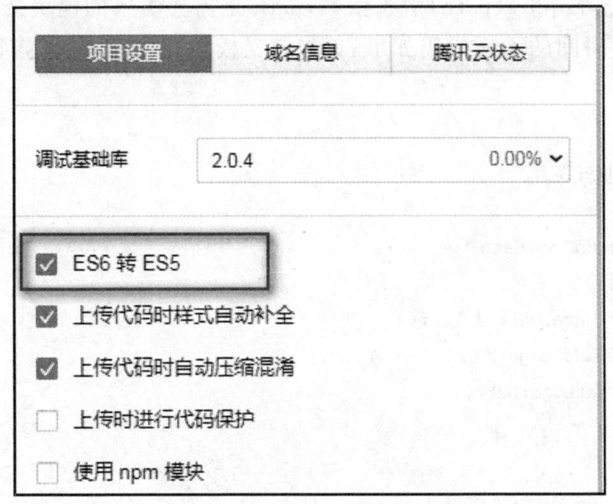

图 3-11　ES6 转 ES5

同步调用是一种阻塞式调用，调用另一段代码时，必须等待这段代码执行结束并返回结果后，代码才能继续执行下去。同步执行的情况下，应用程序每次的修改都会执行完成，较为安全，不会出现后续要用的变量为定义的情况。

同步调用的应用场景如下：

➢ 逻辑简单；

➢ 如果多处代码共享资源时需要同步；

➢ 代码中涉及对变量进行修改时需要用同步。

异步调用是一种非阻塞式调用，一段代码还未执行完成，可以执行下一段代码的逻辑，而代码执行完毕后，通过回调返回继续执行相应的逻辑，而不耽误其他代码的执行。异步执行的情况下，应用程序的修改需要在回调函数内进行。

异步调用的应用场景如下：

➢ 请求会耗费大量时间进行处理；

➢ 调用所需数据需要从外部获取。

3.4.3　模块化

“模块化”的由来：最早通过文件拆分对代码进行管理，但这仅是物理上的分离，没有真正地实现作用域的隔离，由于不知道其他文件内已存在的变量名，造成了全局冲突问题。这时我们需要一种新的组织方式，于是诞生了“模块化”。

小程序内采用的是 commonJS 来进行模块的管理和处理的，故而我们在进行封装模块时，也应该遵循 commonJS 规范。

小程序内模块接口的暴露和引入十分简单：

第一步，定义模块化代码，并通过 exports 暴露接口；

第二步，通过 require(path)引入依赖，path 是需要引入的模块文件的相对路径。

模块化处理时间的示例代码如下，首先定义模块化内容，并暴露接口：

```
//util.js
/**
 * 处理具体业务逻辑
 */
function formatTime(date) {
   //获取年月日
   var year = date.getFullYear()
   var month = date.getMonth() + 1
   var day = date.getDate()

   //获取时分秒
   var hour = date.getHours()
   var minute = date.getMinutes()
   var second = date.getSeconds();

   //格式化日期
   return [year, month, day].map(formatNumber).join('/') + ' ' + [hour, minute, sec-
ond].map(formatNumber).join(':')
}

function formatNumber(n) {
   n = n.toString()
   return n[1] ? n : '0' + n
}

/**
 * 模块化导出暴露接口
 */
module.exports = {
  formatTime: formatTime
}
```

使用该模块，代码如下：

```
//导入模块化方式
var util = require('../../utils/util.js')
Page({
  data: {
    logs: []
  },
```

```
  onLoad: function () {
    this.setData({
      logs: (wx.getStorageSync('logs') || []).map(function (log) {
        // 通过暴露的接口调用模块化方法
        return util.formatTime(new Date(log))
      })
    })
  }
})
```

小程序开发高级篇

第 4 章

小程序框架组件

微信小程序框架为开发者提供了一系列完备的 UI 组件，方便开发者快速构建小程序 UI 界面。借助这些 UI 组件开发者可以像搭积木一样快速地拼装出一栋房子的样子，这非常类似于当下建筑行业比较流行的装配式建筑，UI 组件就好比预先定制好的建筑构件，只需要按照设计图纸即可快速组装各个组件，便捷迅速地完成界面布局和渲染工作。

本章来为读者大致讲解下小程序框架的 UI 组件，由于涉及的知识点较多，此处给出一个学习方法建议：首先，读者需要对小程序框架组件分类有一个基本的认知，然后需要知道哪些是常见且重要的组件，最好能有个基本印象，最后需要知道这部分常见组件的使用场景。本章建议快速阅读，形成整体大概印象即可，以后在需要使用某具体组件的时候再来精读对应的组件知识内容。精读的时候建议将官方的详细文档还有其他技术大神所撰写的实战 DEMO 结合起来研究，最后自己再手敲代码实践这部分内容，这样慢慢的即可对小程序组件的运用达到熟能生巧的程度。

小程序框架组件思维导图，如图 4－1 所示。

4.1 视图容器组件

4.1.1 视图容器

视图容器(View)是小程序框架组件中最常见的基础组件，它的作用跟 HTML 中的 DIV 功能非常相似，用来布局 WXML 界面。view 视图容器的属性如表 4－1 所列。

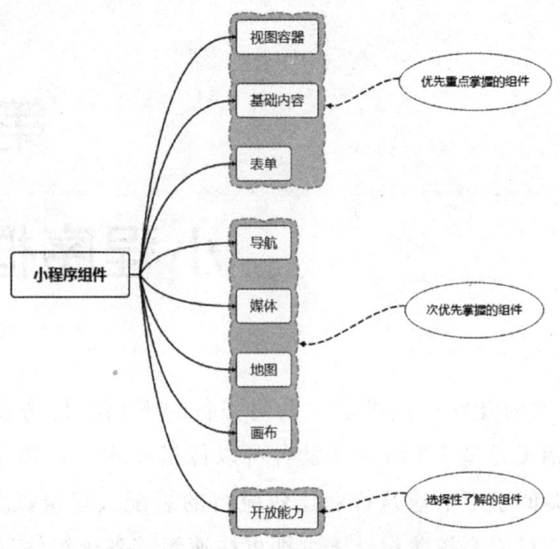

图 4-1 小程序组件思维导图

表 4-1 view 视图容器属性

属性名	类型	默认值	说明	最低版本
hover-class	String	none	指定按下去的样式类。当 hover-class="none" 时，没有点击态效果	
hover-stop-propagation	Boolean	false	指定是否阻止本节点的祖先节点出现点击态	1.5.0
hover-start-time	Number	50	按住后多久出现点击态，单位 ms	
hover-stay-time	Number	400	手指松开后点击态保留时间，单位 ms	

代码示例如下：

```
< view class = "section" >
< view class = "section__title" > flex - direction: row < /view >
< view class = "flex - wrp" style = "flex - direction:row;" >
< view class = "flex - item bc_green" > 1 < /view >
< view class = "flex - item bc_red" > 2 < /view >
< view class = "flex - item bc_blue" > 3 < /view >
< /view >
< /view >
< view class = "section" >
< view class = "section__title" > flex - direction: column < /view >
< view class = "flex - wrp" style = "height: 300px;flex - direction:column;" >
```

```
< view class = "flex - item bc_green" > 1 < /view >
< view class = "flex - item bc_red" > 2 < /view >
< view class = "flex - item bc_blue" > 3 < /view >
< /view >
< /view >
```

示例效果图如图 4 - 2 所示。

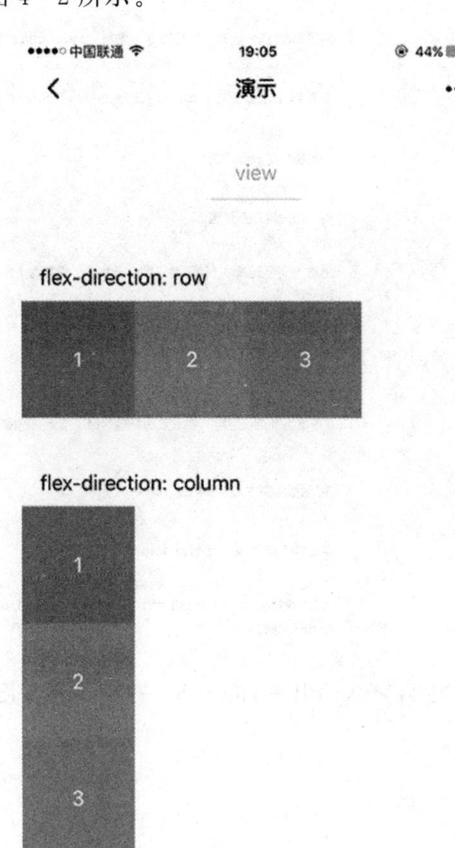

图 4 - 2　View 视图容器示意图

4.1.2　可滚动视图区域

可滚动视图区域(scroll - view),视图区域中的内容如同浏览器横向滚动或纵向滚动一样可以自由滚动。scroll - view 的属性如表 4 - 2 所列。

表 4-2 scroll-view 属性

属性名	类型	默认值	说明
scroll-x	Boolean	false	允许横向滚动
scroll-y	Boolean	false	允许纵向滚动
upper-threshold	Number	50	距顶部/左边多远时（单位px），触发 scrolltoupper 事件
lower-threshold	Number	50	距底部/右边多远时（单位px），触发 scrolltolower 事件
scroll-top	Number		设置竖向滚动条位置
scroll-left	Number		设置横向滚动条位置
scroll-into-view	String		值应为某子元素id（id不能以数字开头）。设置哪个方向可滚动，则在哪个方向滚动到该元素
scroll-with-animation	Boolean	false	在设置滚动条位置时使用动画过渡
enable-back-to-top	Boolean	false	iOS单击顶部状态栏、安卓双击标题栏时，滚动条返回顶部，只支持竖向
bindscrolltoupper	EventHandle		滚动到顶部/左边，会触发 scrolltoupper 事件
bindscrolltolower	EventHandle		滚动到底部/右边，会触发 scrolltolower 事件
bindscroll	EventHandle		滚动时触发，event.detail = {scrollLeft, scrollTop, scrollHeight, scrollWidth, deltaX, deltaY}

使用竖向滚动时，需要给 <scroll-view/> 一个固定高度，通过 WXSS 设置 height。

示例代码如下：

```
//wxml 代码
<view class="section">
<view class="section__title">vertical scroll</view>
<scroll-view scroll-y style="height: 200px;" bindscrolltoupper="upper" bindscrolltolower="lower" bindscroll="scroll" scroll-into-view="{{toView}}" scroll-top="{{scrollTop}}">
<view id="green" class="scroll-view-item bc_green"></view>
<view id="red"  class="scroll-view-item bc_red"></view>
<view id="yellow" class="scroll-view-item bc_yellow"></view>
<view id="blue" class="scroll-view-item bc_blue"></view>
</scroll-view>

<view class="btn-area">
```

```
< button size = "mini" bindtap = "tap" > click me to scroll into view < /button >
< button size = "mini" bindtap = "tapMove" > click me to scroll < /button >
< /view >
< /view >
< view class = "section section_gap" >
< view class = "section__title" > horizontal scroll < /view >
< scroll - view class = "scroll - view_H" scroll - x style = "width: 100 % " >
< view id = "green" class = "scroll - view - item_H bc_green" > < /view >
< view id = "red"   class = "scroll - view - item_H bc_red" > < /view >
< view id = "yellow" class = "scroll - view - item_H bc_yellow" > < /view >
< view id = "blue" class = "scroll - view - item_H bc_blue" > < /view >
< /scroll - view >
< /view >

//js 代码
var order = ['red', 'yellow', 'blue', 'green', 'red']
Page({
  data: {
    toView: 'red',
    scrollTop: 100
  },
  upper: function(e) {
    console.log(e)
  },
  lower: function(e) {
    console.log(e)
  },
  scroll: function(e) {
    console.log(e)
  },
  tap: function(e) {
    for (var i = 0; i < order.length; + + i) {
      if (order[i] === this.data.toView) {
        this.setData({
          toView: order[i + 1]
        })
        break
      }
    }
  },
  tapMove: function(e) {
    this.setData({
```

```
        scrollTop: this.data.scrollTop + 10
      })
    }
  })
```

示例效果图如图 4 - 3 所示。

图 4 - 3　scroll - view 示例图

scroll - view 使用场景:可以用来实现锚点滑动等效果。需要监听滚动的时候也可以用 scroll - view。

4.1.3　滑块视图容器

滑块视图容器(swiper),用来在指定区域内切换内容显示,一般用来控制一组轮播图或页签内容切换。在这个 swiper 容器里需要存放的是它的组件,也就是 swiper-item。swiper 组件的属性列表如表 4 - 3 所列。

表 4-3 swiper 组件属性

属性名	类型	默认值	说明	最低版本
indicator-dots	Boolean	false	是否显示面板指示点	
indicator-color	Color	rgba(0, 0, 0, .3)	指示点颜色	1.1.0
indicator-active-color	Color	#000000	当前选中的指示点颜色	1.1.0
autoplay	Boolean	false	是否自动切换	
current	Number	0	当前所在滑块的 index	
current-item-id	String	""	当前所在滑块的 item-id ，不能与 current 被同时指定	1.9.0
interval	Number	5000	自动切换时间间隔	
duration	Number	500	滑动动画时长	
circular	Boolean	false	是否采用衔接滑动	
vertical	Boolean	false	滑动方向是否为纵向	
previous-margin	String	"0px"	前边距，可用于露出前一项的一小部分，接受 px 和 rpx 值	1.9.0
next-margin	String	"0px"	后边距，可用于露出后一项的一小部分，接受 px 和 rpx 值	1.9.0
display-multiple-items	Number	1	同时显示的滑块数量	1.9.0
skip-hidden-item-layout	Boolean	false	是否跳过未显示的滑块布局，设为 true 可优化复杂情况下的滑动性能，但会丢失隐藏状态滑块的布局信息	1.9.0
bindchange	EventHandle		current 改变时会触发 change 事件，event.detail = {current: current, source: source}	
bindanimationfinish	EventHandle		动画结束时会触发 animationfinish 事件，event.detail 同上	1.9.0

swiper-item 仅可放置在 <swiper/> 组件中，宽、高自动设置为 100%。

swiper 组件代码示例如下：

```
//wxml 代码
<swiper indicator-dots="{{indicatorDots}}"
  autoplay="{{autoplay}}" interval="{{interval}}" duration="{{duration}}">
<block wx:for="{{imgUrls}}">
<swiper-item>
<image src="{{item}}" class="slide-image" width="355" height="150"/>
```

```
    </swiper-item>
    </block>
    </swiper>
    <button bindtap="changeIndicatorDots"> indicator-dots </button>
    <button bindtap="changeAutoplay"> autoplay </button>
    <slider bindchange="intervalChange" show-value min="500" max="2000"/> inter-
val
    <slider bindchange="durationChange" show-value min="1000" max="10000"/> dura-
tion

    //js 代码
    Page({
      data: {
        imgUrls: [
          'http://img02.tooopen.com/images/20150928/tooopen_sy_143912755726.jpg',
          'http://img06.tooopen.com/images/20160818/tooopen_sy_175866434296.jpg',
          'http://img06.tooopen.com/images/20160818/tooopen_sy_175833047715.jpg'
        ],
        indicatorDots: false,
        autoplay: false,
        interval: 5000,
        duration: 1000
      },
      changeIndicatorDots: function(e) {
        this.setData({
          indicatorDots: ! this.data.indicatorDots
        })
      },
      changeAutoplay: function(e) {
        this.setData({
          autoplay: ! this.data.autoplay
        })
      },
      intervalChange: function(e) {
        this.setData({
          interval: e.detail.value
        })
      },
```

```
  durationChange: function(e) {
    this.setData({
      duration: e.detail.value
    })
  }
})
```

示例效果图如图 4－4 所示。

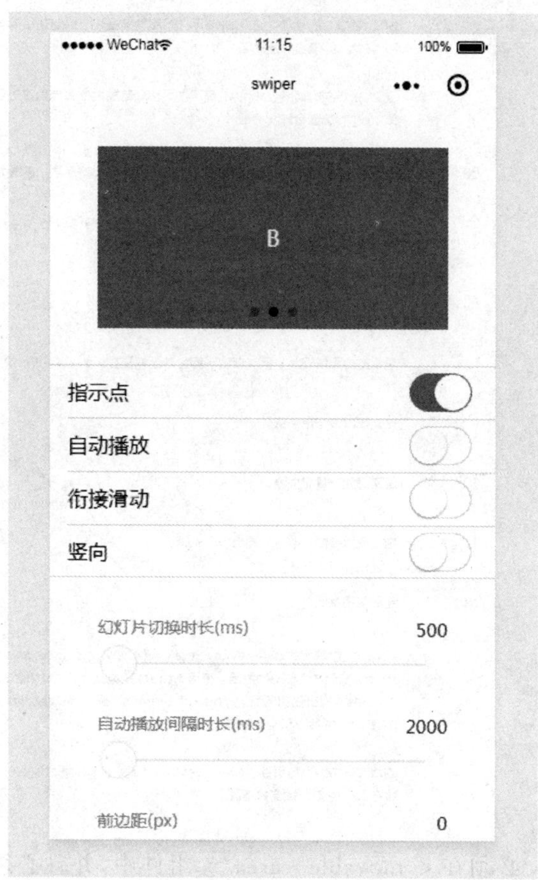

图 4－4　swiper 示例图

4.1.4　可移动视图容器

可移动的视图容器(movable－view),可以在页面中拖拽滑动。该组件从基础库 1.2.0 开始支持,低版本需做兼容处理。该组件的属性如表 4－4 所列。

表 4-4 movable-view 组件属性

属性名	类型	默认值	说明	最低版本
direction	String	none	movable-view的移动方向，属性值有all、vertical、horizontal、none	
inertia	Boolean	false	movable-view是否带有惯性	
out-of-bounds	Boolean	false	超过可移动区域后，movable-view是否还可以移动	
x	Number / String		定义x轴方向的偏移，如果x的值不在可移动范围内，会自动移动到可移动范围；改变x的值会触发动画	
y	Number / String		定义y轴方向的偏移，如果y的值不在可移动范围内，会自动移动到可移动范围；改变y的值会触发动画	
damping	Number	20	阻尼系数，用于控制x或y改变时的动画和过界回弹的动画，值越大移动越快	
friction	Number	2	摩擦系数，用于控制惯性滑动的动画，值越大摩擦力越大，滑动越快停止；其值必须大于0，否则会被设置成默认值	
disabled	Boolean	false	是否禁用	1.9.90
scale	Boolean	false	是否支持双指缩放，默认缩放手势生效区域是在movable-view内	1.9.90
scale-min	Number	0.5	定义缩放倍数最小值	1.9.90
scale-max	Number	10	定义缩放倍数最大值	1.9.90
scale-value	Number	1	定义缩放倍数，取值范围为 0.5 - 10	1.9.90
animation	Boolean	true	是否使用动画	2.1.0
bindchange	EventHandle		拖动过程中触发的事件，event.detail = {x: x, y: y, source: source}，其中source表示产生移动的原因，值可为touch（拖动）、touch-out-of-bounds（超出移动范围）、out-of-bounds（超出移动范围后的回弹）、friction（惯性）和空字符串（setData）	1.9.90
bindscale	EventHandle		缩放过程中触发的事件，event.detail = {x: x, y: y, scale: scale}，其中x和y字段在2.1.0之后开始支持返回	1.9.90

movable-view 必须在 <movable-area/> 组件中，并且必须是直接子节点，否则不能移动。movable-area 从基础库 1.2.0 开始支持，低版本需做兼容处理。组件代码示例如下：

```
//wxml 代码
<view class="section">
<view class="section__title">movable-view 区域小于 movable-area </view>
<movable-area style="height: 200px; width: 200px; background: red;">
<movable-view style="height: 50px; width: 50px; background: blue;" x="{{x}}" y
="{{y}}" direction="all">
```

```
</movable-view>
</movable-area>
<view class="btn-area">
<button size="mini" bindtap="tap">click me to move to (30px, 30px)</button>
</view>
<view class="section__title">movable-view区域大于movable-area</view>
<movable-area style="height: 100px; width: 100px; background: red;">
<movable-view style="height: 200px; width: 200px; background: blue;" direction="all">
</movable-view>
</movable-area>
<view class="section__title">可放缩</view>
<movable-area style="height: 200px; width: 200px; background: red;" scale-area>
<movable-view style="height: 50px; width: 50px; background: blue;" direction="all" bindchange="onChange" bindscale="onScale" scale scale-min="0.5" scale-max="4" scale-value="2">
</movable-view>
</movable-area>
</view>

//js代码
Page({
  data: {
    x: 0,
    y: 0
  },
  tap: function(e) {
    this.setData({
      x: 30,
      y: 30
    });
  },
  onChange: function(e) {
    console.log(e.detail)
  },
  onScale: function(e) {
    console.log(e.detail)
  }
})
```

示例效果图如图4-5所示。

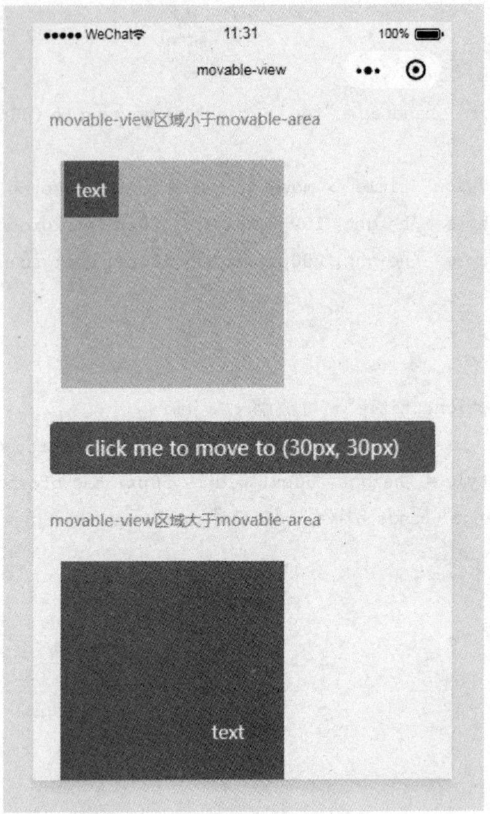

图 4-5　movable-view 示例

4.1.5　覆盖在原生组件上的文本视图

覆盖在原生组件之上的文本视图(cover-view),可覆盖的原生组件包括 map、video、canvas、camera、live-player 和 live-pusher,只支持嵌套 cover-view 和 cover-image,可在 cover-view 中使用 button。从基础库 1.4.0 开始支持 cover-view,低版本需做兼容处理,该组件属性如表 4-5 所列。

表 4-5　cover-view 组件属性

属性名	类型	默认值	说明	最低版本
scroll-top	Number / String		设置顶部滚动偏移量，仅在设置了 overflow-y: scroll 成为滚动元素后生效（单位px或rpx，默认为px）	2.1.0

4.1.6　覆盖在原生组件上的图片视图

覆盖在原生组件之上的图片视图(cover-image),可覆盖的原生组件同 cover-view,支持嵌套在 cover-view 里。从基础库 1.4.0 开始支持 cover-image,低版本需做兼容处理，该组件属性如表 4-6 所列。

表 4-6 cover-image 组件属性

属性名	类型	默认值	说明	最低版本
src	String		图标路径，支持临时路径、网络地址（1.6.0起支持）、云文件ID（2.2.3起支持）。暂不支持base64格式。	
bindload	EventHandle		图片加载成功时触发	2.1.0
binderror	EventHandle		图片加载失败时触发	2.1.0

cover-view 和 cover-image 两个组件在使用时需要注意的事项比较多，也是两个比较特殊的新组件。注意事项如下：

- 基础库 2.2.4 起支持 touch 相关事件，也可使用 hover-class 设置点击态；
- 基础库 2.1.0 起支持设置 scale rotate 的 css 样式，包括 transition 动画；
- 基础库 1.9.90 起 cover-view 支持 overflow: scroll，但不支持动态更新 overflow；
- 基础库 1.9.90 起最外层 cover-view 支持 position: fixed；
- 基础库 1.9.0 起支持插在 view 等标签下，在此之前只可嵌套在原生组件 map、video、canvas 和 camera 内，避免嵌套在其他组件内；
- 基础库 1.6.0 起支持 css transition 动画，transition-property 只支持 transform (translateX, translateY)与 opacity；
- 基础库 1.6.0 起支持 css opacity；
- 事件模型遵循冒泡模型，但不会冒泡到原生组件；
- 文本建议都套上 cover-view 标签，避免排版错误；
- 只支持基本的定位、布局、文本样式，不支持设置单边的 border、background-image、shadow 和 overflow: visible 等；
- 建议子节点不要溢出父节点；
- 默认设置的样式有：white-space: nowrap、line-height: 1.2 和 display: block。
- 自定义组件嵌套 cover-view 时，自定义组件的 slot 及其父节点暂不支持通过 wx:if 控制显隐，否则会导致 cover-view 不显示。

这两个组件的示例代码如下：

```
//wxml 代码
<video id = "myVideo"
src = "http://wxsnsdy.tc.qq.com/test video controls = "{{false}}" event - model = "bub-
ble" >
<cover - view class = "controls" >
<cover - view class = "play" bindtap = "play" >
```

```
< cover - image class = "img" src = "/path/to/icon_play" / >
< /cover - view >
< cover - view class = "pause" bindtap = "pause" >
< cover - image class = "img" src = "/path/to/icon_pause" / >
< /cover - view >
< cover - view class = "time" > 00:00 < /cover - view >
< /cover - view >
< /video >

//wxss 代码
.controls {
  position: relative;
  top: 50 % ;
  height: 50px;
  margin - top: - 25px;
  display: flex;
}
.play,.pause,.time {
  flex: 1;
  height: 100 % ;
}
.time {
  text - align: center;
  background - color: rgba(0, 0, 0, .5);
  color: white;
  line - height: 50px;
}
.img {
  width: 40px;
  height: 40px;
  margin: 5px auto;
}

//js 代码
Page({
  onReady() {
    this.videoCtx = wx.createVideoContext('myVideo')
  },
  play() {
    this.videoCtx.play()
  },
  pause() {
```

```
    this.videoCtx.pause()
  }
})
```

示例效果图如图 4－6 所示。

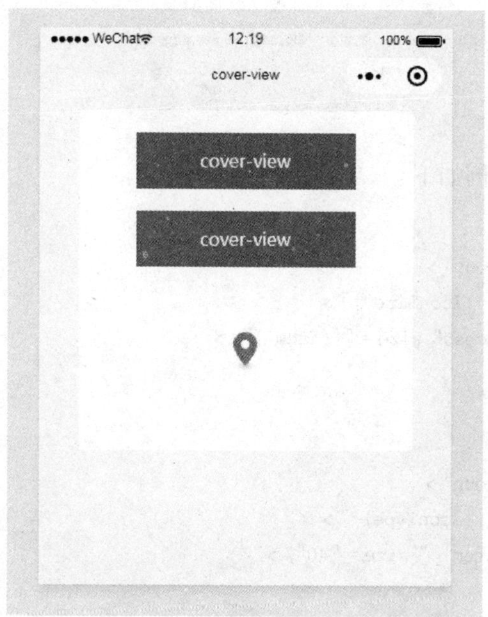

图 4－6 cover－view 和 cover－image 示例图

组件使用场景：用 cover－view 和 cover－image 可以在原生的 video 组件上实现一个自定义的播放控件。已经支持实现的功能包括播放/暂停、全屏/退出全屏、显示播放进度/拖拽播放、设置播放倍速等。

4.2 基础内容组件

4.2.1 图 标

图标(icon)，其意义在于给予用户见名知义的视觉用户体验。该组件的属性如表 4－7 所列。

表 4-7 icon 图标

属性名	类型	默认值	说明
type	String		icon的类型，有效值：success、success_no_circle、info、warn、waiting、cancel、download、search 和 clear
size	Number / String	23px	icon的大小，单位px或rpx，默认为px
color	Color		icon的颜色，同css的color

该组件的代码示例如下：

```
//wxml 代码
<view class="group">
<block wx:for="{{iconSize}}">
<icon type="success" size="{{item}}"/>
</block>
</view>

<view class="group">
<block wx:for="{{iconType}}">
<icon type="{{item}}" size="40"/>
</block>
</view>

<view class="group">
<block wx:for="{{iconColor}}">
<icon type="success" size="40" color="{{item}}"/>
</block>
</view>

//js 代码
Page({
  data: {
    iconSize: [20, 30, 40, 50, 60, 70],
    iconColor: [
      'red', 'orange', 'yellow', 'green', 'rgb(0,255,255)', 'blue', 'purple'
    ],
    iconType: [
      'success', 'success_no_circle', 'info', 'warn', 'waiting', 'cancel', 'download', 'search', 'clear'
    ]
  }
```

```
})
```

示例效果图如图 4－7 所示。

图 4－7　图标组件示例

4.2.2 文　本

文本组件(text),是一个比较常用的基础内容组件,支持转义符“\”。< text > < /text >标签和 HTML 中的< span >标签基本类似。需要注意的是,< text/ >组件内只支持 < text/ > 嵌套,再就是除了文本节点以外的其他节点都无法长按选中。该组件的属性如表 4－8 所列。

表 4－8　text 组件属性

属性名	类型	默认值	说明	最低版本
selectable	Boolean	false	文本是否可选	1.1.0
space	String	false	显示连续空格	1.4.0
decode	Boolean	false	是否解码	1.4.0

代码示例如下:

```
//wxml 代码
 < view class = "btn - area" >
 < view class = "body - view" >
 < text > {{text}} < /text >
 < button bindtap = "add" > add line < /button >
 < button bindtap = "remove" > remove line < /button >
```

```
  </view>
</view>

//js 代码
var initData = 'this is first line\nthis is second line'
var extraLine = [];
Page({
  data: {
    text: initData
  },
  add: function(e) {
    extraLine.push('other line')
    this.setData({
      text: initData + '\n' + extraLine.join('\n')
    })
  },
  remove: function(e) {
    if (extraLine.length > 0) {
      extraLine.pop()
      this.setData({
        text: initData + '\n' + extraLine.join('\n')
      })
    }
  }
})
```

组件效果图如图 4－8 所示。

图 4－8　text 组件示例图

4.2.3 富文本

富文本(rich - text)是一个新增的微信小程序组件，从基础库 1.4.0 开始，低版本需要做兼容处理。这个组件一般会结合富文本编辑器来配合使用，该组件的属性如表 4 - 9 所列。

表 4 - 9　rich - text 属性

属性	类型	默认值	说明	最低版本
nodes	Array / String	[]	HTML String / 节点列表	1.4.0

示例代码如下：

```
<!-- rich-text.wxml -->
<rich-text nodes="{{nodes}}" bindtap="tap"></rich-text>

// rich-text.js
Page({
  data: {
    nodes: [{
      name: 'div',
      attrs: {
        class: 'div_class',
        style: 'line-height: 60px; color: red;'
      },
      children: [{
        type: 'text',
        text: 'Hello World! '
      }]
    }]
  },
  tap() {
    console.log('tap')
  }
})
```

示例效果图如图 4 - 9 所示。

图 4-9 rich-text 示例图

4.2.4 进度条

进度条(progress)是一种使用户查看播放进度、完成进度比较直观的组件,提高了用户体验,该组件的属性如表 4-10 所列。

表 4-10 progress 属性

属性名	类型	默认值	说明	最低版本
percent	Float	无	百分比0~100	
show-info	Boolean	false	在进度条右侧显示百分比	
border-radius	Number / String	0	圆角大小,单位px或rpx,默认为px	2.3.1
font-size	Number / String	16	右侧百分比字体大小,单位px或rpx,默认为px	2.3.1
stroke-width	Number / String	6	进度条线的宽度,单位px或rpx,默认为px	
color	Color	#09BB07	进度条颜色(请使用 activeColor)	
activeColor	Color		已选择的进度条的颜色	
backgroundColor	Color		未选择的进度条的颜色	
active	Boolean	false	进度条从左往右的动画	
active-mode	String	backwards	backwards: 动画从头播;forwards:动画从上次结束点接着播	1.7.0

代码示例如下：

```
//wxml 代码
<progress percent="20" show-info />
<progress percent="40" stroke-width="12" />
<progress percent="60" color="pink" />
<progress percent="80" active />
```

示例效果图如图 4-10 所示。

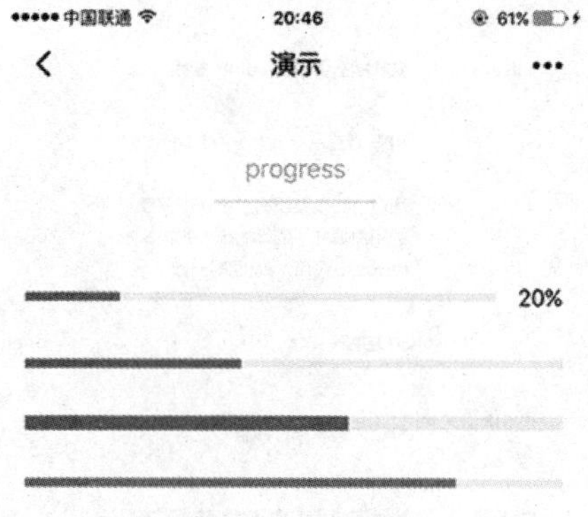

图 4-10 进度条示例

4.3 表单组件

表单组件由一系列组件组成，包括：button、checkbox、form、input、label、picker、picker-view、radio、slider、switch 和 textarea。熟悉 WEB 开发的朋友应该知道表单在 WEB 开发中应用非常广泛，表单主要用来构建与用户进行的交互，因而建议大家务必重点掌握表单组件的应用。

4.3.1 按 钮

按钮(button)提供了 3 种类型的按钮：默认类型(属性 type 为 default)、基本类型(属性 type 为 primary)和警告类型(属性 type 为 warn)。按钮大小分两种：default 默认大小和 mini 小尺寸，该组件的属性如表 4-11 所列。

表 4-11 button 属性

属性名	类型	默认值	说明	生效时机	最低版本
size	String	default	按钮的大小		
type	String	default	按钮的样式类型		
plain	Boolean	false	按钮是否镂空，背景色透明		
disabled	Boolean	false	是否禁用		
loading	Boolean	false	名称前是否带 loading 图标		
			用于 <form/> 组件，点击分别会		
bindgetuserinfo	Handler		用户点击该按钮时，会返回获取到的用户信息，回调的detail数据与wx.getUserInfo返回的一致	open-type="getUserInfo"	1.3.0
session-from	String		会话来源	open-type="contact"	1.4.0
send-message-title	String	当前标题	会话内消息卡片标题	open-type="contact"	1.5.0
send-message-path	String	当前分享路径	会话内消息卡片点击跳转小程序路径	open-type="contact"	1.5.0
send-message-img	String	截图	会话内消息卡片图片	open-type="contact"	1.5.0
show-message-card	Boolean	false	显示会话内消息卡片	open-type="contact"	1.5.0
bindcontact	Handler		客服消息回调	open-type="contact"	1.5.0
bindgetphonenumber	Handler		获取用户手机号回调	open-type="getPhoneNumber"	1.2.0
app-parameter	String		打开 APP 时，向 APP 传递的参数	open-type="launchApp"	1.9.5
binderror	Handler		当使用开放能力时，发生错误的回调	open-type="launchApp"	1.9.5
bindopensetting	Handler		在打开授权设置页后回调	open-type="openSetting"	2.0.7

组件示例代码如下：

```
//wxml 代码
<button type="default" size="{{defaultSize}}" loading="{{loading}}" plain="{{plain}}"
        disabled="{{disabled}}" bindtap="default" hover-class="other-button-
        hover"> default </button>
<button type="primary" size="{{primarySize}}" loading="{{loading}}" plain="{{plain}}"
```

```
        disabled = "{{disabled}}" bindtap = "primary" > primary </button >
    < button type = " warn" size = " {{warnSize}}" loading = " {{loading}}" plain = "
{{plain}}"
        disabled = "{{disabled}}" bindtap = "warn" > warn </button >
    < button bindtap = "setDisabled" >点击设置以上按钮 disabled 属性</button >
    < button bindtap = "setPlain" >点击设置以上按钮 plain 属性</button >
    < button bindtap = "setLoading" >点击设置以上按钮 loading 属性</button >
    < button open-type = "contact" >进入客服会话</button >
    < button open-type = "getUserInfo" lang = "zh_CN" bindgetuserinfo = "onGotUserInfo" >
获取用户信息</button >
    < button open-type = "openSetting" >打开授权设置页</button >

    /** wxss **/
    /** 修改 button 默认的点击态样式类 **/
    .button-hover {
      background-color: red;
    }
    /** 添加自定义 button 点击态样式类 **/
    .other-button-hover {
      background-color: blue;
    }

    //js 代码
    var types = ['default', 'primary', 'warn']
    var pageObject = {
      data: {
        defaultSize: 'default',
        primarySize: 'default',
        warnSize: 'default',
        disabled: false,
        plain: false,
        loading: false
      },
      setDisabled: function(e) {
        this.setData({
          disabled: ! this.data.disabled
        })
      },
      setPlain: function(e) {
        this.setData({
          plain: ! this.data.plain
        })
      },
      setLoading: function(e) {
        this.setData({
          loading: ! this.data.loading
        })
```

```
  },
  onGotUserInfo: function(e) {
    console.log(e.detail.errMsg)
    console.log(e.detail.userInfo)
    console.log(e.detail.rawData)
  },
}

for (var i = 0; i < types.length; ++i) {
  (function(type) {
    pageObject[type] = function(e) {
      var key = type + 'Size'
      var changedData = {}
      changedData[key] =
        this.data[key] === 'default' ? 'mini' : 'default'
      this.setData(changedData)
    }
  })(types[i])
}

Page(pageObject)
```

示例效果图如图 4 - 11 所示。

图 4 - 11　button 示例

4.3.2 多项选择器

多项选择器(checkbox - group),内部由多个 checkbox 组成。checkbox 多项选择器组件,和 HTML 中的多选框是一样的,用来进行多项选择。checkbox 组件的属性如表 4 - 12 所列。

表 4 - 12 多选项目属性

属性名	类型	默认值	说明
value	String		<checkbox/> 标识,选中时触发 <checkbox-group/> 的 change 事件,并携带 <checkbox/> 的 value
disabled	Boolean	false	是否禁用
checked	Boolean	false	当前是否选中,可用来设置默认选中
color	Color		checkbox的颜色,同css的color

示例代码如下:

```
//wxml 代码
< checkbox - group bindchange = "checkboxChange" >
< label class = "checkbox" wx:for = "{{items}}" >
< checkbox value = "{{item.name}}" checked = "{{item.checked}}"/ > {{item.value}}
< /label >
< /checkbox - group >

//js 代码
Page({
  data: {
    items: [
      {name: 'USA', value: '美国'},
      {name: 'CHN', value: '中国', checked: 'true'},
      {name: 'BRA', value: '巴西'},
      {name: 'JPN', value: '日本'},
      {name: 'ENG', value: '英国'},
      {name: 'TUR', value: '法国'},
    ]
  },
  checkboxChange: function(e) {
    console.log('checkbox 发生 change 事件,携带 value 值为:', e.detail.value)
  }
})
```

示例效果图如图 4 - 12 所示。

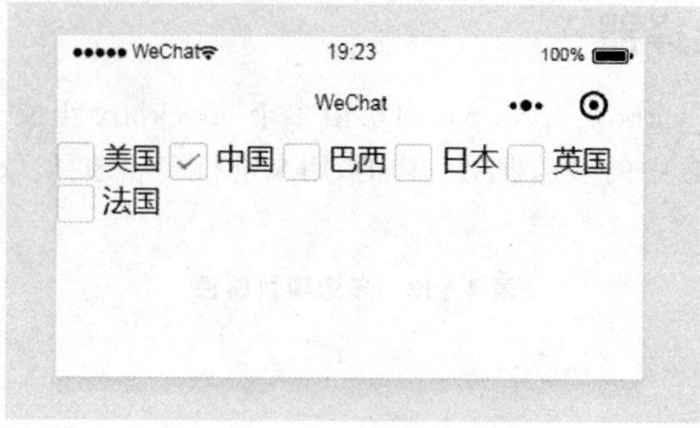

图 4-12 多项选择器示例

4.3.3 表 单

表单(form)组件会将组件内用户输入的 < switch/ > < input/ > < checkbox/ > < slider/ > < radio/ > < picker/ > 提交。当点击 < form/ > 表单中 formType 为 submit 的 < button/ > 组件时,会将表单组件中的 value 值进行提交,需要在表单组件中加上 name 作为 key。该组件的属性如表 4-13 所列。

表 4-13 form 表单属性

属性名	类型	说明	最低版本
report-submit	Boolean	是否返回 formId 用于发送模板消息	
bindsubmit	EventHandle	携带 form 中的数据触发 submit 事件，event.detail = {value : {'name': 'value'}, formId: ''}	
bindreset	EventHandle	表单重置时会触发 reset 事件	

示例代码如下：

```
//wxml 代码
<form bindsubmit = "formSubmit" bindreset = "formReset" >
  <view class = "section section_gap" >
    <view class = "section__title" > switch </view >
    <switch name = "switch"/ >
  </view >
  <view class = "section section_gap" >
    <view class = "section__title" > slider </view >
    <slider name = "slider" show - value > </slider >
  </view >
  <view class = "section" >
```

```
    < view class = "section__title" > input </view >
    < input name = "input" placeholder = "please input here" / >
  </view >
  < view class = "section section_gap" >
    < view class = "section__title" > radio </view >
    < radio - group name = "radio - group" >
      < label > < radio value = "radio1"/ > radio1 </label >
      < label > < radio value = "radio2"/ > radio2 </label >
    </radio - group >
  </view >
  < view class = "section section_gap" >
    < view class = "section__title" > checkbox </view >
    < checkbox - group name = "checkbox" >
      < label > < checkbox value = "checkbox1"/ > checkbox1 </label >
      < label > < checkbox value = "checkbox2"/ > checkbox2 </label >
    </checkbox - group >
  </view >
  < view class = "btn - area" >
    < button formType = "submit" > Submit </button >
    < button formType = "reset" > Reset </button >
  </view >
</form >

//js 代码
Page({
  formSubmit: function(e) {
    console.log('form 发生了 submit 事件,携带数据为:', e.detail.value)
  },
  formReset: function() {
    console.log('form 发生了 reset 事件')
  }
})
```

示例效果图如图 4 - 13 所示。

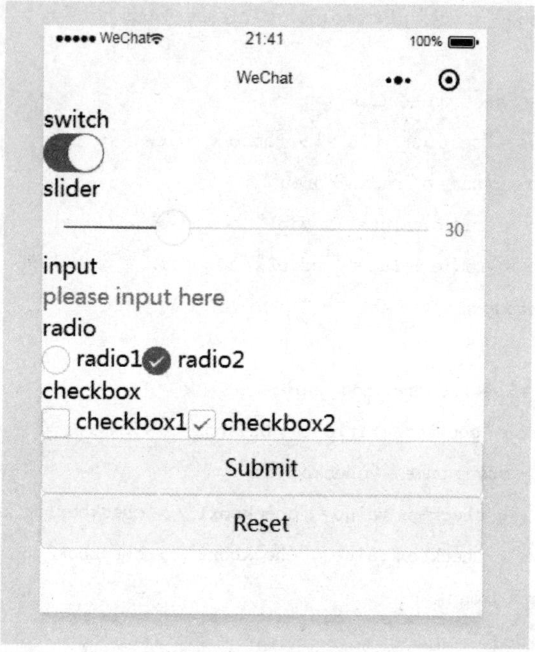

图 4-13　form 示例

4.3.4　输入框

输入框(input),该组件是原生组件,需要注意的是:

- 请注意原生组件使用限制;
- 微信版本 6.3.30,focus 属性设置无效;
- 微信版本 6.3.30,placeholder 在聚焦时出现重影问题;
- input 组件是一个原生组件,字体是系统字体,所以无法设置 font-family;
- 在 input 聚焦期间,避免使用 css 动画;
- 对于将 input 封装在自定义组件中、而 from 在自定义组件外的情况,form 将不能获得这个自定义组件中 input 的值。此时需要使用自定义组件的 内置 behaviors wx://form-field。

该组件的属性如表 4-14 所列。

表 4-14 input 属性

属性名	类型	默认值	说明	最低版本
value	String		输入框的初始内容	
type	String	"text"	input 的类型	
password	Boolean	false	是否是密码类型	
placeholder	String		输入框为空时占位符	
placeholder-style	String		指定 placeholder 的样式	
placeholder-class	String	"input-placeholder"	指定 placeholder 的样式类	
disabled	Boolean	false	是否禁用	
maxlength	Number	140	最大输入长度，设置为“-1”的时候不限制最大长度	
cursor-spacing	Number	0	指定光标与键盘的距离，单位 px 。取 input 距离底部的距离和 cursor-spacing 指定的距离的最小值作为光标与键盘的距离	
auto-focus	Boolean	false	(即将废弃，请直接使用 focus)自动聚焦，拉起键盘	
focus	Boolean	false	获取焦点	
confirm-type	String	"done"	设置键盘右下角按钮的文字，仅在type='text'时生效	1.1.0
confirm-hold	Boolean	false	点击键盘右下角按钮时是否保持键盘不收起	1.1.0
cursor	Number		指定focus时的光标位置	1.5.0
selection-start	Number	-1	光标起始位置，自动聚集时有效，需与selection-end搭配使用	1.9.0
selection-end	Number	-1	光标结束位置，自动聚集时有效，需与selection-start搭配使用	1.9.0
adjust-position	Boolean	true	键盘弹起时，是否自动上推页面	1.9.90
bindinput	EventHandle		键盘输入时触发，event.detail = {value, cursor, keyCode}，keyCode 为键值，2.1.0 起支持，处理函数可以直接 return 一个字符串，将替换输入框的内容。	
bindfocus	EventHandle		输入框聚焦时触发，event.detail = { value, height }，height 为键盘高度，在基础库 1.9.90 起支持	
bindblur	EventHandle		输入框失去焦点时触发，event.detail = {value: value}	
bindconfirm	EventHandle		点击完成按钮时触发，event.detail = {value: value}	

示例代码如下：

```
<!-- input.wxml -->
<view class="section">
  <input placeholder="这是一个可以自动聚焦的 input" auto-focus/>
</view>
<view class="section">
  <input placeholder="这个只有在按钮点击的时候才聚焦" focus="{{focus}}" />
  <view class="btn-area">
    <button bindtap="bindButtonTap">使得输入框获取焦点</button>
  </view>
</view>
<view class="section">
  <input  maxlength="10" placeholder="最大输入长度 10" />
</view>
<view class="section">
  <view class="section__title">你输入的是:{{inputValue}}</view>
  <input  bindinput="bindKeyInput" placeholder="输入同步到 view 中"/>
</view>
<view class="section">
  <input  bindinput="bindReplaceInput" placeholder="连续的两个 1 会变成 2" />
</view>
<view class="section">
  <input password type="number" />
</view>
<view class="section">
  <input password type="text" />
</view>
<view class="section">
  <input type="digit" placeholder="带小数点的数字键盘"/>
</view>
<view class="section">
  <input type="idcard" placeholder="身份证输入键盘" />
</view>
<view class="section">
  <input placeholder-style="color:red" placeholder="占位符字体是红色的" />
</view>
```

```
//input.js
Page({
  data: {
    focus: false,
    inputValue: ''
  },
  bindButtonTap: function() {
    this.setData({
      focus: true
    })
  },
  bindKeyInput: function(e) {
    this.setData({
      inputValue: e.detail.value
    })
  },
  bindReplaceInput: function(e) {
    var value = e.detail.value
    var pos = e.detail.cursor
    if(pos != -1){
      //光标在中间
      var left = e.detail.value.slice(0,pos)
      //计算光标的位置
      pos = left.replace(/11/g,'2').length
    }

    //直接返回对象,可以对输入进行过滤处理,同时可以控制光标的位置
    return {
      value: value.replace(/11/g,'2'),
      cursor: pos
    }

    //或者直接返回字符串,光标在最后边
    //return value.replace(/11/g,'2'),
  }
})
```

示例效果图如图 4-14 所示。

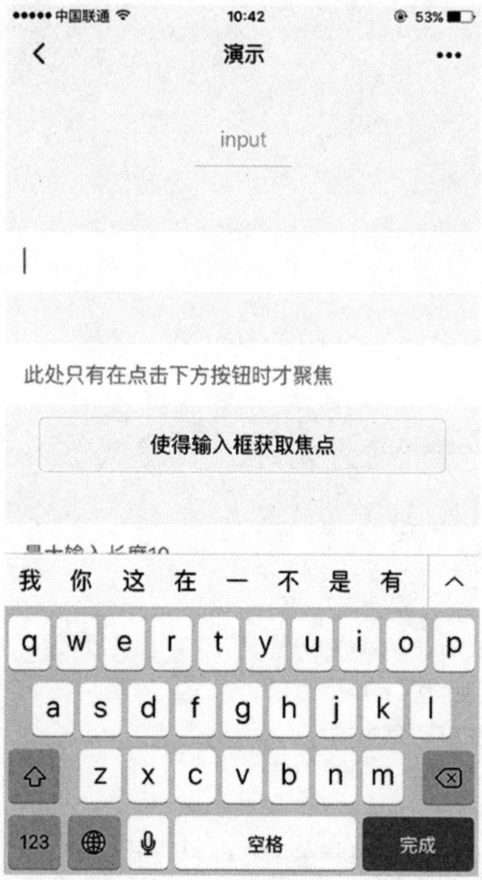

图 4－14 input 示例图

4.3.5 标 签

标签(label)组件用来改进表单组件的可用性，使用 for 属性找到对应的 id，或者将控件放在该标签下，点击时就会触发对应的控件。for 优先级高于内部控件，内部有多个控件的时候默认触发第一个控件。目前可以绑定的控件有：< button/ > < checkbox/ > < radio/ > 和 < switch/ >，该组件属性如表 4－15 所列。

表 4－15 Label 组件属性

属性名	类型	说明
for	String	绑定控件的 id

示例代码如下：

```
//wxml 代码
```

```
< view class = "section section_gap" >
< view class = "section__title" > 表单组件在 label 内 < /view >
< checkbox - group class = "group" bindchange = "checkboxChange" >
  < view class = "label - 1" wx:for = "{{checkboxItems}}" >
    < label >
      < checkbox hidden value = "{{item.name}}" checked = "{{item.checked}}" > < /
checkbox >
      < view class = "label - 1__icon" >
        < view class = "label - 1__icon - checked" style = "opacity:{{item.checked ? 1:
0}}" > < /view >
      < /view >
      < text class = "label - 1__text" > {{item.value}} < /text >
    < /label >
  < /view >
< /checkbox - group >
< /view >

< view class = "section section_gap" >
< view class = "section__title" > label 用 for 标识表单组件 < /view >
< radio - group class = "group" bindchange = "radioChange" >
  < view class = "label - 2" wx:for = "{{radioItems}}" >
    < radio id = "{{item.name}}" hidden value = "{{item.name}}" checked = "{{item.
checked}}" > < /radio >
    < view class = "label - 2__icon" >
      < view class = "label - 2__icon - checked" style = "opacity:{{item.checked ? 1:
0}}" > < /view >
    < /view >
    < label class = "label - 2__text" for = "{{item.name}}" > < text > {{item.name}}
< /text > < /label >
  < /view >
< /radio - group >
< /view >

//js 代码
Page({
  data: {
    checkboxItems: [
      {name: 'USA', value: '美国'},
      {name: 'CHN', value: '中国', checked: 'true'},
      {name: 'BRA', value: '巴西'},
      {name: 'JPN', value: '日本', checked: 'true'},
      {name: 'ENG', value: '英国'},
```

```
      {name: 'TUR', value: '法国'},
    ],
    radioItems: [
      {name: 'USA', value: '美国'},
      {name: 'CHN', value: '中国', checked: 'true'},
      {name: 'BRA', value: '巴西'},
      {name: 'JPN', value: '日本'},
      {name: 'ENG', value: '英国'},
      {name: 'TUR', value: '法国'},
    ],
    hidden: false
  },
  checkboxChange: function(e) {
    var checked = e.detail.value
    var changed = {}
    for (var i = 0; i < this.data.checkboxItems.length; i ++) {
      if (checked.indexOf(this.data.checkboxItems[i].name) !== -1) {
        changed['checkboxItems['+i+'].checked'] = true
      } else {
        changed['checkboxItems['+i+'].checked'] = false
      }
    }
    this.setData(changed)
  },
  radioChange: function(e) {
    var checked = e.detail.value
    var changed = {}
    for (var i = 0; i < this.data.radioItems.length; i ++) {
      if (checked.indexOf(this.data.radioItems[i].name) !== -1) {
        changed['radioItems['+i+'].checked'] = true
      } else {
        changed['radioItems['+i+'].checked'] = false
      }
    }
    this.setData(changed)
  }
})

//wxss 代码
.label-1, .label-2{
    margin-bottom: 15px;
}
```

```
.label-1__text, .label-2__text {
    display: inline-block;
    vertical-align: middle;
}

.label-1__icon {
    position: relative;
    margin-right: 10px;
    display: inline-block;
    vertical-align: middle;
    width: 18px;
    height: 18px;
    background: #fcfff4;
}

.label-1__icon-checked {
    position: absolute;
    top: 3px;
    left: 3px;
    width: 12px;
    height: 12px;
    background: #1aad19;
}

.label-2__icon {
    position: relative;
    display: inline-block;
    vertical-align: middle;
    margin-right: 10px;
    width: 18px;
    height: 18px;
    background: #fcfff4;
    border-radius: 50px;
}

.label-2__icon-checked {
    position: absolute;
    left: 3px;
    top: 3px;
    width: 12px;
    height: 12px;
    background: #1aad19;
```

```
    border-radius: 50%;
}

.label-4_text{
    text-align: center;
    margin-top: 15px;
}
```

示例效果图如图 4-15 所示。

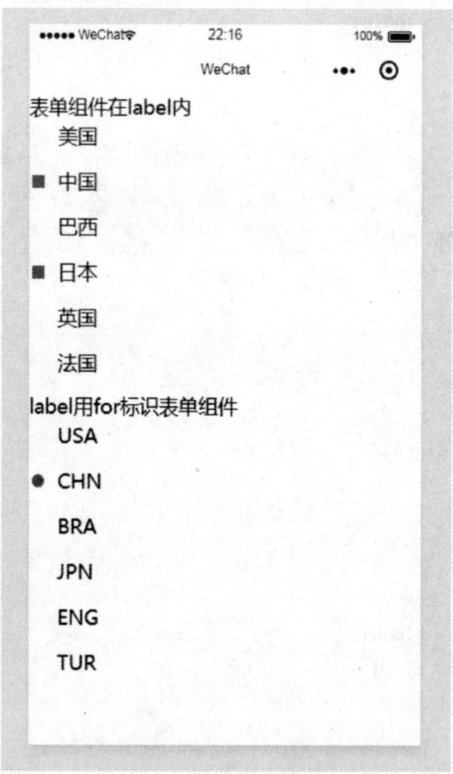

图 4-15　label 示例

4.3.6　滚动选择器

从底部弹起的滚动选择器(picker)组件，其通过 mode 来区分，分别是普通选择器、多列选择器、时间选择器和日期选择器和省市区选择器，默认为普通选择器。Picker 滚动选择器极大地提升了用户体验，非常方便用户进行时间、省市区等常用选择操作。该组件应用比较频繁，建议读者重点学习并掌握。

普通选择器：mode = selector，属性如表 4-16 所列。

表 4-16 普通选择器属性

属性名	类型	默认值	说明	最低版本
range	Array / Object Array	[]	mode为 selector 或 multiSelector 时，range 有效	
range-key	String		当 range 是一个 Object Array 时，通过 range-key 来指定 Object 中 key 的值作为选择器显示内容	
value	Number	0	value 的值表示选择了 range 中的第几个（下标从 0 开始）	
bindchange	EventHandle		value 改变时触发 change 事件，event.detail = {value: value}	
disabled	Boolean	false	是否禁用	
bindcancel	EventHandle		取消选择或点遮罩层收起 picker 时触发	1.9.90

多列选择器：mode = multiSelector（最低版本为 1.4.0），属性如表 4-17 所列。

表 4-17 多列选择器属性

属性名	类型	默认值	说明	最低版本
range	二维Array / 二维Object Array	[]	mode为 selector 或 multiSelector 时，range 有效。二维数组，长度表示多少列，数组的每项表示每列的数据，如 [["a","b"], ["c","d"]]	
range-key	String		当 range 是一个 二维Object Array 时，通过 range-key 来指定 Object 中 key 的值作为选择器显示内容	
value	Array	[]	value 每一项的值表示选择了 range 对应项中的第几个（下标从 0 开始）	
bindchange	EventHandle		value 改变时触发 change 事件，event.detail = {value: value}	
bindcolumnchange	EventHandle		某一列的值改变时触发 columnchange 事件，event.detail = {column: column, value: value}，column 的值表示改变了第几列（下标从0开始），value 的值表示变更值的下标	
bindcancel	EventHandle		取消选择时触发	1.9.90
disabled	Boolean	false	是否禁用	

时间选择器：mode = time，属性如表 4-18 所列。

表 4－18　时间选择器属性

属性名	类型	默认值	说明	最低版本
value	String		表示选中的时间，格式为"hh:mm"	
start	String		表示有效时间范围的开始，字符串格式为"hh:mm"	
end	String		表示有效时间范围的结束，字符串格式为"hh:mm"	
bindchange	EventHandle		value 改变时触发 change 事件，event.detail = {value: value}	
bindcancel	EventHandle		取消选择时触发	1.9.90
disabled	Boolean	false	是否禁用	

日期选择器：mode ＝ date，属性如表 4－19 所列。

表 4－19　日期选择器

属性名	类型	默认值	说明	最低版本
value	String	0	表示选中的日期，格式为"YYYY-MM-DD"	
start	String		表示有效日期范围的开始，字符串格式为"YYYY-MM-DD"	
end	String		表示有效日期范围的结束，字符串格式为"YYYY-MM-DD"	
fields	String	day	有效值 year,month,day，表示选择器的粒度	
bindchange	EventHandle		value 改变时触发 change 事件，event.detail = {value: value}	
bindcancel	EventHandle		取消选择时触发	1.9.90
disabled	Boolean	false	是否禁用	

省市区选择器：mode ＝ region（最低版本为 1.4.0），属性如表 4－20 所列。

表 4－20　省市区选择器

属性名	类型	默认值	说明	最低版本
value	Array	[]	表示选中的省市区，默认选中每一列的第一个值	
custom-item	String		可为每一列的顶部添加一个自定义的项	1.5.0
bindchange	EventHandle		value 改变时触发 change 事件，event.detail = {value: value, code: code, postcode: postcode}，其中字段code是统计用区划代码，postcode是邮政编码	
bindcancel	EventHandle		取消选择时触发	1.9.90
disabled	Boolean	false	是否禁用	

示例代码如下：

```
//wxml 代码
  < view class = "section" >
    < view class = "section__title" > 普通选择器 < /view >
    < picker bindchange = "bindPickerChange" value = "{{index}}" range = "{{array}}" >
      < view class = "picker" >
        当前选择:{{array[index]}}
      < /view >
    < /picker >
  < /view >
  < view class = "section" >
    < view class = "section__title" > 多列选择器 < /view >
    < picker mode = "multiSelector" bindchange = "bindMultiPickerChange" bindcolumn-
change = "bindMultiPickerColumnChange" value = "{{multiIndex}}" range = "{{multiArray}}" >
      < view class = "picker" >
        当前选择:{{multiArray[0][multiIndex[0]]}},{{multiArray[1][multiIndex[1]]}},
{{multiArray[2][multiIndex[2]]}}
      < /view >
    < /picker >
  < /view >
  < view class = "section" >
    < view class = "section__title" > 时间选择器 < /view >
    < picker mode = "time" value = "{{time}}" start = "09:01" end = "21:01" bindchange = "
bindTimeChange" >
      < view class = "picker" >
        当前选择: {{time}}
      < /view >
    < /picker >
  < /view >

  < view class = "section" >
    < view class = "section__title" > 日期选择器 < /view >
    < picker mode = "date" value = "{{date}}" start = "2015 - 09 - 01" end = "2017 - 09 -
01" bindchange = "bindDateChange" >
      < view class = "picker" >
        当前选择: {{date}}
      < /view >
    < /picker >
  < /view >
  < view class = "section" >
    < view class = "section__title" > 省市区选择器 < /view >
```

```
<picker mode="region" bindchange="bindRegionChange" value="{{region}}" custom-item="{{customItem}}">
    <view class="picker">
      当前选择：{{region[0]}}，{{region[1]}}，{{region[2]}}
    </view>
  </picker>
</view>

//js 代码
Page({
  data: {
    array: ['美国', '中国', '巴西', '日本'],
    objectArray: [
      {
        id: 0,
        name: '美国'
      },
      {
        id: 1,
        name: '中国'
      },
      {
        id: 2,
        name: '巴西'
      },
      {
        id: 3,
        name: '日本'
      }
    ],
    index: 0,
    multiArray: [['无脊柱动物', '脊柱动物'], ['扁性动物', '线形动物', '环节动物', '软体动物', '节肢动物'], ['猪肉绦虫', '吸血虫']],
    objectMultiArray: [
      [
        {
          id: 0,
          name: '无脊柱动物'
        },
        {
          id: 1,
          name: '脊柱动物'
```

```
        }
      ],[
        {
          id: 0,
          name: '扁性动物'
        },
        {
          id: 1,
          name: '线形动物'
        },
        {
          id: 2,
          name: '环节动物'
        },
        {
          id: 3,
          name: '软体动物'
        },
        {
          id: 3,
          name: '节肢动物'
        }
      ],[
        {
          id: 0,
          name: '猪肉绦虫'
        },
        {
          id: 1,
          name: '吸血虫'
        }
      ]
    ],
    multiIndex: [0, 0, 0],
    date: '2016-09-01',
    time: '12:01',
    region: ['广东省', '广州市', '海珠区'],
    customItem: '全部'
  },
  bindPickerChange: function(e) {
    console.log('picker发送选择改变,携带值为', e.detail.value)
    this.setData({
```

```
        index: e.detail.value
      })
    },
    bindMultiPickerChange: function (e) {
      console.log('picker 发送选择改变,携带值为', e.detail.value)
      this.setData({
        multiIndex: e.detail.value
      })
    },
    bindMultiPickerColumnChange: function (e) {
      console.log('修改的列为', e.detail.column, ',值为', e.detail.value);
      var data = {
        multiArray: this.data.multiArray,
        multiIndex: this.data.multiIndex
      };
      data.multiIndex[e.detail.column] = e.detail.value;
      switch (e.detail.column) {
        case 0:
          switch (data.multiIndex[0]) {
            case 0:
              data.multiArray[1] = ['扁性动物','线形动物','环节动物','软体动物',
                                    '节肢动物'];
              data.multiArray[2] = ['猪肉绦虫','吸血虫'];
              break;
            case 1:
              data.multiArray[1] = ['鱼','两栖动物','爬行动物'];
              data.multiArray[2] = ['鲫鱼','带鱼'];
              break;
          }
          data.multiIndex[1] = 0;
          data.multiIndex[2] = 0;
          break;
        case 1:
          switch (data.multiIndex[0]) {
            case 0:
              switch (data.multiIndex[1]) {
                case 0:
                  data.multiArray[2] = ['猪肉绦虫','吸血虫'];
                  break;
                case 1:
                  data.multiArray[2] = ['蛔虫'];
                  break;
```

```
          case 2:
            data.multiArray[2] = ['蚂蚁','蚂蟥'];
            break;
          case 3:
            data.multiArray[2] = ['河蚌','蜗牛','蛞蝓'];
            break;
          case 4:
            data.multiArray[2] = ['昆虫','甲壳动物','蛛形动物','多足动物
'];
            break;
        }
        break;
      case 1:
        switch (data.multiIndex[1]) {
          case 0:
            data.multiArray[2] = ['鲫鱼','带鱼'];
            break;
          case 1:
            data.multiArray[2] = ['青蛙','娃娃鱼'];
            break;
          case 2:
            data.multiArray[2] = ['蜥蜴','龟','壁虎'];
            break;
        }
        break;
    }
    data.multiIndex[2] = 0;
    break;
  }
  console.log(data.multiIndex);
  this.setData(data);
},
bindDateChange: function(e) {
  console.log('picker发送选择改变,携带值为', e.detail.value)
  this.setData({
    date: e.detail.value
  })
},
bindTimeChange: function(e) {
  console.log('picker发送选择改变,携带值为', e.detail.value)
  this.setData({
    time: e.detail.value
```

```
    })
  },
  bindRegionChange: function (e) {
    console.log('picker 发送选择改变,携带值为 ', e.detail.value)
    this.setData({
      region: e.detail.value
    })
  }
})
```

普通选择器示例图如图 4 - 16 所示。

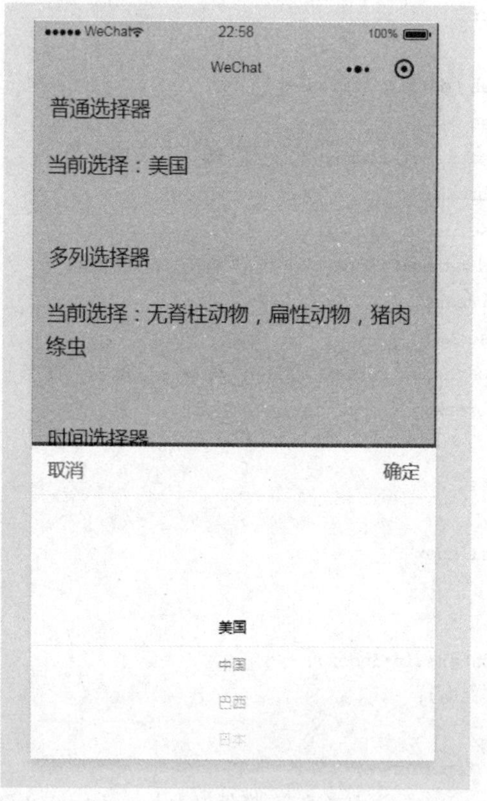

图 4 - 16　普通选择器示例图

多列选择器示例图如图 4 - 17 所示。
时间选择器示例图如图 4 - 18 所示。
日期选择器示例图如图 4 - 19 所示。
省市区选择器示例图如图 4 - 20 所示。

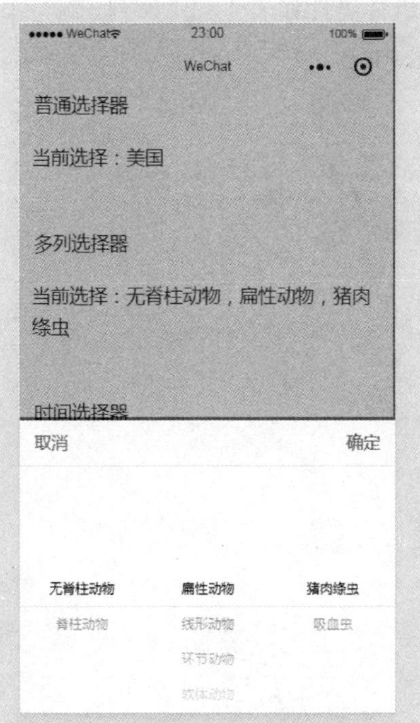

图 4－17　多列选择器示例图

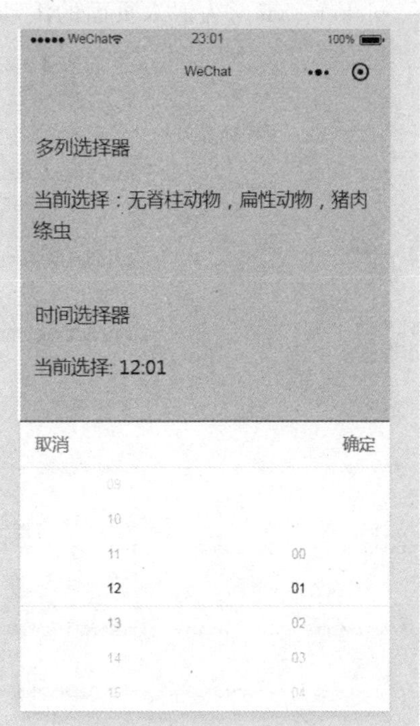

图 4－18　时间选择器示例图

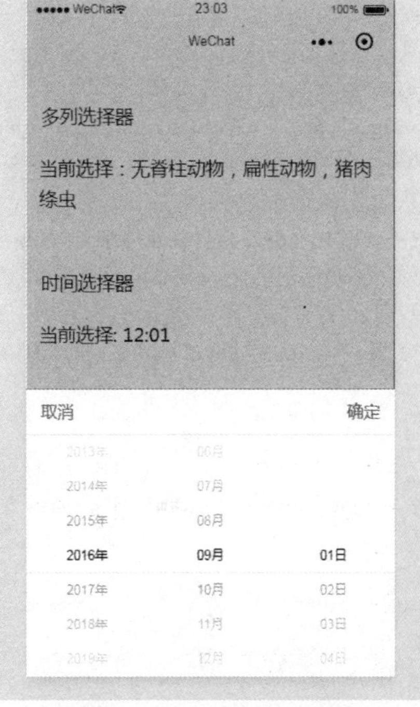

图 4－19　日期选择器示例图

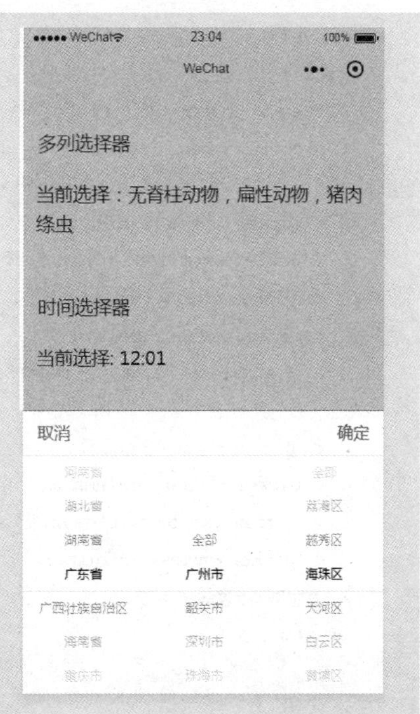

图 4－20　省市区选择器示例图

picker - view 为嵌入页面的滚动选择器,该组件属性如表 4 - 21 所列。

表 4 - 21 picker - view 属性

属性名	类型	说明	最低版本
value	NumberArray	数组中的数字依次表示 picker-view 内的 picker-view-column 选择的第几项(下标从 0 开始),数字大于 picker-view-column 可选项长度时,选择最后一项	
indicator-style	String	设置选择器中间选中框的样式	
indicator-class	String	设置选择器中间选中框的类名	1.1.0
mask-style	String	设置蒙层的样式	1.5.0
mask-class	String	设置蒙层的类名	1.5.0
bindchange	EventHandle	当滚动选择,value 改变时触发 change 事件,event.detail = {value: value};value为数组,表示 picker-view 内的 picker-view-column 当前选择的是第几项(下标从 0 开始)	
bindpickstart	EventHandle	当滚动选择开始时触发事件	2.3.1
bindpickend	EventHandle	当滚动选择结束时触发事件	2.3.1

示例代码如下:

```
//wxml 代码
 < view >
   < view >{{year}}年{{month}}月{{day}}日 </view >
   < picker - view indicator - style = "height: 50px;" style = "width: 100 %; height: 300px;" value = "{{value}}" bindchange = "bindChange" >
     < picker - view - column >
       < view wx:for = "{{years}}" style = "line - height: 50px" > {{item}}年 </view >
     </picker - view - column >
     < picker - view - column >
       < view wx: for = "{{months}}" style = "line - height: 50px" > {{item}} 月 </view >
     </picker - view - column >
     < picker - view - column >
       < view wx:for = "{{days}}" style = "line - height: 50px" > {{item}}日 </view >
     </picker - view - column >
   </picker - view >
 </view >

//js 代码
const date = new Date()
```

```
const years = []
const months = []
const days = []

for (let i = 1990; i < = date.getFullYear(); i++) {
  years.push(i)
}

for (let i = 1 ; i < = 12; i++) {
  months.push(i)
}

for (let i = 1 ; i < = 31; i++) {
  days.push(i)
}

Page({
  data: {
    years: years,
    year: date.getFullYear(),
    months: months,
    month: 2,
    days: days,
    day: 2,
    value: [9999, 1, 1],
  },
  bindChange: function(e) {
    const val = e.detail.value
    this.setData({
      year: this.data.years[val[0]],
      month: this.data.months[val[1]],
      day: this.data.days[val[2]]
    })
  }
})
```

示例效果图如图 4-21 所示。

图 4－21　picker－view 示例图

4.3.7　单项选择器

单项选择器(radio－group),内部由多个 < radio/ > 组成。radio 为单选项目,其属性如表 4－22 所列。

表 4－22　radio 属性

属性名	类型	默认值	说明
value	String		<radio/> 标识。当该 <radio/> 选中时，<radio-group/> 的 change 事件会携带 <radio/> 的 value
checked	Boolean	false	当前是否选中
disabled	Boolean	false	是否禁用
color	Color		radio的颜色，同css的color

示例代码如下：

```
//wxml 代码
< radio - group class = "radio - group" bindchange = "radioChange" >
```

```
    < label class = "radio" wx:for = "{{items}}" >
      < radio value = "{{item.name}}" checked = "{{item.checked}}"/ > {{item.value}}
    </label >
  </radio - group >

//js 代码
Page({
  data: {
    items: [
      {name: 'USA', value: '美国'},
      {name: 'CHN', value: '中国', checked: 'true'},
      {name: 'BRA', value: '巴西'},
      {name: 'JPN', value: '日本'},
      {name: 'ENG', value: '英国'},
      {name: 'TUR', value: '法国'},
    ]
  },
  radioChange: function(e) {
    console.log('radio 发生 change 事件,携带 value 值为:', e.detail.value)
  }
})
```

示例效果图如图 4 - 22 所示。

图 4 - 22　radio 示例

4.3.8 滑动选择器

滑动选择器(slider)组件常用来作为控制按钮使用,比如控制声音大小、设置特定值等场景。该组件的属性如表 4-23 所列。

表 4-23 slider 属性

属性名	类型	默认值	说明	最低版本
min	Number	0	最小值	
max	Number	100	最大值	
step	Number	1	步长,取值必须大于 0,并且可被(max - min)整除	
disabled	Boolean	false	是否禁用	
value	Number	0	当前取值	
color	Color	#e9e9e9	背景条的颜色(请使用 backgroundColor)	
selected-color	Color	#1aad19	已选择的颜色(请使用 activeColor)	
activeColor	Color	#1aad19	已选择的颜色	
backgroundColor	Color	#e9e9e9	背景条的颜色	
block-size	Number	28	滑块的大小,取值范围为 12~28	1.9.0
block-color	Color	#ffffff	滑块的颜色	1.9.0
show-value	Boolean	false	是否显示当前 value	
bindchange	EventHandle		完成一次拖动后触发的事件,event.detail = {value: value}	
bindchanging	EventHandle		拖动过程中触发的事件,event.detail = {value: value}	1.7.0

示例代码如下:

```
//wxml 代码
 < view class = "section section_gap" >
   < text class = "section__title" > 设置 step < /text >
   < view class = "body - view" >
     < slider bindchange = "slider2change" step = "5"/ >
   < /view >
 < /view >
```

```
< view class = "section section_gap" >
  < text class = "section__title" > 显示当前 value < /text >
  < view class = "body - view" >
    < slider bindchange = "slider3change" show - value/ >
  < /view >
< /view >

< view class = "section section_gap" >
  < text class = "section__title" > 设置最小/最大值 < /text >
  < view class = "body - view" >
    < slider bindchange = "slider4change" min = "50" max = "200" show - value/ >
  < /view >
< /view >

//js 代码
var pageData = {}
for (var i = 1; i < 5; i++) {
  (function (index) {
    pageData['slider' + index + 'change'] = function(e) {
      console.log('slider' + 'index' + '发生 change 事件,携带值为', e.detail.value)
    }
  })(i)
}
Page(pageData)
```

示例效果图如图 4 - 23 所示。

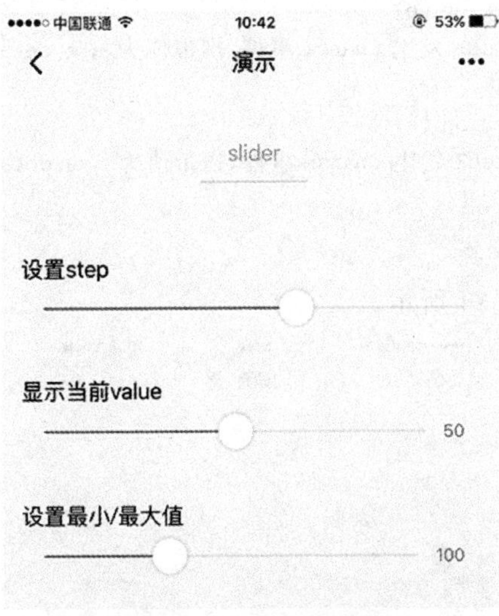

图 4 - 23 slider 示例

4.3.9 开关选择器

开关选择器(switch)组件,只有开和关两个状态,但是它的应用非常广泛。比如以下场景:是否接收、是否开启、是否震动等。该组件经常用来作为应用的个性化定制功能来使用,该组件属性如表 4-24 所列。

表 4-24 switch 属性

属性名	类型	默认值	说明
checked	Boolean	false	是否选中
disabled	Boolean	false	是否禁用
type	String	switch	样式,有效值:switch, checkbox
bindchange	EventHandle		checked 改变时触发 change 事件,event.detail={ value:checked}
color	Color		switch 的颜色,同 css 的 color

示例代码如下:

```
//wxml 代码
<view class="body-view">
    <switch checked bindchange="switch1Change"/>
    <switch bindchange="switch2Change"/>
</view>

//js 代码
Page({
  switch1Change: function (e){
    console.log('switch1 发生 change 事件,携带值为', e.detail.value)
  },
  switch2Change: function (e){
    console.log('switch2 发生 change 事件,携带值为', e.detail.value)
  }
})
```

示例效果图如图 4-24 所示。

图 4-24 switch 示例

4.3.10 多行输入框

多行输入框(textarea)组件用来输入多行文本内容,它与 input 单行输入框的功能基本类似。Textarea 组件的属性如表 4-25 所列。

表 4-25 textarea 属性

属性名	类型	默认值	说明	最低版本
value	String		输入框的内容	
placeholder	String		输入框为空时占位符	
placeholder-style	String		指定 placeholder 的样式	
placeholder-class	String	textarea-placeholder	指定 placeholder 的样式类	
disabled	Boolean	false	是否禁用	
maxlength	Number	140	最大输入长度,设置为"-1"的时候不限制最大长度	
auto-focus	Boolean	false	自动聚焦,拉起键盘	
focus	Boolean	false	获取焦点	
auto-height	Boolean	false	是否自动增高,设置auto-height时,style.height不生效	
fixed	Boolean	false	如果 textarea 是在一个 `position:fixed` 的区域,需要显示指定属性 fixed 为 true	
cursor-spacing	Number	0	指定光标与键盘的距离,单位 px 。取 textarea 距离底部的距离和 cursor-spacing 指定的距离的最小值作为光标与键盘的距离	
cursor	Number		指定focus时的光标位置	1.5.0
show-confirm-bar	Boolean	true	是否显示键盘上方带有"完成"按钮那一栏	1.6.0
selection-start	Number	-1	光标起始位置,自动聚集时有效,需与selection-end搭配使用	1.9.0
selection-end	Number	-1	光标结束位置,自动聚集时有效,需与selection-start搭配使用	1.9.0
adjust-position	Boolean	true	键盘弹起时,是否自动上推页面	1.9.90
bindfocus	EventHandle		输入框聚焦时触发,event.detail = { value, height },height 为键盘高度,从基础库 1.9.90 起支持	
bindblur	EventHandle		输入框失去焦点时触发,event.detail = {value, cursor}	
bindlinechange	EventHandle		输入框行数变化时调用,event.detail = {height: 0, heightRpx: 0, lineCount: 0}	
bindinput	EventHandle		当键盘输入时,触发 input 事件,event.detail = {value, cursor},**bindinput 处理函数的返回值并不会反映到 textarea 上**	
bindconfirm	EventHandle		点击完成时,触发 confirm 事件,event.detail = {value: value}	

示例代码如下：

```
<!-- textarea.wxml -->
<view class="section">
  <textarea bindblur="bindTextAreaBlur" auto-height placeholder="自动变高" />
</view>
<view class="section">
  <textarea placeholder="placeholder颜色是红色的" placeholder-style="color:red;" />
</view>
<view class="section">
  <textarea placeholder="这是一个可以自动聚焦的textarea" auto-focus />
</view>
<view class="section">
  <textarea placeholder="这个只有在按钮点击的时候才聚焦" focus="{{focus}}" />
  <view class="btn-area">
    <button bindtap="bindButtonTap">使得输入框获取焦点</button>
  </view>
</view>
<view class="section">
  <form bindsubmit="bindFormSubmit">
    <textarea placeholder="form 中的 textarea" name="textarea"/>
    <button form-type="submit"> 提交 </button>
  </form>
</view>
```

```
//textarea.js
Page({
  data: {
    height: 20,
    focus: false
  },
  bindButtonTap: function() {
    this.setData({
      focus: true
    })
  },
```

```
    bindTextAreaBlur: function(e) {
      console.log(e.detail.value)
    },
    bindFormSubmit: function(e) {
      console.log(e.detail.value.textarea)
    }
  })
```

示例效果图如图 4-25 所示。

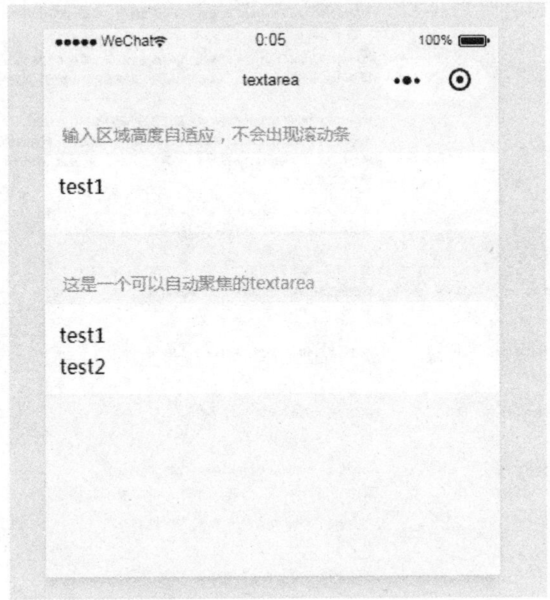

图 4-25　textarea 示例

4.4　导航组件

4.4.1　页面导航

页面导航(navigator)组件，它类似于 HTML 中的 a 标签，该组件的属性如表 4-26 所示。

表 4-26 navigator 属性

属性名	类型	默认值	说明	最低版本
target	String	self	在哪个目标上发生跳转，默认当前小程序，可选值self/miniProgram	2.0.7
url	String		当前小程序内的跳转链接	
open-type	String	navigate	跳转方式	
delta	Number		当 open-type 为 'navigateBack' 时有效，表示回退的层数	
app-id	String		当target="miniProgram"时有效，要打开小程序的 appId	2.0.7
path	String		当target="miniProgram"时有效，打开页面路径，如果为空则打开首页	2.0.7
extra-data	Object		当target="miniProgram"时有效，需要传递给目标小程序的数据，目标小程序可在 App.onLaunch()，App.onShow() 中获取到这份数据。	2.0.7
version	version	release	当target="miniProgram"时有效，要打开的小程序版本，有效值 develop（开发版），trial（体验版），release（正式版），仅在当前小程序为开发版或体验版时此参数有效；如果当前小程序是正式版，则打开的小程序必定是正式版。	2.0.7
hover-class	String	navigator-hover	指定点击时的样式类，当 hover-class="none" 时，没有点击态效果	
hover-stop-propagation	Boolean	false	指定是否阻止本节点的祖先节点出现点击态	1.5.0
hover-start-time	Number	50	按住后多久出现点击态，单位 ms	
hover-stay-time	Number	600	手指松开后点击态保留时间，单位 ms	
bindsuccess	String		当target="miniProgram"时有效，跳转小程序成功	2.0.7
bindfail	String		当target="miniProgram"时有效，跳转小程序失败	2.0.7
bindcomplete	String		当target="miniProgram"时有效，跳转小程序完成	2.0.7

其中属性 open-type 有效值如表 4-27 所列。

表 4-27 open-type 有效值

值	说明	最低版本
navigate	对应 wx.navigateTo 或 wx.navigateToMiniProgram 的功能	
redirect	对应 wx.redirectTo 的功能	
switchTab	对应 wx.switchTab 的功能	
reLaunch	对应 wx.reLaunch 的功能	1.1.0
navigateBack	对应 wx.navigateBack 的功能	1.1.0
exit	退出小程序，target="miniProgram"时生效	2.1.0

示例代码如下：

```
/ ** wxss **/
/ ** 修改默认的 navigator 点击态 **/
.navigator - hover {
  color:blue;
}
/ ** 自定义其他点击态样式类 **/
.other - navigator - hover {
  color:red;
}

< !-- sample.wxml -->
< view class = "btn - area" >
  < navigator url = "/page/navigate/navigate? title = navigate" hover - class = "navi-
gator - hover" > 跳转到新页面 < /navigator >
  < navigator url = "../../redirect/redirect/redirect? title = redirect" open - type
= "redirect" hover - class = "other - navigator - hover" > 在当前页打开 < /navigator >
  < navigator url = "/page/index/index" open - type = "switchTab" hover - class = "other
- navigator - hover" > 切换 Tab < /navigator >
  < navigator target = "miniProgram" open - type = "navigate" app - id = "" path = "" ex-
tra - data = "" version = "release" > 打开绑定的小程序 < /navigator >
< /view >

< !-- navigator.wxml -->
< view style = "text - align:center" > {{title}} < /view >
< view > 点击左上角返回回到之前页面 < /view >

< !-- redirect.wxml -->
< view style = "text - align:center" > {{title}} < /view >
< view > 点击左上角返回回到上级页面 < /view >

// redirect.js navigator.js
Page({
  onLoad: function(options) {
    this.setData({
      title: options.title
    })
  }
})
```

4.4.2 功能页导航

功能页导航(functional - page - navigator)这个组件从小程序基础库版本 2.1.0

开始支持，但仅在插件的自定义组件中有效，用于跳转到插件功能页。该组件的属性如表 4－28 所列。

表 4－28　functional－page－navigator 属性

属性名	类型	默认值	说明	最低版本
version	String	release	跳转到的小程序版本，有效值 develop（开发版），trial（体验版），release（正式版）；线上版本必须设置为 release	2.1.0
name	String		要跳转到的功能页	2.1.0
args	Object	null	功能页参数，参数格式与具体功能页相关	2.1.0
bindsuccess	EventHandler		功能页返回，且操作成功时触发，detail 格式与具体功能页相关	2.1.0
bindfail	EventHandler		功能页返回，且操作失败时触发，detail 格式与具体功能页相关	2.1.0

目前支持的功能页和 name 有效值，如表 4－29 所列。

表 4－29　name 属性有效值

值	功能页	最低版本
loginAndGetUserInfo	用户信息功能页	2.1.0
requestPayment	支付功能页	2.1.0

示例代码如下：

```
<!-- sample.wxml -->
<functional-page-navigator name="loginAndGetUserInfo" bind:success="loginSuccess">
  <button>登录到插件</button>
</functional-page-navigator>

// redirect.js navigator.js
Component({
  methods: {
    loginSuccess: function(e) {
      console.log(e.detail.code) // wx.login 的 code
      console.log(e.detail.userInfo) // wx.getUserInfo 的 userInfo
    }
  }
})
```

4.5 媒体组件

媒体组件主要用来显示图片，播放音视频以及直播，在当下自媒体、短视频如此火爆的情况下，图片和音视频比文本具备更强大的视觉冲击力，这类应用也非常受当下的互联网用户所喜爱。一般来说，一家公司的某一个应用会主打多媒体中的一个分支，比如主打音频的喜马拉雅，主打视频播放的腾讯视频和主打短视频的微视等。故而学习媒体组件，此处建议读者首先对各种媒体组件有一个大致认知即可，当以后具体开发某一个分支的时候再来深入研究具体相对应的媒体组件。

4.5.1 音　频

音频(audio)组件，从 1.6.0 版本开始，不再维护该组件。建议使用能力更强的 wx.createInnerAudioContext 接口。该组件的属性如表 4-30 所列。

表 4-30　audio 属性

属性名	类型	默认值	说明
id	String		audio 组件的唯一标识符
src	String		要播放音频的资源地址
loop	Boolean	false	是否循环播放
controls	Boolean	false	是否显示默认控件
poster	String		默认控件上的音频封面的图片资源地址，如果 controls 属性值为 false 则设置 poster 无效
name	String	未知音频	默认控件上的音频名字，如果 controls 属性值为 false 则设置 name 无效
author	String	未知作者	默认控件上的作者名字，如果 controls 属性值为 false 则设置 author 无效
binderror	EventHandle		当发生错误时触发 error 事件，detail = {errMsg: MediaError.code}
bindplay	EventHandle		当开始/继续播放时触发play事件
bindpause	EventHandle		当暂停播放时触发 pause 事件
bindtimeupdate	EventHandle		当播放进度改变时触发 timeupdate 事件，detail = {currentTime, duration}
bindended	EventHandle		当播放到末尾时触发 ended 事件

返回错误码如表 4-31 所列。

表 4-31　返回错误码

返回错误码	描述
1	获取资源被用户禁止
2	网络错误
3	解码错误
4	不合适资源

示例代码如下：

```
<!-- audio.wxml -->
<!-- id 为该组件的唯一标识 -->
<audio poster="{{poster}}" name="{{name}}" author="{{author}}" src="{{src}}"
id="myAudio" controls loop></audio>
<!-- bindtap 属性这里标识绑定了音频播放事件 -->
<button type="primary" bindtap="audioPlay">播放</button>
<button type="primary" bindtap="audioPause">暂停</button>
<button type="primary" bindtap="audio14">设置当前播放时间为 14 s</button>
<button type="primary" bindtap="audioStart">回到开头</button>

// audio.js
Page({
  onReady: function (e) {
    // 使用 wx.createAudioContext 获取 audio 上下文 context
    this.audioCtx = wx.createAudioContext('myAudio')
  },
  data: {
    poster: 'http://y.gtimg.cn/music/photo_new/T002R300x300M000003rsKF44GyaSk.jpg?
max_age=2592000',
    name: '此时此刻',
    author: '许巍',
      src: 'http://ws.stream.qqmusic.qq.com/M500001VfvsJ21xFqb.mp3?guid=
      ffffffff82def4af4b12b3cd9337d5e7&uin=346897220&vkey=6292F51E1E384E06-DCB-
      DC9AB7C49FD713D632D313AC4858BACB8DDD29067D3C601481D36E62053BF8DFEAF74C0A5C-
      CFADD6471160CAF3E6A&fromtag=46',
  },
      //绑定的播放事件
  audioPlay: function () {
    this.audioCtx.play()
```

```
  },
  audioPause: function () {
    this.audioCtx.pause()
  },
  audio14: function () {
    this.audioCtx.seek(14)
  },
  audioStart: function () {
    this.audioCtx.seek(0)
  }
})
```

示例效果图如图 4－26 所示。

图 4－26　Audio 示例

4.5.2　图　片

图片(image)组件，有两种模式：裁剪和缩放。需要注意：image 组件默认宽度 300px、高度 225px；image 组件中二维码/小程序码图片不支持长按识别，仅在 wx.previewImage 中支持长按识别。Image 组件的属性如表 4－32 所列。

表 4－32　image 属性

属性名	类型	默认值	说明	最低版本
src	String		图片资源地址，支持云文件ID（2.2.3起）	
mode	String	'scaleToFill'	图片裁剪、缩放的模式	
lazy-load	Boolean	false	图片懒加载。只针对page与scroll-view下的image有效	1.5.0
binderror	HandleEvent		当错误发生时，发布到 AppService 的事件名，事件对象event.detail = {errMsg: 'something wrong'}	
bindload	HandleEvent		当图片载入完毕时，发布到 AppService 的事件名，事件对象event.detail = {height:'图片高度px', width:'图片宽度px'}	

mode 有 13 种模式，其中 4 种是缩放模式，9 种是裁剪模式。详细分类说明如表 4－33 所列。

表 4－33　mode 模式

模式	值	说明
缩放	scaleToFill	不保持纵横比缩放图片，使图片的宽高完全拉伸至填满 image 元素
缩放	aspectFit	保持纵横比缩放图片，使图片的长边能完全显示出来。也就是说，可以完整地将图片显示出来。
缩放	aspectFill	保持纵横比缩放图片，只保证图片的短边能完全显示出来。也就是说，图片通常只在水平或垂直方向是完整的，另一个方向将会发生截取。
缩放	widthFix	宽度不变，高度自动变化，保持原图宽高比不变
裁剪	top	不缩放图片，只显示图片的顶部区域
裁剪	bottom	不缩放图片，只显示图片的底部区域
裁剪	center	不缩放图片，只显示图片的中间区域
裁剪	left	不缩放图片，只显示图片的左边区域
裁剪	right	不缩放图片，只显示图片的右边区域
裁剪	top left	不缩放图片，只显示图片的左上边区域
裁剪	top right	不缩放图片，只显示图片的右上边区域
裁剪	bottom left	不缩放图片，只显示图片的左下边区域
裁剪	bottom right	不缩放图片，只显示图片的右下边区域

示例代码如下：

```
//wxml 代码
 <view class="page">
   <view class="page__hd">
     <text class="page__title">image</text>
     <text class="page__desc">图片</text>
   </view>
   <view class="page__bd">
     <view class="section section_gap" wx:for="{{array}}" wx:for-item="item">
       <view class="section__title">{{item.text}}</view>
       <view class="section__ctn">
         <image style="width: 200px; height: 200px; background-color: #eeeeee;"
```

```
mode="{{item.mode}}" src="{{src}}"></image>
        </view>
      </view>
    </view>
  </view>

  //js代码
  Page({
    data: {
      array: [{
        mode: 'scaleToFill',
        text: 'scaleToFill:不保持纵横比缩放图片,使图片完全适应'
      }, {
        mode: 'aspectFit',
        text: 'aspectFit:保持纵横比缩放图片,使图片的长边能完全显示出来'
      }, {
        mode: 'aspectFill',
        text: 'aspectFill:保持纵横比缩放图片,只保证图片的短边能完全显示出来'
      }, {
        mode: 'top',
        text: 'top:不缩放图片,只显示图片的顶部区域'
      }, {
        mode: 'bottom',
        text: 'bottom:不缩放图片,只显示图片的底部区域'
      }, {
        mode: 'center',
        text: 'center:不缩放图片,只显示图片的中间区域'
      }, {
        mode: 'left',
        text: 'left:不缩放图片,只显示图片的左边区域'
      }, {
        mode: 'right',
        text: 'right:不缩放图片,只显示图片的右边边区域'
      }, {
        mode: 'top left',
        text: 'top left:不缩放图片,只显示图片的左上边区域'
      }, {
```

```
      mode: 'top right',
      text: 'top right:不缩放图片,只显示图片的右上边区域'
    }, {
      mode: 'bottom left',
      text: 'bottom left:不缩放图片,只显示图片的左下边区域'
    }, {
      mode: 'bottom right',
      text: 'bottom right:不缩放图片,只显示图片的右下边区域'
    }],
    src: '../../resources/cat.jpg'
  },
  imageError: function(e) {
    console.log('image3 发生 error 事件,携带值为', e.detail.errMsg)
  }
})
```

此处以一图片作为示例,原图如图 4-27 所示。

图 4-27　原图

scaleToFill 缩放模式,不保持纵横比缩放图片,使图片完全适应,如图 4-28 所示。

aspectFit 缩放模式,保持纵横比缩放图片,使图片的长边能完全显示出来,如图 4-29 所示。

aspectFill 缩放模式,保持纵横比缩放图片,只保证图片的短边能完全显示出来,如图 4-30 所示。

top 裁剪模式,不缩放图片,只显示图片的顶部区域,如图 4-31 所示。

图 4-28 scaleToFill 缩放模式

图 4-29 aspectFit 缩放模式

图 4-30 aspectFill 缩放模式

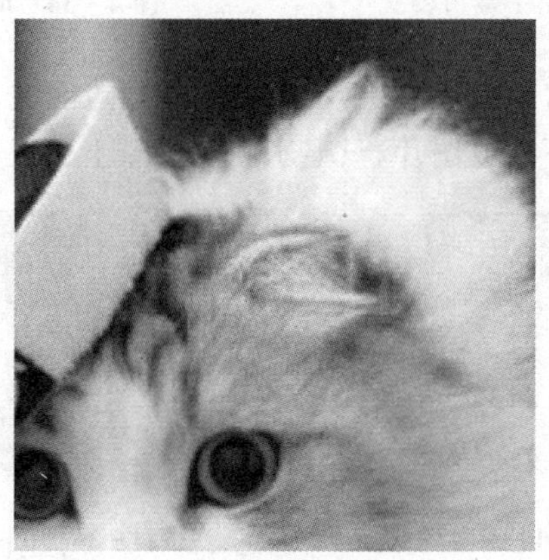

图 4-31 top 裁剪模式

bottom 裁剪模式，不缩放图片，只显示图片的底部区域，如图 4-32 所示。

图 4－32　bottom 裁剪模式

center 裁剪模式，不缩放图片，只显示图片的中间区域，如图 4－33 所示。

left 裁剪模式，不缩放图片，只显示图片的左边区域，如图 4－34 所示。

图 4－33　center 裁剪模式

图 4－34　left 裁剪模式

right 裁剪模式，不缩放图片，只显示图片的右边边区域，如图 4－35 所示。

top left 裁剪模式，不缩放图片，只显示图片的左上边区域，如图 4－36 所示。

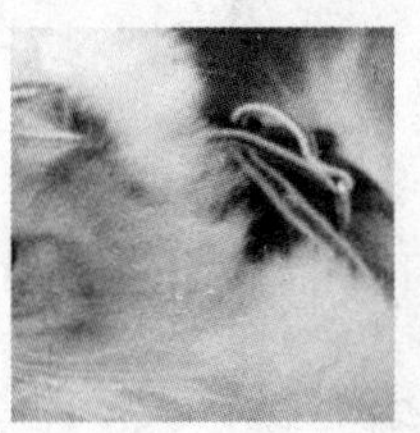

图 4－35　right 裁剪模式

图 4－36　top left 裁剪模式

top right 裁剪模式，不缩放图片，只显示图片的右上边区域，如图 4－37 所示。

bottom left 裁剪模式，不缩放图片，只显示图片的左下边区域，如图 4－38 所示。

bottom right 裁剪模式，不缩放图片，只显示图片的右下边区域，如图 4－39 所示。

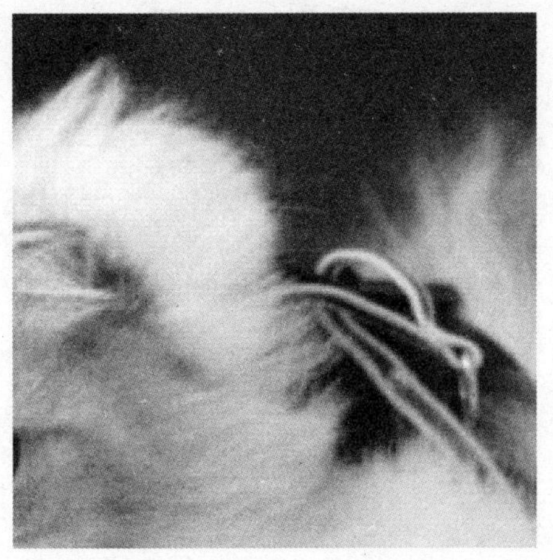

图 4-37　top right 裁剪模式

图 4-38　bottom left 裁剪模式

图 4-39　bottom right 裁剪模式

4.5.3 视　频

视频(video),该组件是原生组件,使用时请注意相关限制。视频组件默认宽度 300px、高度 225px,可通过 wxss 设置宽、高。该组件的属性如表 4-34 所列。

表 4-34　video 属性

属性名	类型	默认值	说明	最低版本
src	String		要播放视频的资源地址，支持云文件ID（2.2.3起）	
initial-time	Number		指定视频初始播放位置	1.6.0
duration	Number		指定视频时长	1.1.0
controls	Boolean	true	是否显示默认播放控件（播放/暂停按钮、播放进度、时间）	
danmu-list	Object Array		弹幕列表	
danmu-btn	Boolean	false	是否显示弹幕按钮，只在初始化时有效，不能动态变更	
enable-danmu	Boolean	false	是否展示弹幕，只在初始化时有效，不能动态变更	
autoplay	Boolean	false	是否自动播放	
loop	Boolean	false	是否循环播放	1.4.0
muted	Boolean	false	是否静音播放	1.4.0
page-gesture	Boolean	false	在非全屏模式下，是否开启亮度与音量调节手势	1.6.0
direction	Number		设置全屏时视频的方向，不指定则根据宽高比自动判断。有效值为 0（正常竖向），90（屏幕逆时针90度），-90（屏幕顺时针90度）	1.7.0
show-progress	Boolean	true	若不设置，宽度大于240时才会显示	1.9.0
show-fullscreen-btn	Boolean	true	是否显示全屏按钮	1.9.0
show-play-btn	Boolean	true	是否显示视频底部控制栏的播放按钮	1.9.0
show-center-play-btn	Boolean	true	是否显示视频中间的播放按钮	1.9.0
enable-progress-gesture	Boolean	true	是否开启控制进度的手势	1.9.0
objectFit	String	contain	当视频大小与 video 容器大小不一致时，视频的表现形式；contain：包含，fill：填充，cover：覆盖	
poster	String		视频封面的图片网络资源地址或云文件ID（2.2.3起支持）如果 controls 属性值为 false 则设置 poster 无效	
bindplay	EventHandle		当开始/继续播放时触发play事件	
bindpause	EventHandle		当暂停播放时触发 pause 事件	
bindended	EventHandle		当播放到末尾时触发 ended 事件	
bindtimeupdate	EventHandle		播放进度变化时触发，event.detail = {currentTime, duration}。触发频率 250ms 一次	
bindfullscreenchange	EventHandle		视频进入和退出全屏时触发，event.detail = {fullScreen, direction}，direction取为 vertical 或 horizontal	1.4.0
bindwaiting	EventHandle		视频出现缓冲时触发	1.7.0
binderror	EventHandle		视频播放出错时触发	1.7.0
bindprogress	EventHandle		加载进度变化时触发，只支持一段加载。event.detail = {buffered}，百分比	2.4.0

示例代码如下：

```
//wxml 代码
<view class="section tc">
  <video src="{{src}}"  controls ></video>
  <view class="btn-area">
    <button bindtap="bindButtonTap">获取视频</button>
  </view>
</view>

<view class="section tc">
  <video id="myVideo" src="http://wxsnsdy.tc.qq.com/105/20210/snsdyvideodownload?filekey=30280201010421301f0201690402534804102ca905ce620b1241b726bc41dcff44e00204012882540400&bizid=1023&hy=SH&fileparam=302c020101042530230204136ffd93020457e3c4ff02024ef202031e8d7f02030f42400204045a320a0201000400" danmu-list="{{danmuList}}" enable-danmu danmu-btn controls ></video>
  <view class="btn-area">
    <button bindtap="bindButtonTap">获取视频</button>
    <input bindblur="bindInputBlur"/>
    <button bindtap="bindSendDanmu">发送弹幕</button>
  </view>
</view>

//js 代码
function getRandomColor () {
  let rgb = []
  for (let i = 0 ; i < 3; ++i){
    let color = Math.floor(Math.random() * 256).toString(16)
    color = color.length == 1 ? '0' + color : color
    rgb.push(color)
  }
  return '#' + rgb.join('')
}

Page({
  onReady: function (res) {
    this.videoContext = wx.createVideoContext('myVideo')
  },
  inputValue: '',
```

```
data: {
  src: '',
  danmuList: [
    {
      text: '第 1s 出现的弹幕',
      color: '#ff0000',
      time: 1
    },
    {
      text: '第 3s 出现的弹幕',
      color: '#ff00ff',
      time: 3
  }]
},
bindInputBlur: function(e) {
  this.inputValue = e.detail.value
},
bindButtonTap: function() {
  var that = this
  wx.chooseVideo({
    sourceType: ['album', 'camera'],
    maxDuration: 60,
    camera: ['front','back'],
    success: function(res) {
      that.setData({
        src: res.tempFilePath
      })
    }
  })
},
bindSendDanmu: function () {
  this.videoContext.sendDanmu({
    text: this.inputValue,
    color: getRandomColor()
  })
}
})
```

示例效果图如图 4－40 所示。

图 4－40 Video 示例

4.5.4 相 机

相机(camera)组件，从基础库 1.6.0 开始支持，低版本需做兼容处理。该组件是原生组件，使用时请注意相关限制。扫二维码功能，需升级微信客户端至 6.7.3。需要注意的是：相机组件使用时需要用户授权 scope.camera，再就是同一页面只能插入一个 camera 组件。该组件的属性如表 4－35 所列。

表 4－35 camera 属性

属性名	类型	默认值	说明	最低版本
mode	String	normal	有效值为 normal, scanCode	2.1.0
device-position	String	back	前置或后置，值为front, back	
flash	String	auto	闪光灯，值为auto, on, off	
bindstop	EventHandle		摄像头在非正常终止时触发，如退出后台等情况	
binderror	EventHandle		用户不允许使用摄像头时触发	
bindscancode	EventHandle		在扫码识别成功时触发，仅在 mode="scanCode" 时生效	2.1.0支持一维码，2.4.0支持二维码

示例代码如下：

```
<!-- camera.wxml -->
<camera device-position="back" flash="off" binderror="error" style="width:
100%; height: 300px;"></camera>
<button type="primary" bindtap="takePhoto">拍照</button>
<view>预览</view>
<image mode="widthFix" src="{{src}}"></image>

// camera.js
Page({
  takePhoto() {
    const ctx = wx.createCameraContext()
    ctx.takePhoto({
      quality: 'high',
      success: (res) => {
        this.setData({
          src: res.tempImagePath
        })
      }
    })
  },
  error(e) {
    console.log(e.detail)
  }
})
```

示例效果图如图 4-41 所示。

4.5.5 实时音视频播放

实时音视频播放(live-player)组件是原生组件,使用时请注意相关限制。该组件从基础库 1.7.0 开始支持,低版本需做兼容处理。这两年风靡的直播应用慢慢地也可以用小程序来实现,该组件即可用来实现直播流媒体的实时播放。直播功能组件目前需要进行相关的审核才可以开通该组件的权限。该组件的属性如表 4-36 所列。

图 4－41　Camera 示例

表 4－36　live－player 属性

属性名	类型	默认值	说明	最低版本
src	String		音视频地址。目前仅支持 flv, rtmp 格式	
mode	String	live	live（直播），RTC（实时通话，该模式时延更低）	
autoplay	Boolean	false	自动播放	
muted	Boolean	false	是否静音	
orientation	String	vertical	画面方向，可选值有 vertical，horizontal	
object-fit	String	contain	填充模式，可选值有 contain，fillCrop	
background-mute	Boolean	false	进入后台时是否静音（已废弃，默认退台静音）	
min-cache	Number	1	最小缓冲区，单位s（RTC 模式推荐 0.2s）	
max-cache	Number	3	最大缓冲区，单位s（RTC 模式推荐 0.8s）	
bindstatechange	EventHandle		播放状态变化事件，detail = {code}	
bindfullscreenchange	EventHandle		全屏变化事件，detail = {direction, fullScreen}	
bindnetstatus	EventHandle		网络状态通知，detail = {info}	1.9.0

需要注意：live-player组件默认宽度300px、高度225px，可通过wxss设置宽、高。目前开发者工具上暂不支持该组件的动态调试。

示例代码如下：

```
//wxml 代码
< live-player src = "https://domain/pull_stream" mode = "RTC" autoplay bindstate-
change = "statechange" binderror = "error" style = "width: 300px; height: 225px;" />

//js 代码
Page({
  statechange(e) {
    console.log('live-player code:', e.detail.code)
  },
  error(e) {
    console.error('live-player error:', e.detail.errMsg)
  }
})
```

示例效果图如图4-42所示。

图4-42 live-player示例

4.5.6 实时音视频录制

实时音视频录制(live - pusher),该组件是原生组件,使用时请注意相关限制。该组件从基础库 1.7.0 开始支持,低版本需做兼容处理。使用该组件需要用户授权 scope. camera 和 scope. record。该组件可用来实现直播推流。该组件的属性如表 4 - 37 所列。

表 4 - 37 live - pusher 属性

属性名	类型	默认值	说明	最低版本
url	String		推流地址。目前仅支持 flv, rtmp 格式	
mode	String	RTC	SD(标清), HD(高清), FHD(超清), RTC(实时通话)	
autopush	Boolean	false	自动推流	
muted	Boolean	false	是否静音	
enable-camera	Boolean	true	开启摄像头	
auto-focus	Boolean	true	自动聚集	
orientation	String	vertical	vertical, horizontal	
beauty	Number	0	美颜,取值范围 0-9 ,0 表示关闭	
min-bitrate	Number	200	最小码率	
max-bitrate	Number	1000	最大码率	
waiting-image	String		进入后台时推流的等待画面	
waiting-image-hash	String		等待画面资源的MD5值	
zoom	Boolean	false	调整焦距	2.1.0
device-position	String	front	前置或后置,值为front, back	2.3.0
background-mute	Boolean	false	进入后台时是否静音	
bindstatechange	EventHandle		状态变化事件,detail = {code}	
bindnetstatus	EventHandle		网络状态通知,detail = {info}	1.9.0
binderror	EventHandle		渲染错误事件,detail = {errMsg, errCode}	1.7.4
bindbgmstart	EventHandle		背景音开始播放时触发	2.4.0
bindbgmprogress	EventHandle		背景音进度变化时触发,detail = {progress, duration}	2.4.0
bindbgmcomplete	EventHandle		背景音播放完成时触发	2.4.0

需要注意的是:live - pusher 组件的默认宽度为 100%,无默认高度,请通过 wxss 设置宽、高。目前开发者工具上暂不支持该组件的动态调试。示例代码如下:

```
//wxml 代码
< live - pusher url = "https://domain/push_stream" mode = "RTC" autopush bindstatechange = "statechange" style = "width: 300px; height: 225px;" />

//js 代码
Page({
  statechange(e) {
    console.log('live - pusher code:', e.detail.code)
  }
})
```

示例效果图如图 4 - 43 所示。

图 4 - 43　live - pusher 示例

4.6 地图组件

地图(map)组件是原生组件,使用时请注意相关限制。该组件的使用场景在LBS(基于位置的服务)应用的非常广泛,比如我们常用的摩拜小程序就是基于map地图组件开发的。如果需要个性化的地图能力,可以在小程序后台"设置→开发者工具→腾讯位置服务"申请开通。设置subkey(如果没有key,需要先创建,这个key用于配置在小程序的map组件中,也就是subkey,个性样式就是绑定在key上的)后,小程序内的地图组件均会使用该地图效果,地图场景的切换会在后续版本提供,详见《小程序个性地图使用指南》,网址为:https://lbs.qq.com/product/miniapp/guide/。注意,个性化地图暂不支持在工具中调试,请先使用微信客户端进行测试。地图组件的属性如表4-38所列。

表4-38 map属性

属性名	类型	默认值	说明	最低版本
longitude	Number		中心经度	
latitude	Number		中心纬度	
scale	Number	16	缩放级别,取值范围为5~18	
markers	Array		标记点	
covers	Array		**即将移除,请使用 markers**	
polyline	Array		路线	
polygons	Array		多边形	2.3.0
circles	Array		圆	
controls	Array		控件(即将废弃,建议使用 cover-view 代替)	
include-points	Array		缩放视野以便包含所有给下的坐标点	
show-location	Boolean		显示带有方向的当前定位点	
subkey	String	''	个性化地图使用的key,仅初始化地图时有效	2.3.0
enable-3D	Boolean	false	展示3D楼块(工具暂不支持)	2.3.0
show-compass	Boolean	false	显示指南针	2.3.0

续表 4-38

属性名	类型	默认值	说明	最低版本
enable-overlooking	Boolean	false	开启俯视	2.3.0
enable-zoom	Boolean	true	是否支持缩放	2.3.0
enable-scroll	Boolean	true	是否支持拖动	2.3.0
enable-rotate	Boolean	false	是否支持旋转	2.3.0
bindmarkertap	EventHandle		点击标记点时触发，会返回marker的id	
bindcallouttap	EventHandle		点击标记点对应的气泡时触发，会返回marker的id	1.2.0
bindcontroltap	EventHandle		点击控件时触发，会返回control的id	
bindregionchange	EventHandle		视野发生变化时触发	2.3.0起增加 causedBy 参数区分拖动、缩放和调用接口等来源
bindtap	EventHandle		点击地图时触发	
bindupdated	EventHandle		在地图渲染更新完成时触发	1.6.0
bindpoitap	EventHandle		点击地图poi点时触发	2.3.0

示例代码如下：

```
<!-- map.wxml -->
< map id = "map" longitude = "113.324520" latitude = "23.099994" scale = "14" controls = "{{controls}} " bindcontroltap = "controltap" markers = "{{markers}}" bindmarkertap = " markertap" polyline = "{{polyline}}" bindregionchange = "regionchange" show - location style = "width: 100 % ; height: 300px;" > </map >

// map.js
Page({
  data: {
    markers: [{
      iconPath: "/resources/others.png",
      id: 0,
      latitude: 23.099994,
      longitude: 113.324520,
      width: 50,
      height: 50
    }],
```

```
    polyline：[{
      points：[{
        longitude：113.3245211，
        latitude：23.10229
      }，{
        longitude：113.324520，
        latitude：23.21229
      }]，
      color："#FF0000DD"，
      width：2，
      dottedLine：true
    }]，
    controls：[{
      id：1，
      iconPath：'/resources/location.png'，
      position：{
        left：0，
        top：300 - 50，
        width：50，
        height：50
      }，
      clickable：true
    }]
  }，
  regionchange(e) {
    console.log(e.type)
  }，
  markertap(e) {
    console.log(e.markerId)
  }，
  controltap(e) {
    console.log(e.controlId)
  }
})
```

示例效果图如图 4-44 所示。

图 4－44　map 示例

4.7　画布组件

画布(canvas)组件是原生组件,使用时请注意相关限制。需要注意的是:canvas 标签默认宽度 300px、高度 225px;同一页面中的 canvas－id 不可重复,如果使用一个已经出现过的 canvas－id,该 canvas 标签对应的画布将被隐藏并不再正常工作。该组件的属性如表 4－39 所列。

表 4-39 canvas 属性

属性名	类型	默认值	说明
canvas-id	String		canvas 组件的唯一标识符
disable-scroll	Boolean	false	当在 canvas 中移动时且有绑定手势事件时，禁止屏幕滚动以及下拉刷新
bindtouchstart	EventHandle		手指触摸动作开始
bindtouchmove	EventHandle		手指触摸后移动
bindtouchend	EventHandle		手指触摸动作结束
bindtouchcancel	EventHandle		手指触摸动作被打断，如来电提醒，弹窗
bindlongtap	EventHandle		手指长按 500ms 之后触发，触发了长按事件后进行移动不会触发屏幕的滚动
binderror	EventHandle		当发生错误时触发 error 事件，detail = {errMsg: 'something wrong'}

示例代码如下：

```
<!-- canvas.wxml -->
<canvas style="width: 300px; height: 200px;" canvas-id="firstCanvas"></canvas>
<!-- 当使用绝对定位时,文档流后边的 canvas 的显示层级高于前边的 canvas -->
<canvas style="width: 400px; height: 500px;" canvas-id="secondCanvas"></canvas>
<!-- 因为 canvas-id 与前一个 canvas 重复,该 canvas 不会显示,并会发送一个错误事件到 AppService -->
<canvas style="width: 400px; height: 500px;" canvas-id="secondCanvas" binderror="canvasIdErrorCallback"></canvas>

// canvas.js
Page({
  canvasIdErrorCallback: function (e) {
    console.error(e.detail.errMsg)
  },
  onReady: function (e) {
    // 使用 wx.createContext 获取绘图上下文 context
    var context = wx.createCanvasContext('firstCanvas')

    context.setStrokeStyle("#00ff00")
    context.setLineWidth(5)
    context.rect(0, 0, 200, 200)
    context.stroke()
```

```
    context.setStrokeStyle("#ff0000")
    context.setLineWidth(2)
    context.moveTo(160, 100)
    context.arc(100, 100, 60, 0, 2 * Math.PI, true)
    context.moveTo(140, 100)
    context.arc(100, 100, 40, 0, Math.PI, false)
    context.moveTo(85, 80)
    context.arc(80, 80, 5, 0, 2 * Math.PI, true)
    context.moveTo(125, 80)
    context.arc(120, 80, 5, 0, 2 * Math.PI, true)
    context.stroke()
    context.draw()
  }
})
```

示例效果图如图 4-45 所示。

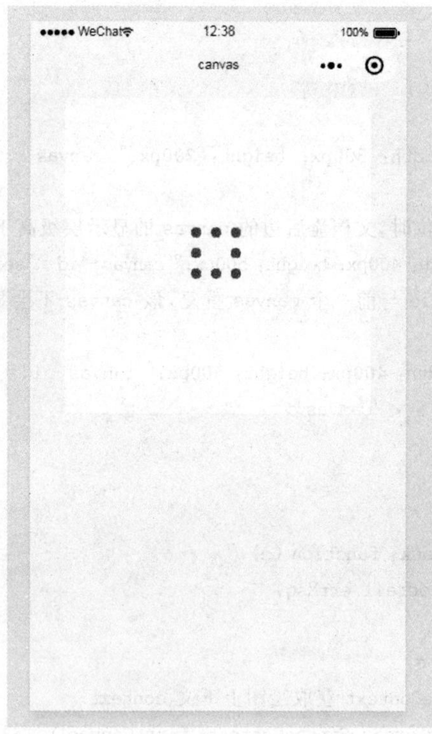

图 4-45 canvas 示例

4.8 开放能力组件

开放能力组件大多为新增组件，且使用频率较低，属于相对比较冷门的组件。读者可以先简单了解一下。

4.8.1 开放数据

开放数据(open - data)组件用于展示微信开放的数据。该组件从基础库 1.4.0 开始支持，低版本需做兼容处理，该组件属性如表 4 - 40 所列。

表 4 - 40 open - data 属性

属性名	类型	默认值	说明
type	String		开放数据类型
open-gid	String		当 type="groupName" 时生效, 群id
lang	String	en	当 type="user*" 时生效，以哪种语言展示 userInfo，有效值有：en、zh_CN、zh_TW

示例代码如下：

```
< open - data type = "groupName" open - gid = "xxxxxx" > < /open - data >
 < open - data type = "userAvatarUrl" > < /open - data >
 < open - data type = "userGender" lang = "zh_CN" > < /open - data >
```

4.8.2 网页容器

网页容器(web - view)组件是一个可以用来承载网页的容器，会自动铺满整个小程序页面。个人类型与海外类型的小程序暂不支持使用。从基础库 1.6.4 开始支持，低版本需做兼容处理，该组件属性如表 4 - 41 所列。

表 4 - 41 web - view 属性

属性名	类型	默认值	说明
src	String		webview 指向网页的链接。可打开关联的公众号的文章，其他网页需登录小程序管理后台配置业务域名
bindmessage	EventHandler		网页向小程序 postMessage 时，会在特定时机（小程序后退、组件销毁、分享）触发并收到消息。e.detail = { data }
bindload	EventHandler		网页加载成功时候触发此事件。e.detail = { src }
binderror	EventHandler		网页加载失败的时候触发此事件。e.detail = { src }

示例代码如下：

```
<!-- wxml -->
<!-- 指向微信公众平台首页的 web-view -->
<web-view src="https://mp.weixin.qq.com/"></web-view>

<!-- html -->
<script type="text/javascript" src="https://res.wx.qq.com/open/js/jweixin-1.3.2.js"></script>

// javascript
wx.miniProgram.navigateTo({url: '/path/to/page'})
wx.miniProgram.postMessage({ data: 'foo' })
wx.miniProgram.postMessage({ data: {foo: 'bar'} })
wx.miniProgram.getEnv(function(res) { console.log(res.miniprogram) // true })
```

示例效果图如图 4-46 所示。

图 4-46 web-view 示例

4.8.3 广 告

广告(ad)组件从基础库 1.9.94 开始支持，低版本需做兼容处理。目前暂时以邀请制开放申请。

4.8.4 公众号关注

公众号关注(official-account)组件,从基础库 2.3.0 开始支持,低版本需做兼容处理。用户扫码打开小程序时,开发者可在小程序内配置公众号关注组件,方便用户快捷地关注公众号,可嵌套在原生组件内。该组件非常适合用来引流用户关注公众号,然后再做相关的营销推广活动。

需要注意的是:

(1) 使用组件前,需前往小程序后台,在“设置→接口设置→公众号关注组件”中设置要展示的公众号。设置的公众号需与小程序主体一致。

(2) 在一个小程序的生命周期内,只有从以下场景进入小程序,才具有展示引导关注公众号组件的能力:当小程序从扫二维码场景(场景值 1011)打开时;当小程序从扫小程序码场景(场景值 1047)打开时;当小程序从聊天顶部场景(场景值 1089)中的“最近使用”内打开时,若小程序之前未被销毁,则该组件保持上一次打开小程序时的状态;当从其他小程序返回小程序(场景值 1038)时,若小程序之前未被销毁,则该组件保持上一次打开小程序时的状态。

(3) 每个页面只能配置一个该组件。

示例代码如下:

```
<official-account></official-account>
```

第 5 章

小程序框架 API

微信小程序框架所提供的 API 接口也是相当完备的，如果说小程序组件是用来构建小程序的视图层，那么小程序 API 则在小程序逻辑层担当重任。随着小程序版本的更新迭代，目前小程序框架 API 分类已经达到了 15 个大类，如图 5-1 所示。

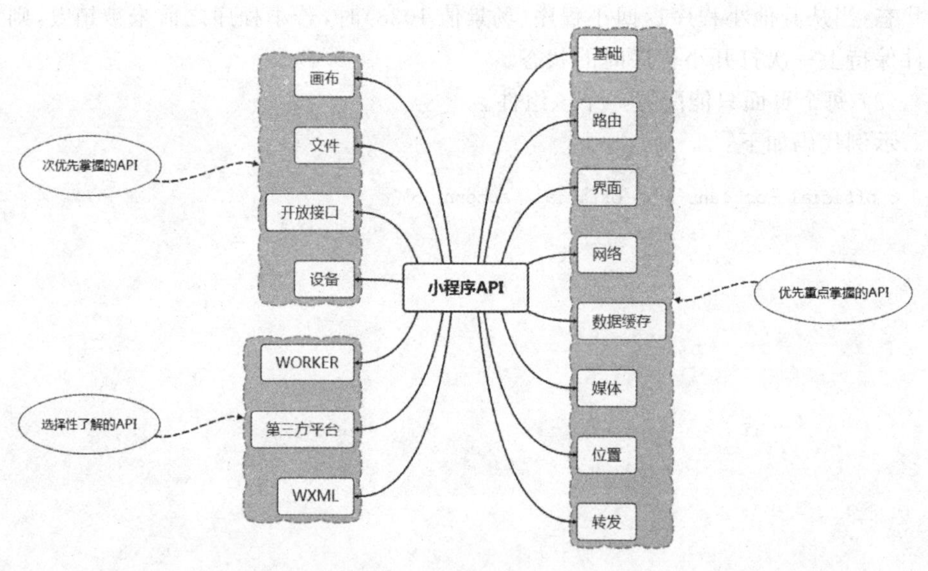

图 5-1　小程序 API 分类

当然这么多 API 并不需要读者一个个去详细研究，而是建议大家重点掌握其中常用的 API 如何查阅和使用，然后即可触类旁通使用其他的 API。其实查阅 API 有点类似于查词典，我们只需要知道如何去查阅的方法即可，而没必要将整部词典学会，故而掌握查阅 API 定义以及如何使用才是至关重要的。

接下来，就开始重点讲解小程序框架中一些常用的 API，通过这些 API 的学习和理解，希望读者能够对小程序 API 如何查阅和使用形成一定的思维认知。

5.1 网络 API

网络 API 在开发需要进行后端服务接口调用、文件上传下载、长连接 WebSocket 和监听 mDNS 服务等功能时非常有用。发起请求和上传下载这类 API 在开发中经常使用，属于高频 API，建议读者重点掌握。

5.1.1 发起请求

wx.request(Object object)

该接口用于发起 HTTPS 网络请求。接口参数如表 5-1 所列。

表 5-1 wx.request 接口参数

属性	类型	默认值	必填	说明	最低版本
url	string		是	开发者服务器接口地址	
data	string/object/ArrayBuffer		否	请求的参数	
header	Object		否	设置请求的 header，header 中不能设置 Referer。content-type 默认为 application/json	
method	string	GET	否	HTTP 请求方法	
dataType	string	json	否	返回的数据格式	
responseType	string	text	否	响应的数据类型	1.7.0
success	function		否	接口调用成功的回调函数	
fail	function		否	接口调用失败的回调函数	
complete	function		否	接口调用结束的回调函数（调用成功、失败都会执行）	

其中属性 method 的合法值有：OPTIONS、GET、HEAD、POST、PUT、DELETE、TRACE 和 CONNECT。示例代码如下：

```
wx.request({
  url: 'https://www.itsoho.cn/test/req', //仅为示例,并非真实接口地址
  data: {name:"tom",password:"123",id:"123"},
  method: 'post', // OPTIONS, GET, HEAD, POST, PUT, DELETE, TRACE, CONNECT
```

```
      header：{
'content-type'：'application/json' // 默认值
},
  // dataType:JSON,//该语句会将服务器端的数据自动转为 string 类型
  success：function(res){
    // success
    console.log("返回数据为：" + res.data.employees[1].firstName);
    console.log('submit success');
  },
  fail：function() {
    // fail
    console.log('submit fail');
  },
  complete：function() {
    // complete
    console.log('submit comlete');
  }
})
```

5.1.2 上传下载

1. wx.uploadFile(Object object)

将本地资源上传到服务器。客户端发起一个 HTTPS POST 请求，其中 content-type 为 multipart/form-data。该接口参数如表 5-2 所列。

表 5-2 wx.uploadFile 接口参数

属性	类型	默认值	必填	说明	最低版本
url	string		是	开发者服务器地址	
filePath	string		是	要上传文件资源的路径	
name	string		是	文件对应的 key，开发者在服务端可以通过这个 key 获取文件的二进制内容	
header	Object		否	HTTP 请求 Header，Header 中不能设置 Referer	
formData	Object		否	HTTP 请求中其他额外的 form data	
success	function		否	接口调用成功的回调函数	
fail	function		否	接口调用失败的回调函数	
complete	function		否	接口调用结束的回调函数（调用成功、失败都会执行）	

示例代码如下：

```
wx.chooseImage({
  success (res) {
    const tempFilePaths = res.tempFilePaths
    wx.uploadFile({
      url: 'https://example.weixin.qq.com/upload', //仅为示例,非真实的接口地址
      filePath: tempFilePaths[0],
      name: 'file',
      formData: {
        'user': 'test'
      },
      success (res){
        const data = res.data
        //do something
      }
    })
  }
})
```

2. wx.downloadFile(Object object)

下载文件资源到本地。客户端直接发起一个 HTTPS GET 请求，返回文件的本地临时路径。需要注意：请在服务端响应的 header 中指定合理的 Content-Type 字段，以保证客户端正确处理文件类型。该接口的参数如表 5-3 所列。

表 5-3 wx.downloadFile 接口参数

属性	类型	默认值	必填	说明	最低版本
url	string		是	下载资源的 url	
header	Object		否	HTTP 请求的 Header，Header 中不能设置 Referer	
filePath	string		否	指定文件下载后存储的路径	1.8.0
success	function		否	接口调用成功的回调函数	
fail	function		否	接口调用失败的回调函数	
complete	function		否	接口调用结束的回调函数（调用成功、失败都会执行）	

示例代码如下：

```
wx.downloadFile({
  url: 'https://example.com/audio/123', //仅为示例,并非真实的资源
```

```
    success (res) {
      // 只要服务器有响应数据，就会把响应内容写入文件并进入 success 回调，业务需要自
         行判断是否下载到了想要的内容
      if (res.statusCode === 200) {
        wx.playVoice({
          filePath: res.tempFilePath
        })
      }
    }
  })
```

5.1.3 长连接 WebSocket

长连接 WebSocket 允许服务端和客户端彼此相互推送信息，它适用于需要高实时性的场景，比如：社交聊天、弹幕、多玩家在线游戏、协同编辑、股票基金实时报价，体育实况更新和视频会议等。长连接 WebSocket 常用 API 列表如下：

- wx.connectSocket：创建一个 WebSocket 连接；
- wx.closeSocket：关闭 WebSocket 连接；
- wx.onSocketOpen：监听 WebSocket 连接打开事件；
- wx.onSocketError：监听 WebSocket 错误事件；
- wx.onSocketClose：监听 WebSocket 连接关闭事件；
- wx.sendSocketMessage：通过 WebSocket 连接发送数据；
- wx.onSocketMessage：监听 WebSocket 接收服务器的消息事件。

长连接 WebSocket 示例代码如下：

```
//wxml 代码
<button bindtap="WebSocket">WebSocket</button>

//js 代码
Page({
  WebSocket:() => {

    wx.connectSocket({
      url: 'wss://echo.websocket.org',
      data: {
        x: '',
        y: ''
      },
      header: {
        'content-type': 'application/json'
      },
```

```
    method: "GET",
    success:(res) = > {
      console.log(" === 请求连接 === ");
      console.log(res)
    },
    fail: (res) = > {
      console.log(" === 请求失败 === ");
      console.log(res)
    },
    complete: (res) = > {
      console.log(" === 请求完成 === ");
      console.log(res)
    }
  });

  wx.onSocketOpen(function (res) {
    console.log('WebSocket 连接已打开! ');
    wx.sendSocketMessage({
      data: "Netease",
      success: (res) = > {
        console.log(" === 信息发送完成 === ");
        console.log(res)
      },
      fail: (res) = > {
        console.log(" === 信息发送失败 === ");
        console.log(res)
      },
      complete: (res) = > {
        console.log(" === 信息发送完成 === ");
        console.log(res)
      }
    })
  });

  wx.onSocketError(function (res) {
    console.log('WebSocket 连接打开失败,请检查! ')
  })

  wx.onSocketMessage(function (res) {
    console.log('收到服务器内容:' + res.data)

    wx.onSocketOpen(function () {
```

```
        wx.closeSocket()
      })

      wx.onSocketClose(function (res) {
        console.log('WebSocket 已关闭！')
      })
    })

  }
})
```

5.2 媒体 API

随着当下互联网短视频、微语音和图片信息流等应用的火爆，多媒体功能的开发变得越来越频繁。在小程序中，开发者只需借助媒体 API 即可轻松完成诸如音视频播放、录音、图片浏览与预览等常用的多媒体功能。

5.2.1 音视频

1. wx.playVoice(Object object)

开始播放语音。只允许一个语音文件正在播放，如果前一个语音文件还没播完，将中断前一个语音播放。从基础库 1.6.0 开始，该接口不再维护，建议使用 wx.createInnerAudioContext 代替。该接口参数如表 5－4 所列。

表 5－4 wx.playVoice 接口参数

属性	类型	默认值	必填	说明	最低版本
filePath	string		是	需要播放的语音文件的文件路径	
duration	number	60	否	指定录音时长，到达指定的录音时长后会自动停止录音，单位：S	1.6.0
success	function		否	接口调用成功的回调函数	
fail	function		否	接口调用失败的回调函数	
complete	function		否	接口调用结束的回调函数（调用成功、失败都会执行）	

示例代码如下：

```
wx.startRecord({
  success (res) {
    const tempFilePath = res.tempFilePath
```

```
    wx.playVoice({
      filePath: tempFilePath,
      complete () { }
    })
  }
})
```

2. wx.pauseVoice(Object object)

暂停正在播放的语音。再次调用 wx.playVoice 播放同一个文件时,会从暂停处开始播放。如果想从头开始播放,需要先调用 wx.stopVoice。从基础库 1.6.0 开始,该接口不再维护,建议使用 wx.createInnerAudioContext 代替。该接口参数如表 5-5 所列。

表 5-5 wx.pauseVoice 接口参数

属性	类型	默认值	必填	说明	最低版本
success	function		否	接口调用成功的回调函数	
fail	function		否	接口调用失败的回调函数	
complete	function		否	接口调用结束的回调函数(调用成功、失败都会执行)	

示例代码如下:

```
wx.startRecord({
  success (res) {
    const tempFilePath = res.tempFilePath
    wx.playVoice({
      filePath: tempFilePath
    })

    setTimeout(() => { wx.pauseVoice() }, 5000)
  }
})
```

3. wx.stopVoice(Object object)

结束播放语音。从基础库 1.6.0 开始,该接口不再维护,建议使用 wx.createInnerAudioContext 代替。该接口参数如表 5-6 所列。

表 5-6　wx.pauseVoice 接口参数

属性	类型	默认值	必填	说明	最低版本
success	function		否	接口调用成功的回调函数	
fail	function		否	接口调用失败的回调函数	
complete	function		否	接口调用结束的回调函数（调用成功、失败都会执行）	

示例代码如下：

```
wx.startRecord({
  success (res) {
    const tempFilePath = res.tempFilePath
    wx.playVoice({
      filePath: tempFilePath,
    })

    setTimeout(() = > { wx.stopVoice() }, 5000)
  }
})
```

4. wx.createInnerAudioContext()

通过该接口可以创建内部 Audio 上下文的 InnerAudioContext 对象，该对象中的一部分属性和方法列表如下所示：

① 属　性

- string src：音频资源的地址，用于直接播放。2.2.3 版本开始支持云文件 ID。
- number startTime：开始播放的位置(单位：s)，默认为 0。
- boolean autoplay：是否自动开始播放，默认为 false。
- boolean loop：是否循环播放，默认为 false。

② 方　法

- InnerAudioContext.play()：播放。
- InnerAudioContext.pause()：暂停。暂停后的音频再播放会从暂停处开始播放。
- InnerAudioContext.stop()：停止。停止后的音频再播放会从头开始播放。
- InnerAudioContext.seek(number position)：跳转到指定位置。
- InnerAudioContext.destroy()：销毁当前实例。
- InnerAudioContext.onCanplay(function callback)：监听音频进入可以播放状态的事件。但不保证后面可以流畅播放。
- InnerAudioContext.offCanplay(function callback)：取消监听音频进入可以播放状态的事件。但不保证后面可以流畅播放。

➢ InnerAudioContext.onPlay(function callback):监听音频播放事件。
➢ InnerAudioContext.offPlay(function callback):取消监听音频播放事件。
➢ InnerAudioContext.onPause(function callback):监听音频暂停事件。
➢ InnerAudioContext.offPause(function callback):取消监听音频暂停事件。
➢ InnerAudioContext.onStop(function callback):监听音频停止事件。
➢ InnerAudioContext.offStop(function callback):取消监听音频停止事件。

示例代码如下:

```
const innerAudioContext = wx.createInnerAudioContext()
innerAudioContext.autoplay = true
innerAudioContext.src = 'http://ws.stream.qqmusic.qq.com/M500001VfvsJ21xFqb.mp3?
  guid = ffffffff82def4af4b12b3cd9337d5e7&uin = 346897220&vkey =
  6292F51E1E384E061F-
  F02C31F716658E5C81F5594D561F2E88B854E81CAAB7806D5E4F103E55D33C16F3FAC506D1AB172DE8600-
  B37E43FAD&fromtag = 46'
innerAudioContext.onPlay(() = > {
  console.log('开始播放')
})
innerAudioContext.onError((res) = > {
  console.log(res.errMsg)
  console.log(res.errCode)
})
```

5. wx.chooseVideo(Object object)

拍摄视频或从手机相册中选视频。该 API 常用在用户上传视频的场景中,还有视频分享等视频类社区等应用场景。该接口参数如表 5-7 所列。

表 5-7 wx.chooseVideo 接口参数

属性	类型	默认值	必填	说明	最低版本
sourceType	Array.<string>	['album', 'camera']	否	视频选择的来源	
compressed	boolean	true	否	是否压缩所选择的视频文件	1.6.0
maxDuration	number	60	否	拍摄视频最长拍摄时间,单位 s	
camera	string	'back'	否	默认拉起的是前置或者后置摄像头。部分 Android 手机下由于系统 ROM 不支持无法生效	
success	function		否	接口调用成功的回调函数	
fail	function		否	接口调用失败的回调函数	
complete	function		否	接口调用结束的回调函数(调用成功、失败都会执行)	

示例代码如下：

```
wx.chooseVideo({
  sourceType: ['album','camera'],
  maxDuration: 60,
  camera: 'back',
  success(res) {
    console.log(res.tempFilePath)
  }
})
```

6. wx.createVideoContext(string id, Object this)

该接口用来创建 video 上下文 VideoContext 对象。videoContext 通过 id 跟一个 <video/> 组件绑定，操作对应的 <video/> 组件。videoContext 对象中的方法列表如下：

- VideoContext.play()：播放视频；
- VideoContext.pause()：暂停视频；
- VideoContext.stop()：停止视频；
- VideoContext.seek(number position)：跳转到指定位置；
- VideoContext.sendDanmu(Object data)：发送弹幕；
- VideoContext.playbackRate(number rate)：设置倍速播放；
- VideoContext.requestFullScreen(Object object)：进入全屏；
- VideoContext.exitFullScreen()：退出全屏；
- VideoContext.showStatusBar()：显示状态栏，仅在 iOS 全屏下有效；
- VideoContext.hideStatusBar()：隐藏状态栏，仅在 iOS 全屏下有效。

示例代码如下：

```
//wxml 代码
 <view class="section tc">
   <video id="myVideo" src="http://wxsnsdy.tc.qq.com/105/20210/snsdyvideodownload?filekey=30280201010421301f0201690402534804102ca905ce620b1241b726bc41dcff44e00204012882540400&bizid=1023&hy=SH&fileparam=302c020101042530230204136ffd93020457e3c4ff02024ef202031e8d7f02030f42400204045a320a0201000400" enable-danmu danmu-btn controls ></video>
   <view class="btn-area">
     <input bindblur="bindInputBlur"/>
     <button bindtap="bindSendDanmu">发送弹幕</button>
   </view>
 </view>
```

```
//js 代码
function getRandomColor () {
  let rgb = []
  for (let i = 0 ; i < 3; + + i) {
    let color = Math.floor(Math.random() * 256).toString(16)
    color = color.length == 1 ? '0' + color : color
    rgb.push(color)
  }
  return '#' + rgb.join('')
}

Page({
  onReady (res) {
    this.videoContext = wx.createVideoContext('myVideo')
  },
  inputValue: '',
  bindInputBlur (e) {
    this.inputValue = e.detail.value
  },
  bindSendDanmu () {
    this.videoContext.sendDanmu({
      text: this.inputValue,
      color: getRandomColor()
    })
  }
})
```

示例效果图如图 5 - 2 所示。

7. wx.createLivePlayerContext(string id, Object this)

该接口用于创建 live - player 上下文 LivePlayerContext 对象。livePlayerContext 通过 id 跟一个 < live - player/ > 组件绑定,操作对应的 < live - player/ > 组件。

8. wx.createLivePusherContext()

该接口用于创建 live - pusher 上下文 LivePusherContext 对象。livePusherContext 与页面内唯一的 < live - pusher/ > 组件绑定,操作对应的 < live - pusher/ > 组件。

图 5-2 视频 API 示例

5.2.2 图 片

1. wx.chooseImage(Object object)

该接口用于从本地相册选择图片或使用相机拍照。接口参数如表 5-8 所列。

表 5-8 wx.chooseImage 接口参数

属性	类型	默认值	必填	说明	最低版本
count	number	9	否	最多可以选择的图片张数	
sizeType	Array.<string>	['original', 'compressed']	否	所选的图片的尺寸	
sourceType	Array.<string>	['album', 'camera']	否	选择图片的来源	
success	function		否	接口调用成功的回调函数	
fail	function		否	接口调用失败的回调函数	
complete	function		否	接口调用结束的回调函数(调用成功、失败都会执行)	

示例代码如下：

```
wx.chooseImage({
  count: 1,
  sizeType: ['original', 'compressed'],
  sourceType: ['album', 'camera'],
  success (res) {
    // tempFilePath 可以作为 img 标签的 src 属性显示图片
    const tempFilePaths = res.tempFilePaths
  }
})
```

2. wx.compressImage(Object object)

该接口为压缩图片接口，可选压缩质量，该接口从基础库 2.4.0 开始支持，低版本需做兼容处理。接口参数如表 5-9 所列。

表 5-9 wx.compressImage 接口参数

属性	类型	默认值	必填	说明	最低版本
src	string		是	图片路径，图片的路径，可以是相对路径、临时文件路径、存储文件路径	
quality	number	80	否	压缩质量，范围0~100，数值越小，质量越低，压缩率越高（仅对jpg有效）。	
success	function		否	接口调用成功的回调函数	
fail	function		否	接口调用失败的回调函数	
complete	function		否	接口调用结束的回调函数（调用成功、失败都会执行）	

示例代码如下：

```
wx.compressImage({
  src: '', // 图片路径
  quality: 80 // 压缩质量
})
```

3. wx.getImageInfo(Object object)

接口用于获取图片信息。网络图片需先配置 download 域名才能生效。该接口参数如表 5-10 所列。

表 5 - 10 wx. getImageInfo 接口参数

属性	类型	默认值	必填	说明	最低版本
src	string		是	图片的路径，可以是相对路径、临时文件路径、存储文件路径、网络图片路径	
success	function		否	接口调用成功的回调函数	
fail	function		否	接口调用失败的回调函数	
complete	function		否	接口调用结束的回调函数（调用成功、失败都会执行）	

示例代码如下：

```
wx.getImageInfo({
  src: 'images/a.jpg',
  success (res) {
    console.log(res.width)
    console.log(res.height)
  }
})

wx.chooseImage({
  success (res) {
    wx.getImageInfo({
      src: res.tempFilePaths[0],
      success (res) {
        console.log(res.width)
        console.log(res.height)
      }
    })
  }
})
```

示例效果图如图 5 - 3 所示。

4. wx.previewImage(Object object)

该接口用于在新页面中全屏预览图片。预览的过程中用户可以进行保存图片、发送给朋友等操作。该接口参数如表 5 - 11 所列。

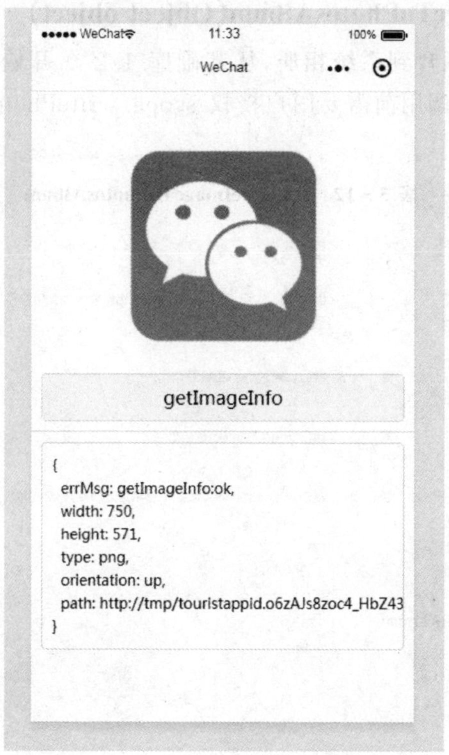

图 5－3　wx. getImageInfo 示例

表 5－11　wx. previewImage 接口参数

属性	类型	默认值	必填	说明	最低版本
urls	Array.<string>		是	需要预览的图片链接列表。2.2.3 起支持云文件ID	
current	string	urls 的第一张	否	当前显示图片的链接	
success	function		否	接口调用成功的回调函数	
fail	function		否	接口调用失败的回调函数	
complete	function		否	接口调用结束的回调函数（调用成功、失败都会执行）	

示例代码如下：

```
wx.previewImage({
  current: '', // 当前显示图片的 http 链接
  urls: [] // 需要预览的图片 http 链接列表
})
```

5. wx.saveImageToPhotosAlbum(Object object)

该接口用于保存图片到系统相册，从基础库 1.2.0 开始支持，低版本需做兼容处理。需要注意的是，调用前需要用户授权 scope.writePhotosAlbum。该接口参数如表 5-12 所列。

表 5-12　wx.saveImageToPhotosAlbum

属性	类型	默认值	必填	说明	最低版本
filePath	string		是	图片文件路径，可以是临时文件路径或永久文件路径，不支持网络图片路径	
success	function		否	接口调用成功的回调函数	
fail	function		否	接口调用失败的回调函数	
complete	function		否	接口调用结束的回调函数（调用成功、失败都会执行）	

示例代码如下：

```
wx.saveImageToPhotosAlbum({
  success(res) { }
})
```

5.2.3　录　音

1. wx.startRecord(Object object)

开始录音。当主动调用 wx.stopRecord，或者录音超过 1min 自动结束录音。当用户离开小程序时，此接口无法调用。从基础库 1.6.0 开始，该接口不再维护，建议使用 wx.getRecorderManager 代替。该接口参数如表 5-13 所列。

表 5-13　wx.startRecord 接口参数

属性	类型	默认值	必填	说明	最低版本
success	function		否	接口调用成功的回调函数	
fail	function		否	接口调用失败的回调函数	
complete	function		否	接口调用结束的回调函数（调用成功、失败都会执行）	

示例代码如下：

```
wx.startRecord({
  success (res) {
    const tempFilePath = res.tempFilePath
  }
```

```
})
setTimeout(function () {
  wx.stopRecord() // 结束录音
}, 10000)
```

2. wx.stopRecord()

停止录音。从基础库 1.6.0 开始,该接口不再维护,建议使用 wx.getRecorderManager 代替。

3. wx.getRecorderManager()

该接口用于获取全局唯一的录音管理器 RecorderManager 对象。从基础库 1.6.0 开始支持,低版本需做兼容处理。全局唯一录音管理器 RecorderManager 对象的一部分常用方法如下:

- RecorderManager.start(Object object):开始录音;
- RecorderManager.pause():暂停录音;
- RecorderManager.resume():继续录音;
- RecorderManager.stop():停止录音;
- RecorderManager.onStart(function callback):监听录音开始事件;
- RecorderManager.onResume(function callback):监听录音继续事件;
- RecorderManager.onPause(function callback):监听录音暂停事件;
- RecorderManager.onStop(function callback):监听录音结束事件;
- RecorderManager.onFrameRecorded(function callback):监听已录制完指定帧大小的文件事件。如果设置了 frameSize,则会回调此事件。

示例代码如下:

```
const recorderManager = wx.getRecorderManager()

recorderManager.onStart(() => {
  console.log('recorder start')
})
recorderManager.onPause(() => {
  console.log('recorder pause')
})
recorderManager.onStop((res) => {
  console.log('recorder stop', res)
  const { tempFilePath } = res
})
recorderManager.onFrameRecorded((res) => {
  const { frameBuffer } = res
  console.log('frameBuffer.byteLength', frameBuffer.byteLength)
```

```
})

const options = {
  duration: 10000,
  sampleRate: 44100,
  numberOfChannels: 1,
  encodeBitRate: 192000,
  format: 'aac',
  frameSize: 50
}

recorderManager.start(options)
```

5.3 文件 API

1. wx.saveFile(Object object)

该接口用于保存文件到本地。注意，saveFile 会移动临时文件，因此调用成功后传入的 tempFilePath 将不可用。小程序本地文件存储的大小限制为 10M。该接口参数如表 5-14 所列。

表 5-14 wx.saveFile 接口参数

属性	类型	默认值	必填	说明	最低版本
tempFilePath	string		是	需要保存的文件的临时路径	
success	function		否	接口调用成功的回调函数	
fail	function		否	接口调用失败的回调函数	
complete	function		否	接口调用结束的回调函数（调用成功、失败都会执行）	

示例代码如下：

```
wx.chooseImage({
  success: function(res) {
    const tempFilePaths = res.tempFilePaths
    wx.saveFile({
      tempFilePath: tempFilePaths[0],
      success (res) {
        const savedFilePath = res.savedFilePath
      }
```

```
    })
  }
})
```

2. wx. getSavedFileList(Object object)

该接口用于获取该小程序下已保存的本地缓存文件列表。接口参数如表 5 - 15 所列。

表 5 - 15　wx. getSavedFileList 接口参数

属性	类型	默认值	必填	说明	最低版本
success	function		否	接口调用成功的回调函数	
fail	function		否	接口调用失败的回调函数	
complete	function		否	接口调用结束的回调函数（调用成功、失败都会执行）	

示例代码如下：

```
wx.getSavedFileList({
  success (res) {
    console.log(res.fileList)
  }
})
```

3. wx. getFileInfo(Object object)

该接口用于获取文件信息。接口参数如表 5 - 16 所列。

表 5 - 16　wx. getFileInfo 接口参数

属性	类型	默认值	必填	说明	最低版本
filePath	string		是	本地文件路径	
digestAlgorithm	string	'md5'	否	计算文件摘要的算法	
success	function		否	接口调用成功的回调函数	
fail	function		否	接口调用失败的回调函数	
complete	function		否	接口调用结束的回调函数（调用成功、失败都会执行）	

示例代码如下：

```
wx.getFileInfo({
  success (res) {
```

```
      console.log(res.size)
      console.log(res.digest)
    }
  })
```

4. wx.getSavedFileInfo(Object object)

该接口用于获取本地文件的文件信息。此接口只能用于获取已保存到本地的文件,若需要获取临时文件信息,请使用 wx.getFileInfo() 接口。接口参数如表 5-17 所列。

表 5-17　wx.getSavedFileInfo

属性	类型	默认值	必填	说明	最低版本
filePath	string		是	文件路径	
success	function		否	接口调用成功的回调函数	
fail	function		否	接口调用失败的回调函数	
complete	function		否	接口调用结束的回调函数（调用成功、失败都会执行）	

示例代码如下：

```
wx.getSavedFileList({
  success (res) {
    console.log(res.fileList)
  }
})
```

5. wx.openDocument(Object object)

该接口用于新开页面打开文档。接口参数如表 5-18 所列。

表 5-18　wx.openDocument

属性	类型	默认值	必填	说明	最低版本
filePath	string		是	文件路径，可通过 downloadFile 获得	
fileType	string		否	文件类型，指定文件类型打开文件	1.4.0
success	function		否	接口调用成功的回调函数	
fail	function		否	接口调用失败的回调函数	
complete	function		否	接口调用结束的回调函数（调用成功、失败都会执行）	

示例代码如下：

```
wx.downloadFile({
    // 示例 url,并非真实存在
    url: 'http://example.com/somefile.pdf',
    success: function(res) {
        const filePath = res.tempFilePath
        wx.openDocument({
            filePath: filePath,
            success: function(res) {
                console.log('打开文档成功')
            }
        })
    }
})
```

6. wx.removeSavedFile(Object object)

该接口用于删除本地缓存文件。接口参数如表 5-19 所列。

表 5-19 wx.removeSavedFile 接口参数

属性	类型	默认值	必填	说明	最低版本
filePath	string		是	需要删除的文件路径	
success	function		否	接口调用成功的回调函数	
fail	function		否	接口调用失败的回调函数	
complete	function		否	接口调用结束的回调函数（调用成功、失败都会执行）	

示例代码如下：

```
wx.getSavedFileList({
 success (res) {
    if (res.fileList.length > 0){
      wx.removeSavedFile({
        filePath: res.fileList[0].filePath,
        complete (res) {
          console.log(res)
        }
      })
    }
 }
})
```

7. wx.getFileSystemManager()

该接口用于获取全局唯一的文件管理器对象 FileSystemManager，该对象的主要方法如下：

- FileSystemManager.access(Object object)：判断文件/目录是否存在。
- FileSystemManager.appendFile(Object object)：在文件结尾追加内容。
- FileSystemManager.saveFile(Object object)：保存临时文件到本地。此接口会移动临时文件，因此调用成功后，tempFilePath 将不可用。
- FileSystemManager.getSavedFileList(Object object)：获取该小程序下已保存的本地缓存文件列表。
- FileSystemManager.removeSavedFile(Object object)：删除该小程序下已保存的本地缓存文件。
- FileSystemManager.copyFile(Object object)：复制文件。
- FileSystemManager.getFileInfo(Object object)：获取该小程序下的本地临时文件或本地缓存文件信息。
- FileSystemManager.mkdir(Object object)：创建目录。
- FileSystemManager.readFile(Object object)：读取本地文件内容。
- FileSystemManager.readdir(Object object)：读取目录内文件列表。
- FileSystemManager.rename(Object object)：重命名文件。可以把文件从 oldPath 移动到 newPath。
- FileSystemManager.rmdir(Object object)：删除目录。
- FileSystemManager.stat(Object object)：获取文件 Stats 对象。
- FileSystemManager.unlink(Object object)：删除文件。
- FileSystemManager.unzip(Object object)：解压文件。
- FileSystemManager.writeFile(Object object)：写文件。

5.4 数据缓存 API

1. wx.setStorage(Object object)

该接口以异步方式将数据存储在本地缓存中指定的 key 中，原来该 key 对应的内容将被覆盖掉。数据存储生命周期跟小程序本身一致，即除用户主动删除或超过一定时间被自动清理，否则数据都一直可用。单个 key 允许存储的最大数据长度为 1MB，所有数据存储上限为 10MB。接口参数如表 5-20 所列。

表 5－20　wx. setStorage

属性	类型	默认值	必填	说明	最低版本
key	string		是	本地缓存中指定的 key	
data	Object		是	需要存储的内容。只支持原生类型、Date、及能够通过 JSON.stringify 序列化的对象	
success	function		否	接口调用成功的回调函数	
fail	function		否	接口调用失败的回调函数	
complete	function		否	接口调用结束的回调函数（调用成功、失败都会执行）	

wx. setStorageSync(string key，Object data)为 wx. setStorage 的同步版本。示例代码如下：

```
wx.setStorage({
  key:"key",
  data:"value"
})
try {
  wx.setStorageSync('key', 'value')
} catch (e) { }
```

2. wx. getStorage(Object object)

该接口用于以异步方式从本地缓存中异步获取指定 key 的内容。接口参数如表 5－21 所列。

表 5－21　wx. getStorage

属性	类型	默认值	必填	说明	最低版本
key	string		是	本地缓存中指定的 key	
success	function		否	接口调用成功的回调函数	
fail	function		否	接口调用失败的回调函数	
complete	function		否	接口调用结束的回调函数（调用成功、失败都会执行）	

wx. getStorageSync(string key)为 wx. getStorage 的同步版本。示例代码如下：

```
wx.getStorage({
  key: 'key',
  success (res) {
```

```
            console.log(res.data)
        }
    })
    try {
        var value = wx.getStorageSync('key')
        if (value) {
            // 逻辑处理,返回值
        }
    } catch (e) {
        // 捕获异常并处理
    }
```

3. wx. getStorageInfo(Object object)

异步获取当前 storage 的相关信息。接口参数如表 5-22 所列。

表 5-22　wx. getStorageInfo

属性	类型	默认值	必填	说明	最低版本
success	function		否	接口调用成功的回调函数	
fail	function		否	接口调用失败的回调函数	
complete	function		否	接口调用结束的回调函数（调用成功、失败都会执行）	

wx. getStorageInfoSync()为 wx. getStorageInfo 的同步版本。

示例代码如下：

```
wx.getStorageInfo({
    success (res) {
        console.log(res.keys)
        console.log(res.currentSize)
        console.log(res.limitSize)
    }
})
try {
    const res = wx.getStorageInfoSync()
    console.log(res.keys)
    console.log(res.currentSize)
    console.log(res.limitSize)
} catch (e) {
    // 捕获异常并处理
}
```

4. wx.removeStorage(Object object)

该接口以异步方式从本地缓存中移除指定 key。接口参数如表 5－23 所列。

表 5－23 wx.removeStorage

属性	类型	默认值	必填	说明	最低版本
key	string		是	本地缓存中指定的 key	
success	function		否	接口调用成功的回调函数	
fail	function		否	接口调用失败的回调函数	
complete	function		否	接口调用结束的回调函数（调用成功、失败都会执行）	

wx.removeStorageSync(string key)为 wx.removeStorage 的同步版本。示例代码如下：

```
wx.removeStorage({
  key：'key',
  success (res) {
    console.log(res.data)
  }
})
try {
  wx.removeStorageSync('key')
} catch (e) {
  // 捕获异常并处理
}
```

5. wx.clearStorage(Object object)

该接口以异步方式清理本地数据缓存。接口参数如表 5－24 所列。

表 5－24 wx.clearStorage

属性	类型	默认值	必填	说明	最低版本
success	function		否	接口调用成功的回调函数	
fail	function		否	接口调用失败的回调函数	
complete	function		否	接口调用结束的回调函数（调用成功、失败都会执行）	

wx.clearStorageSync()为 wx.clearStorage 的同步版本。示例代码如下：

```
wx.clearStorage()
```

```
try {
  wx.clearStorageSync()
} catch(e) {
  // 捕获异常并处理
}
```

5.5 位置 API

1. wx.chooseLocation(Object object)

该接口用于打开地图选择位置。调用前需要用户授权 scope.userLocation。接口参数如表 5-25 所列。

表 5-25 wx.chooseLocation 接口参数

属性	类型	默认值	必填	说明	最低版本
success	function		否	接口调用成功的回调函数	
fail	function		否	接口调用失败的回调函数	
complete	function		否	接口调用结束的回调函数（调用成功、失败都会执行）	

接口返回值参数如表 5-26 所列。

表 5-26 wx.chooseLocation 返回值参数

属性	类型	说明	最低版本
name	string	位置名称	
address	string	详细地址	
latitude	string	纬度，浮点数，范围为"-90~90"，负数表示南纬。使用 gcj02 国测局坐标系	
longitude	string	经度，浮点数，范围为"-180~180"，负数表示西经。使用 gcj02 国测局坐标系	

2. wx.getLocation(Object object)

该接口用于获取当前的地理位置、速度。当用户离开小程序后，该接口无法调用。调用前需要用户授权 scope.userLocation。接口参数如表 5-27 所列。

表 5－27　wx. getLocation 接口参数

属性	类型	默认值	必填	说明	最低版本
type	string	wgs84	否	wgs84 返回 gps 坐标，gcj02 返回可用于 wx.openLocation 的坐标	
altitude	string	false	否	传入 true 会返回高度信息，由于获取高度需要较高精确度，会减慢接口返回速度	1.6.0
success	function		否	接口调用成功的回调函数	
fail	function		否	接口调用失败的回调函数	
complete	function		否	接口调用结束的回调函数（调用成功、失败都会执行）	

接口返回值如表 5－28 所列。

表 5－28　接口返回值

属性	类型	说明	最低版本
latitude	number	纬度，范围为“-90~90”，负数表示南纬	
longitude	number	经度，范围为“-180~180”，负数表示西经	
speed	number	速度，单位 m/s	
accuracy	number	位置的精确度	
altitude	number	高度，单位 m	1.2.0
verticalAccuracy	number	垂直精度，单位 m（Android 无法获取，返回 0）	1.2.0
horizontalAccuracy	number	水平精度，单位 m	1.2.0

示例代码如下：

```
wx.getLocation({
 type：'wgs84',
 success (res) {
    const latitude = res.latitude
    const longitude = res.longitude
    const speed = res.speed
    const accuracy = res.accuracy
 }
 })
```

3. wx.openLocation(Object object)

该接口会使用微信内置地图查看位置。调用前需要用户授权 scope.userLocation。接口参数如表 5-29 所列。

表 5-29 wx.openLocation

属性	类型	默认值	必填	说明	最低版本
latitude	number		是	纬度，范围为"-90~90"，负数表示南纬。使用 gcj02 国测局坐标系	
longitude	number		是	经度，范围为"-180~180"，负数表示西经。使用 gcj02 国测局坐标系	
scale	number	18	否	缩放比例，范围5~18	
name	string		否	位置名	
address	string		否	地址的详细说明	
success	function		否	接口调用成功的回调函数	
fail	function		否	接口调用失败的回调函数	
complete	function		否	接口调用结束的回调函数（调用成功、失败都会执行）	

示例代码如下：

```
wx.getLocation({
 type: 'gcj02', //返回可以用于 wx.openLocation 的经纬度
 success (res) {
    const latitude = res.latitude
    const longitude = res.longitude
    wx.openLocation({
      latitude,
      longitude,
      scale: 18
    })
 }
})
```

5.6 设备 API

5.6.1 加速计、蓝牙、罗盘

1. wx.onAccelerometerChange(function callback)

该接口用来监听加速度数据事件。频率根据 wx.startAccelerometer() 的 interval 参数，可使用 wx.stopAccelerometer() 停止监听。接口的返回值 x、y 和 z 分

别为 X 轴、Y 轴和 Z 轴。

示例代码如下：

```
wx.onAccelerometerChange(function (res) {
  console.log(res.x)
  console.log(res.y)
  console.log(res.z)
})
```

2. wx.openBluetoothAdapter(Object object)

该接口用来初始化蓝牙模块，从基础库 1.1.0 开始支持，低版本需做兼容处理。接口参数如表 5-30 所列。

表 5-30　wx.openBluetoothAdapter 接口参数

属性	类型	默认值	必填	说明	最低版本
success	function		否	接口调用成功的回调函数	
fail	function		否	接口调用失败的回调函数	
complete	function		否	接口调用结束的回调函数（调用成功、失败都会执行）	

示例代码如下：

```
wx.openBluetoothAdapter({
  success (res) {
    console.log(res)
  }
})
```

3. wx.onBluetoothAdapterStateChange(function callback)

该接口用于监听蓝牙适配器状态变化事件，从基础库 1.1.0 开始支持，低版本需做兼容处理。接口参数中 callback 为蓝牙适配器状态变化事件的回调函数。返回值如表 5-31 所列。

表 5-31　wx.onBluetoothAdapterStateChange 返回值

属性	类型	说明	最低版本
available	boolean	蓝牙适配器是否可用	
discovering	boolean	蓝牙适配器是否处于搜索状态	

4. wx.onCompassChange(function callback)

该接口用于监听罗盘数据变化事件。频率：5n/s(次/秒)，接口调用后会自动开始监听，可使用 wx. stopCompass 停止监听。接口参数中，callback 为罗盘数据变化事件的回调函数。返回值如表 5-32 所列。

表 5-32 wx. onCompassChange 返回值

属性	类型	说明	最低版本
direction	number	面对的方向度数	
accuracy	number/string	精度	2.4.0

示例代码：

```
wx.onCompassChange(function(res) {
    console.log(res.direction)
})
```

5.6.2 电量、性能、屏幕

1. wx. getBatteryInfo(Object object)

该接口用来获取设备电量。同步 API wx. getBatteryInfoSync 在 iOS 上不可用，它的同步版本为 wx. getBatteryInfoSync()。该接口的参数如表 5-33 所列。

表 5-33 wx. getBatteryInfo 接口参数

属性	类型	默认值	必填	说明	最低版本
success	function		否	接口调用成功的回调函数	
fail	function		否	接口调用失败的回调函数	
complete	function		否	接口调用结束的回调函数（调用成功、失败都会执行）	

该接口的返回值如表 5-34 所列。

表 5-34 wx. getBatteryInfo 返回值

属性	类型	说明	最低版本
level	string	设备电量，范围 1~100	
isCharging	boolean	是否正在充电中	

2. wx. onMemoryWarning(function callback)

该接口用来监听内存不足告警事件。从基础库 2.0.2 开始支持，低版本需做兼容处理。接口参数中，callback 为内存不足告警事件的回调函数。返回值只有一个

level 值,为 number 类型,表示内存告警等级,只有 Android 才有,对应系统宏定义如表 5-35 所列。

表 5-35 level 的合法值

值	说明
5	TRIM_MEMORY_RUNNING_MODERATE
10	TRIM_MEMORY_RUNNING_LOW
15	TRIM_MEMORY_RUNNING_CRITICAL

示例代码如下:

```
wx.onMemoryWarning(function () {
  console.log('onMemoryWarningReceive')
})
```

3. wx.onUserCaptureScreen(function callback)

该接口用来监听用户主动截屏事件,用户使用系统截屏按键截屏时触发。从基础库 1.4.0 开始支持,低版本需做兼容处理。接口参数中,callback 为用户主动截屏事件的回调函数。示例代码如下:

```
wx.onUserCaptureScreen(function (res) {
  console.log('用户截屏了')
})
```

4. wx.setKeepScreenOn(Object object)

该接口用于设置是否保持常亮状态。仅在当前小程序生效,离开小程序后设置失效。从基础库 1.4.0 开始支持,低版本需做兼容处理。接口参数如表 5-36 所列。

表 5-36 wx.setKeepScreenOn

属性	类型	默认值	必填	说明	最低版本
keepScreenOn	boolean		是	是否保持屏幕常亮	
success	function		否	接口调用成功的回调函数	
fail	function		否	接口调用失败的回调函数	
complete	function		否	接口调用结束的回调函数(调用成功、失败都会执行)	

示例代码如下:

```
wx.setKeepScreenOn({
```

```
  keepScreenOn: true
})
```

5.6.3 联系人、电话、扫码

1. wx.addPhoneContact(Object object)

该接口用于添加手机通讯录联系人。用户可以选择将该表单以「新增联系人」或「添加到已有联系人」的方式,写入手机系统通讯录。

2. wx.makePhoneCall(Object object)

使用该接口可以调用系统的拨打电话功能,接口参数如表 5-37 所列。

表 5-37 wx.makePhoneCall 接口参数

属性	类型	默认值	必填	说明	最低版本
phoneNumber	string		是	需要拨打的电话号码	
success	function		否	接口调用成功的回调函数	
fail	function		否	接口调用失败的回调函数	
complete	function		否	接口调用结束的回调函数(调用成功、失败都会执行)	

示例代码如下:

```
wx.makePhoneCall({
  phoneNumber: '1340000' //仅为示例,并非真实的电话号码
})
```

3. wx.scanCode(Object object)

该接口用于调用客户端扫码界面进行扫码。该接口参数如表 5-38 所列。

表 5-38 wx.scanCode 接口参数

属性	类型	默认值	必填	说明	最低版本
onlyFromCamera	boolean	false	否	是否只能从相机扫码,不允许从相册选择图片	1.2.0
scanType	Array.<string>	['barCode', 'qrCode']	否	扫码类型	1.7.0
success	function		否	接口调用成功的回调函数	
fail	function		否	接口调用失败的回调函数	
complete	function		否	接口调用结束的回调函数(调用成功、失败都会执行)	

该接口的返回值如表 5－39 所列。

表 5－39　wx.scanCode 返回值

属性	类型	说明	最低版本
result	string	所扫码的内容	
scanType	string	所扫码的类型	
charSet	string	所扫码的字符集	
path	string	当所扫的码为当前小程序的合法二维码时，会返回此字段，内容为二维码携带的 path	
rawData	string	原始数据，base64编码	

示例代码如下：

```
// 允许从相机和相册扫码
wx.scanCode({
  success (res) {
    console.log(res)
  }
})

// 只允许从相机扫码
wx.scanCode({
  onlyFromCamera: true,
  success (res) {
    console.log(res)
  }
})
```

5.7　界面 API

5.7.1　动　画

wx.createAnimation(Object object)

该接口用于创建一个动画实例 animation。用调用实例的方法来描述动画，最后通过动画实例的 export 方法导出动画数据，传递给组件的 animation 属性。接口参数如表 5－40 所列。

表 5-40 wx.createAnimation 接口参数

属性	类型	默认值	必填	说明	最低版本
duration	number	400	否	动画持续时间，单位 ms	
timingFunction	string	'linear'	否	动画的效果	
delay	number	0	否	动画延迟时间，单位 ms	
transformOrigin	string	'50% 50% 0'	否		

该接口返回值为 Animation 动画对象，其常用方法如下：

- Array. < Object > Animation. export()用于导出动画队列。export 方法每次调用后会清掉之前的动画操作。
- Animation. step(Object object) 表示一组动画完成。可以在一组动画中调用任意多个动画方法，一组动画中的所有动画会同时开始，一组动画完成后才会进行下一组动画。
- Animation Animation. rotate(number angle)表示从原点顺时针旋转一个角度。
- Animation Animation. scale(number sx，number sy)表示缩放。

示例代码如下：

```
//wxml 代码
 < view   style = "background: red; height: 100rpx; width: 100rpx" animation = "{{animationData}}" > </view >

//js 代码
Page({
  data: {
    animationData: {}
  },
  onShow: function(){
    var animation = wx.createAnimation({
      duration: 1000,
  timingFunction: 'ease',
    })

    this.animation = animation

    animation.scale(2,2).rotate(45).step()
```

```
    this.setData({
      animationData:animation.export()
    })

    setTimeout(function() {
      animation.translate(30).step()
      this.setData({
        animationData:animation.export()
      })
    }.bind(this), 1000)
  },
  rotateAndScale: function () {
    // 旋转同时放大
    this.animation.rotate(45).scale(2, 2).step()
    this.setData({
      animationData: this.animation.export()
    })
  },
  rotateThenScale: function () {
    // 先旋转后放大
    this.animation.rotate(45).step()
    this.animation.scale(2, 2).step()
    this.setData({
      animationData: this.animation.export()
    })
  },
  rotateAndScaleThenTranslate: function () {
    // 先旋转同时放大,然后平移
    this.animation.rotate(45).scale(2, 2).step()
    this.animation.translate(100, 100).step({ duration: 1000 })
    this.setData({
      animationData: this.animation.export()
    })
  }
})
```

示例效果图如图 5－4 所示。

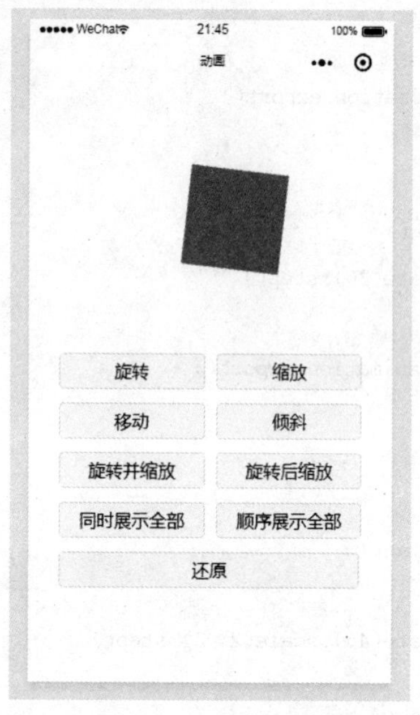

图 5-4 wx.createAnimation 示例

5.7.2 交 互

1. wx.showToast(Object object)

该接口用于显示消息提示框，wx.showToast 常与 wx.hideToast 配对使用。该接口参数如表 5-41 所列。

表 5-41 wx.showToast 接口参数

属性	类型	默认值	必填	说明	最低版本
title	string		是	提示的内容	
icon	string	'success'	否	图标	
image	string		否	自定义图标的本地路径，image 的优先级高于 icon	1.1.0
duration	number	1500	否	提示的延迟时间	
mask	boolean	false	否	是否显示透明蒙层，防止触摸穿透	
success	function		否	接口调用成功的回调函数	
fail	function		否	接口调用失败的回调函数	
complete	function		否	接口调用结束的回调函数（调用成功、失败都会执行）	

示例代码：

```
wx.showToast({
  title: '成功',
  icon: 'success',
  duration: 2000
})
```

2. wx.showLoading(Object object)

该接口用于显示 loading 提示框，需主动调用 wx.hideLoading 才能关闭提示框。wx.showLoading 常与 wx.hideLoading 配对使用。该接口参数如表 5-42 所列。

表 5-42 wx.showLoading 接口参数

属性	类型	默认值	必填	说明	最低版本
title	string		是	提示的内容	
mask	boolean	false	否	是否显示透明蒙层，防止触摸穿透	
success	function		否	接口调用成功的回调函数	
fail	function		否	接口调用失败的回调函数	
complete	function		否	接口调用结束的回调函数（调用成功、失败都会执行）	

示例代码如下：

```
wx.showLoading({
  title: '加载中',
})

setTimeout(function () {
  wx.hideLoading()
}, 2000)
```

需要注意的是，wx.showLoading 和 wx.showToast 同时只能显示一个。

3. wx.showModal(Object object)

该接口用于显示模态对话框，接口参数如表 5-43 所列。

表 5-43　wx.showModal 接口参数

属性	类型	默认值	必填	说明	最低版本
title	string		是	提示的标题	
content	string		是	提示的内容	
showCancel	boolean	true	否	是否显示取消按钮	
cancelText	string	'取消'	否	取消按钮的文字，最多 4 个字符	
cancelColor	string	#000000	否	取消按钮的文字颜色，必须是 16 进制格式的颜色字符串	
confirmText	string	'确定'	否	确认按钮的文字，最多 4 个字符	
confirmColor	string	#3cc51f	否	确认按钮的文字颜色，必须是 16 进制格式的颜色字符串	
success	function		否	接口调用成功的回调函数	
fail	function		否	接口调用失败的回调函数	
complete	function		否	接口调用结束的回调函数（调用成功、失败都会执行）	

接口返回值如表 5-44 所列。

表 5-44　wx.showModal 返回值

属性	类型	说明	最低版本
confirm	boolean	为 true 时，表示用户点击了确定按钮	
cancel	boolean	为 true 时，表示用户点击了取消（用于 Android 系统区分点击蒙层关闭还是点击取消按钮关闭）	1.1.0

示例代码如下：

```
wx.showModal({
  title: '提示',
  content: '这是一个模态弹窗',
  success (res) {
    if (res.confirm) {
      console.log('用户点击确定')
    } else if (res.cancel) {
      console.log('用户点击取消')
    }
  }
})
```

4. wx.showActionSheet(Object object)

该接口用于显示操作菜单,其返回值只有一个值 tapIndex,为 number 类型,为用户点击的按钮序号,从上到下的顺序且从 0 开始。接口参数如表 5-45 所列。

表 5-45 wx.showActionSheet 接口参数

属性	类型	默认值	必填	说明	最低版本
itemList	Array.<string>		是	按钮的文字数组,数组长度最大为 6	
itemColor	string	#000000	否	按钮的文字颜色	
success	function		否	接口调用成功的回调函数	
fail	function		否	接口调用失败的回调函数	
complete	function		否	接口调用结束的回调函数(调用成功、失败都会执行)	

示例代码如下:

```
wx.showActionSheet({
  itemList: ['A', 'B', 'C'],
  success (res) {
    console.log(res.tapIndex)
  },
  fail (res) {
    console.log(res.errMsg)
  }
})
```

5.7.3 导航栏

1. wx.setNavigationBarTitle(Object object)

该接口用于动态设置当前页面的标题。接口参数如表 5-46 所列。

表 5-46 wx.setNavigationBarTitle 接口参数

属性	类型	默认值	必填	说明	最低版本
title	string		是	页面标题	
success	function		否	接口调用成功的回调函数	
fail	function		否	接口调用失败的回调函数	
complete	function		否	接口调用结束的回调函数(调用成功、失败都会执行)	

示例代码如下：

```
wx.setNavigationBarTitle({
  title：'当前页面'
})
```

2. wx.setNavigationBarColor(Object object)

该接口用于设置页面导航条颜色，从基础库 1.4.0 开始支持，低版本需做兼容处理。接口参数如表 5-47 所列。

表 5-47　wx.setNavigationBarColor 接口参数

属性	类型	默认值	必填	说明	最低版本
frontColor	string		是	前景颜色值，包括按钮、标题、状态栏的颜色，仅支持 #ffffff 和 #000000	
backgroundColor	string		是	背景颜色值，有效值为十六进制颜色	
animation	Object		是	动画效果	
success	function		否	接口调用成功的回调函数	
fail	function		否	接口调用失败的回调函数	
complete	function		否	接口调用结束的回调函数（调用成功、失败都会执行）	

示例代码如下：

```
wx.setNavigationBarColor({
    frontColor：'＃ffffff',
    backgroundColor：'＃ff0000',
    animation：{
        duration：400,
        timingFunc：'easeIn'
    }
})
```

5.7.4　置顶、滚动、下拉刷新

1. wx.setTopBarText(Object object)

该接口用于动态设置置顶栏文字内容。只有在当前小程序被置顶时生效，如果当前小程序没有被置顶，也能调用成功，但是不会立即生效，只有在用户将这个小程序置顶后才换上设置的文字内容。从基础库 1.4.3 开始支持该接口，低版本需做兼容处理。接口参数如表 5-48 所列。

表 5 - 48 wx.setTopBarText 接口参数

属性	类型	默认值	必填	说明	最低版本
text	Object		是	置顶栏文字	
success	function		否	接口调用成功的回调函数	
fail	function		否	接口调用失败的回调函数	
complete	function		否	接口调用结束的回调函数（调用成功、失败都会执行）	

示例代码如下：

```
wx.setTopBarText({
  text：'hello，world! '
})
```

2. wx.pageScrollTo(Object object)

该接口会将页面滚动到目标位置，从基础库 1.4.0 开始支持，低版本需做兼容处理。接口参数如表 5 - 49 所列。

表 5 - 49 wx.pageScrollTo 接口参数

属性	类型	默认值	必填	说明	最低版本
scrollTop	number		是	滚动到页面的目标位置，单位 px	
duration	number	300	否	滚动动画的时长，单位 ms	
success	function		否	接口调用成功的回调函数	
fail	function		否	接口调用失败的回调函数	
complete	function		否	接口调用结束的回调函数（调用成功、失败都会执行）	

示例代码如下：

```
wx.pageScrollTo({
  scrollTop：0，
  duration：300
})
```

3. wx.startPullDownRefresh(Object object)

开始下拉刷新。调用后触发下拉刷新动画，效果与用户手动下拉刷新一致。从基础库 1.5.0 开始支持该接口，低版本需做兼容处理。该接口参数如表 5 - 50 所列。

表 5－50 wx.startPullDownRefresh 接口参数

属性	类型	默认值	必填	说明	最低版本
success	function		否	接口调用成功的回调函数	
fail	function		否	接口调用失败的回调函数	
complete	function		否	接口调用结束的回调函数（调用成功、失败都会执行）	

示例代码如下：

```
wx.startPullDownRefresh()
```

4. wx.stopPullDownRefresh(Object object)

停止当前页面下拉刷新。从基础库 1.5.0 开始支持该接口，低版本需做兼容处理。接口参数如表 5－51 所列。

表 5－51 wx.stopPullDownRefresh

属性	类型	默认值	必填	说明	最低版本
success	function		否	接口调用成功的回调函数	
fail	function		否	接口调用失败的回调函数	
complete	function		否	接口调用结束的回调函数（调用成功、失败都会执行）	

示例代码如下：

```
Page({
  onPullDownRefresh () {
    wx.stopPullDownRefresh()
  }
})
```

5.8 开放接口 API

5.8.1 授权、登录、用户信息

1. wx.authorize(Object object)

该接口用于提前向用户发起授权请求，为高频接口，请读者理解透彻。调用后会立刻弹窗询问用户是否同意授权小程序使用某项功能或获取用户的某些数据，但不会实际调用对应接口。如果用户之前已经同意授权，则不会出现弹窗，直接返回成

功。从基础库 1.2.0 开始支持该接口，低版本需做兼容处理。接口参数如表 5 - 52 所列。

表 5 - 52　wx.authorize 接口参数

属性	类型	默认值	必填	说明	最低版本
scope	string		是	需要获取权限的 scope，详见 scope 列表	
success	function		否	接口调用成功的回调函数	
fail	function		否	接口调用失败的回调函数	
complete	function		否	接口调用结束的回调函数（调用成功、失败都会执行）	

属性 scope 所涉及的详细列表如表 5 - 53 所列。需要注意的是，wx.authorize({scope: "scope.userInfo"})无法弹出授权窗口，请使用 < button open - type=" getUserInfo"/ >。

表 5 - 53　scope 列表

scope	对应接口	描述
scope.userInfo	wx.getUserInfo	用户信息
scope.userLocation	wx.getLocation, wx.chooseLocation, wx.openLocation	地理位置
scope.address	wx.chooseAddress	通讯地址
scope.invoiceTitle	wx.chooseInvoiceTitle	发票抬头
scope.invoice	wx.chooseInvoice	获取发票
scope.werun	wx.getWeRunData	微信运动步数
scope.record	wx.startRecord	录音功能
scope.writePhotosAlbum	wx.saveImageToPhotosAlbum, wx.saveVideoToPhotosAlbum	保存到相册
scope.camera	<camera /> 组件	摄像头

示例代码如下：

```
// 可以通过 wx.getSetting 先查询一下用户是否授权了 "scope.record" 这个 scope
wx.getSetting({
  success(res) {
    if (! res.authSetting['scope.record']) {
      wx.authorize({
```

```
          scope: 'scope.record',
          success () {
            // 用户已经同意小程序使用录音功能,后续调用 wx.startRecord 接口不会出现
               询问弹窗
            wx.startRecord()
          }
        })
      }
    }
  })
```

2. wx.login(Object object)

该接口用于调用接口获取登录凭证(code),登录也是一个高频常用接口,请读者务必理解该接口以及相关的接口。通过凭证进而换取用户登录态信息,包括用户的唯一标识(openid)及本次登录的会话密钥(session_key)等。用户数据的加解密通讯需要依赖会话密钥完成。参看如图 5-5 所示的登录时序图可以更好地理解小程序的登录流程。

wx.login 接口参数如表 5-54 所列。

表 5-54　wx.login 接口参数

属性	类型	默认值	必填	说明	最低版本
timeout	number		否	超时时间，单位ms	1.9.90
success	function		否	接口调用成功的回调函数	
fail	function		否	接口调用失败的回调函数	
complete	function		否	接口调用结束的回调函数（调用成功、失败都会执行）	

返回值为用户登录凭证 code,为 string 类型(有效期 5min)。开发者需要在开发者服务器后台调用 code2Session,使用 code 换取 openid 和 session_key 等信息。示例代码如下:

```
wx.login({
  success (res) {
    if (res.code) {
      //发起网络请求
      wx.request({
        url: 'https://test.com/onLogin',
        data: {
          code: res.code
        }
```

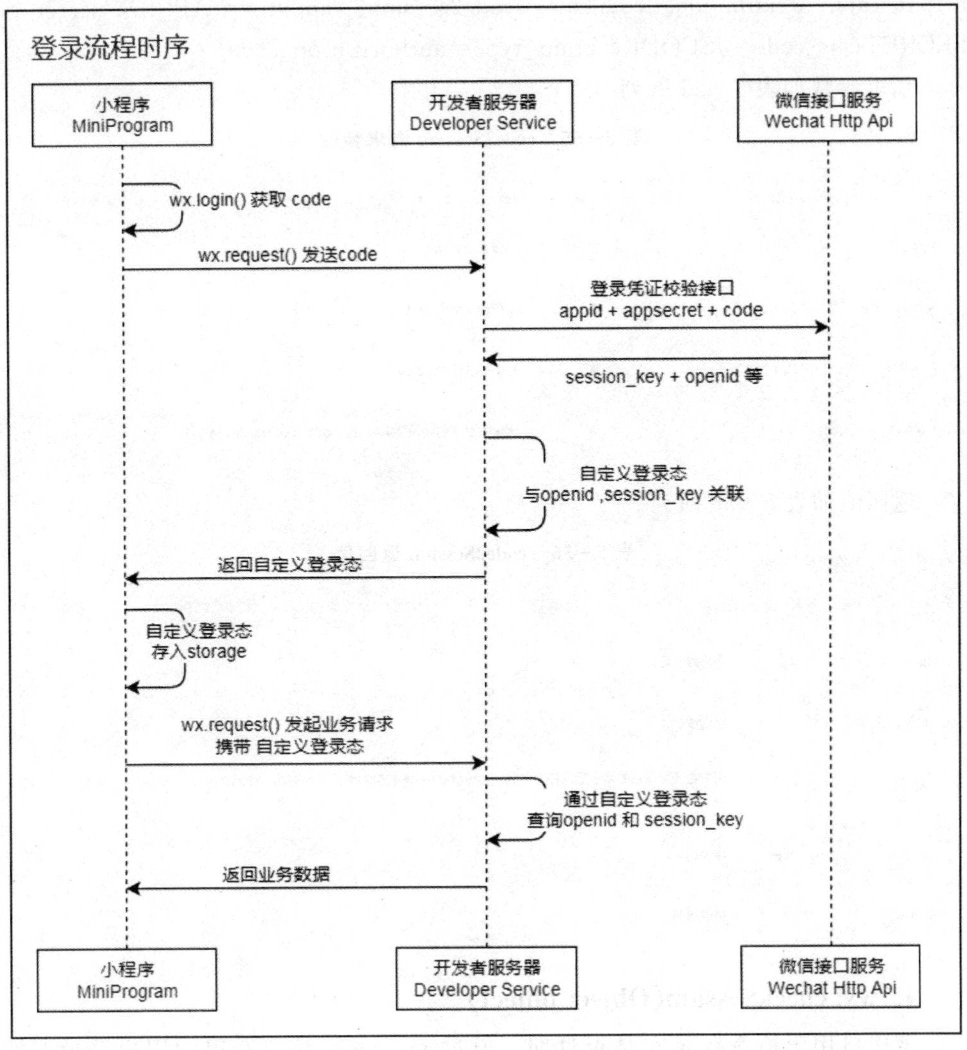

图 5-5　小程序登录流程时序图

```
      })
    } else {
      console.log('登录失败！' + res.errMsg)
    }
  }
})
```

3. code2Session

登录凭证校验。通过 wx.login() 接口获得临时登录凭证 code 后，传到开发者服务器调用此接口完成登录流程。该接口应在后端服务器调用。请求地址为:GET https://api.weixin.qq.com/sns/jscode2session?appid=APPID&secret=SECRET&js_code=JSCODE&grant_type=authorization_code。

请求参数如表 5-55 所列。

表 5-55 code2Session 请求参数

属性	类型	默认值	必填	说明	最低版本
appid	string		是	小程序 appId	
secret	string		是	小程序 appSecret	
js_code	string		是	登录时获取的 code	
grant_type	string		是	授权类型，此处只需填写 authorization_code	

返回值如表 5-56 所列。

表 5-56 code2Session 返回值

属性	类型	说明	最低版本
openid	string	用户唯一标识	
session_key	string	会话密钥	
unionid	string	用户在开放平台的唯一标识符，在满足 UnionID 下发条件的情况下会返回，详见 UnionID 机制说明	
errcode	number	错误码	
errmsg	string	错误信息	

4. wx.checkSession(Object object)

该接口用于检查登录态是否过期。通过 wx.login 接口获得的用户登录态拥有一定的时效性。用户越久未使用小程序，用户登录态越有可能失效。反之如果用户一直在使用小程序，则用户登录态一直保持有效。具体时效逻辑由微信维护，对开发者透明。开发者只需要调用 wx.checkSession 接口检测当前用户登录态是否有效。登录态过期后开发者可以再调用 wx.login 获取新的用户登录态；调用成功说明当前 session_key 未过期，调用失败说明 session_key 已过期。接口参数如表 5-57 所列。

表 5-57 wx.checkSession 接口参数

属性	类型	默认值	必填	说明	最低版本
success	function		否	接口调用成功的回调函数	
fail	function		否	接口调用失败的回调函数	
complete	function		否	接口调用结束的回调函数（调用成功、失败都会执行）	

示例代码如下：

```
wx.checkSession({
  success () {
    //session_key 未过期,并且在本生命周期一直有效
  },
  fail () {
    // session_key 已经失效,需要重新执行登录流程
    wx.login() //重新登录
  }
})
```

5.8.2 微信支付

wx.requestPayment(Object object)

该接口用于发起微信支付。凡涉及需要向用户收费的场景都会用到该接口,电商小程序也基本会用到此接口,微信支付可以很方便地让用户完成交易闭环从而实现社交电商的变现,建议读者重点掌握本接口。小程序支付仅限于认证过的小程序,且需要对小程序进行绑定。接口参数如表 5-58 所列。

表 5-58 wx.requestPayment 接口参数

属性	类型	默认值	必填	说明	最低版本
timeStamp	string		是	时间戳，从 1970 年 1 月 1 日 00:00:00 至今的秒数，即当前的时间	
nonceStr	string		是	随机字符串，长度为32个字符以下	
package	string		是	统一下单接口返回的 prepay_id 参数值，提交格式如：prepay_id=***	
signType	string	MD5	否	签名算法	
paySign	string		是	签名，具体签名方案参见 小程序支付接口文档	
success	function		否	接口调用成功的回调函数	
fail	function		否	接口调用失败的回调函数	
complete	function		否	接口调用结束的回调函数（调用成功、失败都会执行）	

示例代码如下：

```
wx.requestPayment({
  timeStamp: '',
  nonceStr: '',
  package: '',
  signType: 'MD5',
  paySign: '',
  success (res) { },
  fail (res) { }
})
```

小程序支付的业务流程时序图如图 5-6 所示。

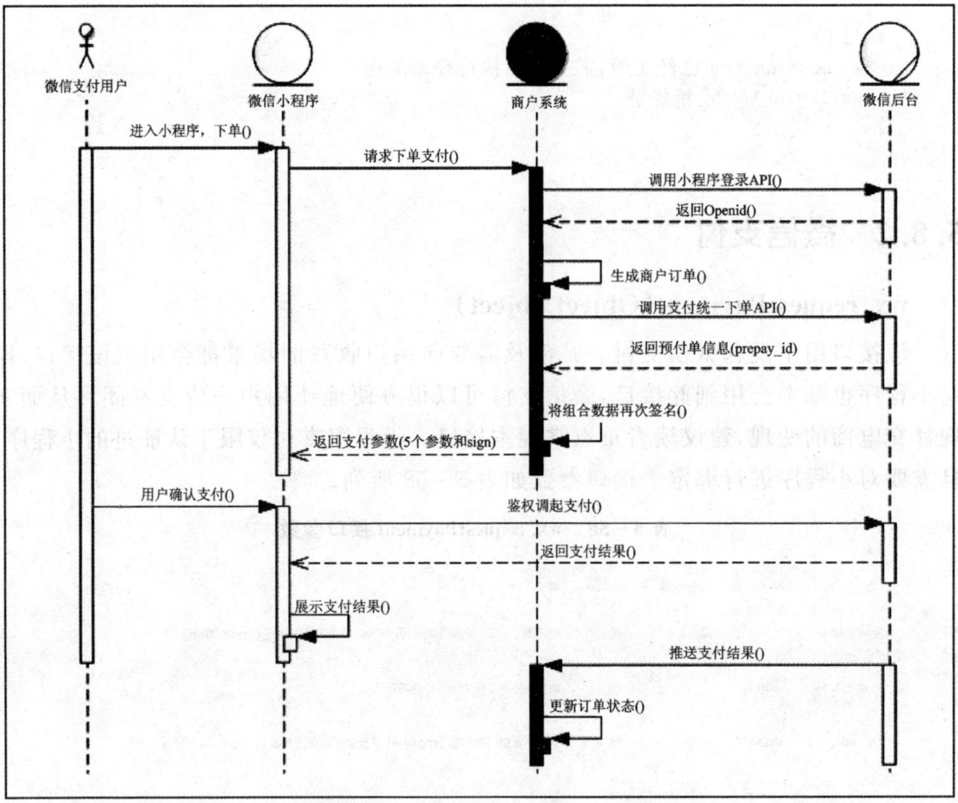

图 5-6　小程序支付的业务流程时序图

从上面的时序图我们可以得知小程序支付大致分以下几步：

- 小程序内调用登录接口；
- 商户服务端调用支付统一下单；
- 商户服务端调用再次签名；

➤ 商户服务端接收支付通知；

➤ 商户服务端查询支付结果。

需要注意的是，小程序支付跟微信 H5 支付不一样，小程序支付没有支付目录和授权域名的限制；用户支付成功事件需要通过小程序端的 success callback 来获取；支付时需要通过统一下单 API 来下单再进行支付。

最后还需要了解一下微信商户系统和微信支付系统两者之间的主要交互所用的 API，有以下几个：

统一下单 API，链接：

https://pay.weixin.qq.com/wiki/doc/api/wxa/wxa_api.php?chapter=9_1&index=1

再次签名 API，链接：

https://pay.weixin.qq.com/wiki/doc/api/wxa/wxa_api.php?chapter=7_7&index=3

支付结果通知 API，链接：

https://pay.weixin.qq.com/wiki/doc/api/wxa/wxa_api.php?chapter=9_7

查询订单 API，链接：

https://pay.weixin.qq.com/wiki/doc/api/wxa/wxa_api.php?chapter=9_2

5.8.3 模板消息、统一服务消息

1. sendTemplateMessage

该接口用于发送模板消息，它需要在后端服务器调用。请求地址如下：

POST https://api.weixin.qq.com/cgi-bin/message/wxopen/template/send?access_token=ACCESS_TOKEN。

请求参数如表 5-59 所列。

表 5-59 sendTemplateMessage 接口参数

属性	类型	默认值	必填	说明	最低版本
access_token	string		是	接口调用凭证	
touser	string		是	接收者（用户）的 openid	
template_id	string		是	所需下发的模板消息的id	
page	string		否	点击模板卡片后的跳转页面，仅限本小程序内的页面。支持带参数，（示例index?foo=bar）。该字段不填则模板无跳转	
form_id	string		是	表单提交场景下，为 submit 事件带上的 formId；支付场景下，为本次支付的 prepay_id	
data	string		否	模板内容，不填则下发空模板	
emphasis_keyword	string		否	模板需要放大的关键词，不填则默认无放大	

接口返回 JSON 数据包，返回值如表 5－60 所列。

表 5－60　sendTemplateMessage 返回值

属性	类型	说明	最低版本
errcode	number	错误码	
errmsg	string	错误信息	

请求数据示例如下：

```
{
  "touser"："OPENID",
  "template_id"："TEMPLATE_ID",
  "page"："index",
  "form_id"："FORMID",
  "data"：{
      "keyword1"：{
          "value"："339208499"
      },
      "keyword2"：{
          "value"："2015 年 01 月 05 日 12:30"
      },
      "keyword3"：{
          "value"："腾讯微信总部"
      },
      "keyword4"：{
          "value"："广州市海珠区新港中路 397 号"
      }
  },
  "emphasis_keyword"："keyword1.DATA"
}
```

返回数据示例如下：

```
{
"errcode"：0,
"errmsg"："ok",
"template_id"："wDYzYZVxobJivW9oMpSCpuvACOfJXQIoKUm0PY397Tc"
}
```

2. sendUniformMessage

该接口用于下发小程序和公众号统一的服务消息，且需要在后端服务器调用。它和模板消息基本类似。请求地址为：

POST https://api.weixin.qq.com/cgi-bin/message/wxopen/template/uniform_send?access_token=ACCESS_TOKEN。

请求参数如表 5-61 所列。

表 5-61 sendUniformMessage 请求参数

属性	类型	默认值	必填	说明	最低版本
access_token	string		是	接口调用凭证	
touser	string		是	用户openid，可以是小程序的openid，也可以是mp_template_msg.appid对应的公众号的openid	
weapp_template_msg	Object		否	小程序模板消息相关的信息，可以参考小程序模板消息接口; 有此节点则优先发送小程序模板消息	
mp_template_msg	Object		是	公众号模板消息相关的信息，可以参考公众号模板消息接口；有此节点并且没有weapp_template_msg节点时，发送公众号模板消息	

接口返回 JSON 数据包，返回值如表 5-62 所列。

表 5-62 sendUniformMessage 返回值

属性	类型	说明	最低版本
errcode	number	错误码	
errmsg	string	错误信息	

请求数据示例如下：

```
{
    "touser":"OPENID",
    "weapp_template_msg":{
        "template_id":"TEMPLATE_ID",
        "page":"page/page/index",
        "form_id":"FORMID",
        "data":{
            "keyword1":{
                "value":"339208499"
            },
            "keyword2":{
                "value":"2015 年 01 月 05 日 12:30"
            },
            "keyword3":{
                "value":"腾讯微信总部"
            },
            "keyword4":{
```

```
                "value":"广州市海珠区新港中路 397 号"
            }
        },
        "emphasis_keyword":"keyword1.DATA"
    },
    "mp_template_msg":{
        "appid":"APPID ",
        "template_id":"TEMPLATE_ID",
        "url":"http://weixin.qq.com/download",
        "miniprogram":{
            "appid":"xiaochengxuappid12345",
            "pagepath":"index? foo = bar"
        },
        "data":{
            "first":{
                "value":"恭喜你购买成功!",
                "color":"#173177"
            },
            "keyword1":{
                "value":"巧克力",
                "color":"#173177"
            },
            "keyword2":{
                "value":"39.8 元",
                "color":"#173177"
            },
            "keyword3":{
                "value":"2014 年 9 月 22 日",
                "color":"#173177"
            },
            "remark":{
                "value":"欢迎再次购买!",
                "color":"#173177"
            }
        }
    }
}
```

返回数据示例如下：

```
{
"errcode": 0,
"errmsg": "ok"
```

```
}
```

5.9 其他 API

5.9.1 基 础

wx.canIUse(string schema)

该接口用来判断小程序的 API、回调、参数和组件等是否在当前版本可用。接口参数使用 ${API}.${method}.${param}.${options} 或者 ${component}.${attribute}.${option} 方式来调用。接口返回值为 boolean，表示当前版本是否可用。

参数说明：

- ${API} 代表 API 名字；
- ${method} 代表调用方式，有效值为 return、success、object 和 callback；
- ${param} 代表参数或者返回值；
- ${options} 代表参数的可选值；
- ${component} 代表组件名字；
- ${attribute} 代表组件属性；
- ${option} 代表组件属性的可选值。

示例代码如下：

```
wx.canIUse('openBluetoothAdapter')
wx.canIUse('getSystemInfoSync.return.screenWidth')
wx.canIUse('getSystemInfo.success.screenWidth')
wx.canIUse('showToast.object.image')
wx.canIUse('onCompassChange.callback.direction')
wx.canIUse('request.object.method.GET')

wx.canIUse('live-player')
wx.canIUse('text.selectable')
wx.canIUse('button.open-type.contact')
```

5.9.2 转 发

1. wx.getShareInfo(Object object)

该接口用于获取转发详细信息，从基础库 1.1.0 开始支持，低版本需做兼容处理。接口参数如表 5-63 所列。

表 5-63　wx.getShareInfo 接口参数

属性	类型	默认值	必填	说明	最低版本
shareTicket	string		是	shareTicket	
timeout	number		否	超时时间，单位 ms	1.9.90
success	function		否	接口调用成功的回调函数	
fail	function		否	接口调用失败的回调函数	
complete	function		否	接口调用结束的回调函数（调用成功、失败都会执行）	

返回值如表 5-64 所列。

表 5-64　wx.getShareInfo 返回值

属性	类型	说明	最低版本
errMsg	string	错误信息	
encryptedData	string	包括敏感数据在内的完整转发信息的加密数据，详细见加密数据解密算法	
iv	string	加密算法的初始向量，详细见加密数据解密算法	

示例代码：

encryptedData 解密后为以下 json 结构，详见加密数据解密算法。其中 openGId 为当前群的唯一标识。

```
{
"openGId"："OPENGID"
}
```

2. wx.showShareMenu(Object object)

该接口用于显示当前页面的转发按钮。从基础库 1.1.0 开始支持，低版本需做兼容处理。接口参数如表 5-65 所列。

表 5-65　wx.showShareMenu 接口参数

属性	类型	默认值	必填	说明	最低版本
withShareTicket	boolean	false	否	是否使用带 shareTicket 的转发详情	
success	function		否	接口调用成功的回调函数	
fail	function		否	接口调用失败的回调函数	
complete	function		否	接口调用结束的回调函数（调用成功、失败都会执行）	

示例代码如下：

```
wx.showShareMenu({
  withShareTicket: true
})
```

5.9.3 系统信息

wx.getSystemInfo(Object object)

该接口用于获取系统信息。wx.getSystemInfoSync()是它的同步版本。接口参数如表 5-66 所列。

表 5-66 wx.getSystemInfo 接口参数

属性	类型	默认值	必填	说明	最低版本
success	function		否	接口调用成功的回调函数	
fail	function		否	接口调用失败的回调函数	
complete	function		否	接口调用结束的回调函数（调用成功、失败都会执行）	

主要的返回值如表 5-67 所列。

表 5-67 wx.getSystemInfo 返回值

属性	类型	说明	最低版本
brand	string	手机品牌	1.5.0
model	string	手机型号	
pixelRatio	number	设备像素比	
screenWidth	number	屏幕宽度	1.1.0
screenHeight	number	屏幕高度	1.1.0
windowWidth	number	可使用窗口宽度	
windowHeight	number	可使用窗口高度	
statusBarHeight	number	状态栏的高度	1.9.0
language	string	微信设置的语言	
version	string	微信版本号	
system	string	操作系统版本	
platform	string	客户端平台	

示例代码如下：

```
wx.getSystemInfo({
  success (res) {
    console.log(res.model)
    console.log(res.pixelRatio)
    console.log(res.windowWidth)
    console.log(res.windowHeight)
    console.log(res.language)
    console.log(res.version)
    console.log(res.platform)
  }
})
try {
  const res = wx.getSystemInfoSync()
  console.log(res.model)
  console.log(res.pixelRatio)
  console.log(res.windowWidth)
  console.log(res.windowHeight)
  console.log(res.language)
  console.log(res.version)
  console.log(res.platform)
} catch (e) {
  // 捕获异常并处理
}
```

第 6 章

小程序服务端开发思路

本章主要讲解小程序服务端的开发思路和相关建议。前面学习的小程序框架基础、框架组件和框架 API 其实大多涉及的是前端技术,对于一个完整的互联网应用而言,后端也是同样重要的一环,因为后端承载的是整个应用的业务逻辑和数据中心。相信读者中有一部分是做前端开发的,一部分是做后端开发的,而随着当下前后端技术的迅猛发展,前后端技术也在相互借鉴和吸收,再加上全栈开发工程师的新理念也使得开发者开始去尝试同时掌握前后端开发技术,这样也非常有利于对整个项目架构设计的理解以及开发效率的提升。

6.1 服务端开发语言和框架选择建议

6.1.1 开发语言和框架的选择

小程序后端开发语言比较主流的有:PHP、Node.js、Java 和.NET,其中前两者相对来说更适合用来作为小程序的后端开发语言。小程序本身就是一个轻量级的应用,功能不是特别复杂,所以后端选择 PHP 和 Node.js 这种适合敏捷开发的脚本语言,在开发效率上非常有优势。当然,如果你本身是其他语言的开发者或手中的项目只是需要增加一个小程序扩展,那么选择自己熟悉的开发在语言即可,这样集成和开发在效率上都不会有太大的延缓。总之,开发语言的选择是一个见仁见智的话题,我们无需在语言选择上纠结太多,适合自己情况即可,最重要的其实依然还是开发思路和解决问题的思维能力。

腾讯云也为开发者提供了非常便捷的小程序服务端配置向导工具,当然它所提供的小程序云服务器是付费的,比较适合追求快捷配置和服务端稳定的用户,如图 6-1 所示。

需要注意的是,目前该云服务器暂不支持 Java 和.NET 环境,可选的只有 PHP 和 Node.js 这两种,此处简单介绍并推荐一下这两种语言的开发框架,供读者参考之用。

关于 PHP 的开发框架,国内采用 ThinkPHP 的应该比较多,做 PHP 开发的想必对该开发框架也是非常熟悉了,我们在后面的实战项目中所采用的后端框架也是

① 设置域名信息 > ② 配置生产环境 > ③ 完成

选择生产环境

开发语言* PHP　Node.js　Java　.NET

操作系统* Centos7.2 64 位

域名配置

域名信息　○ 注册一个新域名　● 使用已有腾讯云域名

请选择

微信小程序暂不支持中文域名

只能选择本账户下的腾讯云域名

小程序服务器域名必须经过ICP备案，所选域名需符合工信部及对应省管局的要求，具体可查看可备案域名公告和各省管局要求

服务器配置

云服务器

	类别	描述	价格
●	小程序开发工具入门版	1核CPU，内存1G，50G本地硬盘	49.00 元/月
○	小程序开发工具基础版	1核CPU，内存2G，50G本地硬盘	109.00 元/月
○	小程序开发工具进阶版	2核CPU，内存4G，50G本地硬盘	229.00 元/月

以上主机均搭载小程序Wafer2 QuickStart PHP Server

图 6-1　微信小程序生产环境初始化向导

基于 ThinkPHP 来开发的。除了 ThinkPHP 外，还有一些比较优秀的框架，比如之前一直在国外非常流行的 laravel 框架，非常的简洁优雅，自动化程度比较高，目前在国内也拥有一大波粉丝。再还有一个笔者本人之前常用的框架 CodeIgniter，简称“CI 框架”，比较小巧但是功能非常强大，且非常容易上手，适合不太懂 PHP 开发的新手学习。

至于 Node.js 的开发框架也非常多，比如：Express 是使用最多的框架，也是大多数新手入门的框架，它的优势在于对应的库完善、社区庞大、文档多和容易上手，比较适合小型项目开发。Koa 框架则比 Express 框架思想更先进，它是 Express 原班人马打造，Koa 框架的核心是 ES6 的 generator；Koa 使用 generator 来实现中间件的流程控制，使用 try/catch 来增强异常处理，同时在 Koa 框架中也看不到复杂的 callback 回调，Koa 框架比较适合用来做大型企业级项目开发。Egg.js 框架是阿里团队推出的基于 Node.js 和 Koa 的一个 Nodejs 的企业级应用开发框架，它可以帮助开发团队及开发人员降低开发和维护成本。Egg.js 是按照约定进行开发的，奉行约定优于配置，具备提供基于 Egg 定制上层框架的能力、高度可扩展的插件机制、内置多进程管理；基于 Koa 开发，性能优异、框架稳定，测试覆盖率高、渐进式开发、开发成本和维护成本低等特点。如果是 Node.js 新手，笔者建议按照 Express→Koa→Egg 这个顺序来进阶学习。

6.1.2 新选择——小程序·云开发

后端采用云服务也是当下一种新兴的后端开发方式。微信在今年也为小程序提供了后端云服务版的解决方案——小程序·云开发，它是基于腾讯云的云开发 TCB 的一站式小程序后端云服务。开发者可以使用云开发开发微信小程序、小游戏，无需搭建服务器，即可使用云端能力。

云开发为开发者提供完整的云端支持，弱化后端和运维概念，无需搭建服务器，使用平台提供的 API 进行核心业务开发，即可实现快速上线和迭代，同时这一能力，同开发者已经使用的云服务相互兼容，并不互斥。

目前云开发提供三大基础能力支持：

- 云函数：在云端运行的代码，微信私有协议天然鉴权，开发者只需编写自身业务逻辑代码；
- 数据库：一个既可在小程序前端操作，也能在云函数中读写的 JSON 数据库；
- 存储：在小程序前端直接上传/下载云端文件，在云开发控制台可视化管理。

云开发控制台如图 6-2 所示。

云开发控制台是管理云开发资源的地方，控制台提供以下能力：

- 概览：查看云开发基础使用数据；
- 用户管理：查看小程序用户信息；
- 数据库：管理数据库，可查看、增加、更新、查找、删除数据、管理索引、管理数据库访问权限等；
- 存储管理：查看和管理存储空间；
- 云函数：查看云函数列表、配置、日志和监控；
- 统计分析：查看云开发资源具体使用统计信息；

小程序云开发文档地址：https://developers.weixin.qq.com/miniprogram/dev/wxcloud/basis/getting-started.html。

借助微信小程序云开发这个解决方案，开发者无需关心太多的框架搭建和服务器运维等问题，重心只需关注在项目业务逻辑开发上即可。云开发这种新模式的优势在于提供了稳定的后端服务器架构，开发者的开发重心只专注于业务逻辑代码，当然缺点就是随着业务的成熟和用户的积累，该软件项目对云开发这套解决方案势必会形成强依赖。因此，自行开发后端和采用云开发这两种方式各有利弊，开发者需要根据自身情况来决定。目前云开发这个产品处于内测阶段，暂不收费。

图 6-2　云开发控制台

6.2　数据库设计思路

6.2.1　小程序项目数据库选型

目前市面上的数据库比较多，大体上分为两大类：关系型数据库和非关系型数据库。

关系型数据库最典型的数据结构是表，由二维表及其之间的联系所组成的一个数据组织。市面上的主流关系型数据库如图 6-3 所示。

优点：

① 易于维护，都是使用表结构，格式一致；

② 使用方便，SQL 语言通用，可用于复杂查询；

③ 复杂操作，支持 SQL，可用于一个表以及多个表之间非常复杂的查询。

缺点：

① 读写性能比较差，尤其是海量数据的高效率读写；

② 固定的表结构，灵活度稍欠；

③ 高并发读写需求，对于传统关系型数据库来说，硬盘 I/O 是一个很大的瓶颈。

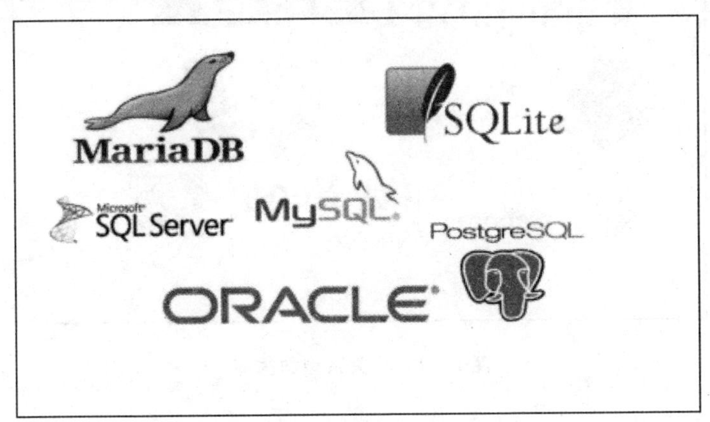

图 6－3　关系型数据库

非关系型数据库严格上不是一种数据库，应该是一种数据结构化存储方法的集合，可以是文档或者键值对（key－value 型）等。市面上主流的非关系型数据库如图 6－4 所示。

优点：

① 格式灵活，存储数据的格式可以是 key－value 形式、文档形式和图片形式等，使用灵活，应用场景广泛，而关系型数据库则只支持基础类型；

② 速度快，nosql 可以使用硬盘或者随机存储器作为载体，而关系型数据库只能使用硬盘；

③ 高扩展性；

④ 成本低，nosql 数据库部署简单，大部分是开源软件。

缺点：

① 不提供 sql 支持，学习和使用成本较高；

② 无事务处理；

③ 数据结构相对复杂，复杂查询方面稍欠。

就小程序项目而言，其所需要的数据库无需太大，保证够用即可，在关系型数据库中，选择 MySQL 数据库即可满足小程序项目后端数据库的存储性能需求。在非关系型数据库中，Key－Value 键值对型的数据库也可以很好地满足小程序的后端数据库存储需求，这是因为前后端是通过 Restful API 接口进行数据交换的，而数据的格式则是 JSON，而 JSON 数据类型正好是这种 Key－Value 键值对结构的数据。Redis 和 Cassandra 这种 Key－Value 键值对型数据库也比较适合用来作为小程序后端数据库存储使用。再者，小程序云开发解决方案中的数据库也是提供的一个 JSON 数据库。

图 6-4　非关系型数据库

6.2.2　数据库设计方法和建议

无论我们设计什么项目的数据库,数据库设计的三范式都需要我们了然于心。数据库三范式如下:

- 第一范式(1NF):数据表中的每一列(每个字段)必须是不可拆分的最小单元,也就是确保每一列的原子性。简记为"行不可重复,列不可再分";
- 第二范式(2NF):满足 1NF 后,要求表中的所有列,都必须依赖于主键,而不能有任何一列与主键没有关系,也就是说一个表只描述一件事情。简记为"非主键列必须依赖于主键列";
- 第三范式(3NF):必须先满足第二范式(2NF),要求表中的每一列只与主键直接相关而不是间接相关,表中的每一列只能依赖于主键。简记为"非主键列之间必须相互独立"。

当然,有时候我们为了追求数据库查询效率,也会采用一些逆范式的数据库设计。逆范式指的是通过增加冗余或重复的数据来提高数据库的读性能。

数据库三大范式只是一般设计数据库的基本理念,可以建立冗余较小、结构合理的数据库。如果有特殊情况,当然要特殊对待,数据库设计最重要的是看需求跟性能,"需求 > 性能 > 表结构",所以不能一味地去追求范式建立数据库。

数据库设计工具 PowerDesigner 是一个挺不错的设计软件,在做大型软件项目的数据库设计时,非常有必要用该软件进行数据库设计。而对于中小型项目为主导的小程序项目而言,我们也可以借助一些比较轻量级的数据库软件来做数据库设计,比如 Navicat 就是一个比较简单易用、适合用来做小型项目数据库设计的软件工具,如图 6-5 所示。一般而言,特定的数据库大多都有一些比较经典的数据库设计工具,这里不再一一列举,读者可以自行搜索做出选择,适合自己使用习惯即可。

关于数据库设计的方法建议,个人建议先从核心基础数据表(主表)开始设计,以

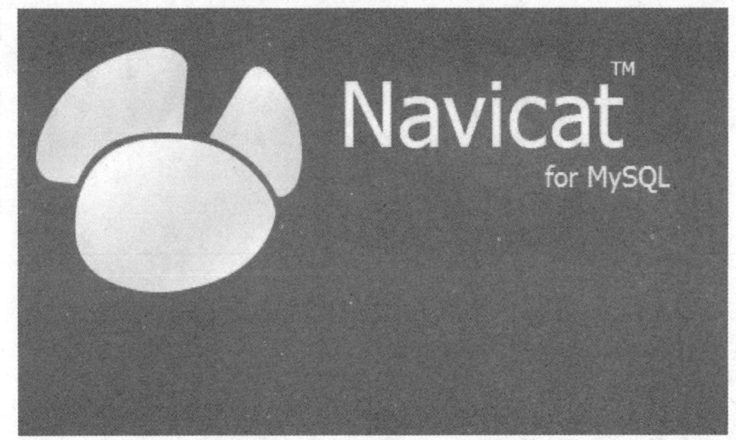

图 6 - 5　Navicat for MySQL

电商小程序为例,可以先设计用户表、产品表、订单表等,有了这些主表,然后再进行思维扩散。设计扩展表(或者叫从表),比如用户收货地址表就是用户表的扩展表,产品品牌表就是产品表的扩展表,订单商品信息表就是订单表的扩展表等。这些扩展表会有一个外键字段指向主表,从而形成主从关联关系。最后其他的一些数据表,可以在开发的过程中进行新增和修改。数据库表结构也不是一次成型的,它会在开发的过程中,不断地迭代修正以适应实际的业务逻辑开发需求。

6.3　服务端接口开发思路

6.3.1　RESTful API 介绍

REST 全称是 Representational State Transfer,即表现层状态转移。它首次出现在 2000 年 Roy Fielding 的博士论文中,Roy Fielding 是 HTTP 规范的主要编写者之一。RESTful API 就是 REST 风格的 API,RESTful 用简单直白的话来描述便是:URL 定位资源,用 HTTP 动词(GET,POST,PUT,DELETE)描述操作。RESTful API 如图 6 - 6 所示。

RESTful 的本质其实是一种软件架构风格,其核心思想是,客户端发出的数据操作指令都是"动词 + 宾语"的结构。比如,GET /articles 这个命令,GET 是动词,/articles 是宾语。

动词通常就是五种 HTTP 方法,对应 CRUD 操作:

- GET:读取(Read)
- POST:新建(Create)
- PUT:更新(Update)

RESTful API

GET PUT POST DELETE

图 6-6 RESTful API

- PATCH:更新(Update),通常是部分更新
- DELETE:删除(Delete)

标准 RESTful API 示例如下:

GET/rest/api/dogs:获取所有小狗狗;

POST/rest/api/dogs:添加一条小狗狗;

PUT/rest/api/dogs/:dog_id:修改一条小狗狗;

DELETE/rest/api/dogs/:dog_id:删除一条小狗狗。

6.3.2 后端 API 开发技巧和建议

RESTful API 只是一种软件架构风格,并不是强制规定,所以在国内,开发者在进行后端 API 开发的时候实际上并不一定是完全遵守 RESTful API 风格的。比如有一种很常见的做法,就是大多只用到了 HTTP 五种方法中的 GET 和 POST 这两种,用 GET 来处理 API 的查询读操作,用 POST 来处理增、删、改这几个写操作,具体是哪个写操作可能会在 URI 中以 add、del 和 update 字符串来区分。

对于小程序这种轻量级移动应用,接口返回值一般以 JSON 数据的形式进行展现。示例如下:

```
HTTP/1.1 200 OK
Content-Type: application/json
Content-Length: xxx

{
  "url" : "/api/categories/1",
  "label" : "Food",
  "items_url" : "/api/items? category=1",
  "brands" : [
        {
          "label" : "可乐",
          "brand_key" : "2073",
          "url" : "/api/brands/2073"
```

```
        }, {
            "label" : "薯片",
            "brand_key" : "6632",
            "url" : "/api/brands/6632"
        }
        ...
    ]
}
```

后端 API 接口开发重心其实在业务逻辑处理上，大部分的代码都有一定的相似之处，不外乎就是 CRUD 这类操作。所以建议读者朋友在开发的时候，最开始以其中某一个模块为标准构建一套标准化的 CRUD 代码模板，方便在后面开发其他接口的时候进行快速参考使用。

关于后端 API 如何记录从而方便开发者共享查阅，目前大致有两种方法：文档或者在线 API 文档系统。这两种方法笔者都曾使用过，个人建议采用在线 API 文档，这样查询 API 效率更高，API 修改之后也可以在线同步更新，十分方便快捷。

6.4 服务端后台管理系统开发思路

6.4.1 后台管理系统页面设计建议

一般大部分的互联网应用都会有一个后台管理系统，这个系统主要是给管理员使用的，用来对系统进行配置以及对应用项目的具体设置。就笔者的开发经验而言，后台管理系统的设计风格尽量遵守简约、醒目和清晰明了的规则，以方便用户操作使用为前提。失败的后台管理设计往往可用性极其糟糕，用户体验也无从谈起。开发者虽然不是设计师出身，但是掌握基本的设计原则或者技巧也可以让后台管理系统变得更加好用、体验更佳。

比如在颜色使用上，尽量采用“三色原则”，后台系统不宜用过多的颜色，最好保证清晰明了即可，主色调控制在三种颜色左右。后台网页颜色太多，容易看起来很凌乱和花哨，这是完全没必要的。

“首页布局导航栏 + dashboard”仪表盘这种设计布局相对来说比较通用。详情页中注意加入一些跳转或返回按钮，用于导向或返回至上一级页面。

统计图表页面需要在数据量增加到一定规模的时候再加上，前期数据量非常小的时候，没有太大的必要去做统计图表页。前期把列表页和查询页的体验做好即可达到查询统计的效果。后台管理系统首页示例图如图 6－7 所示。

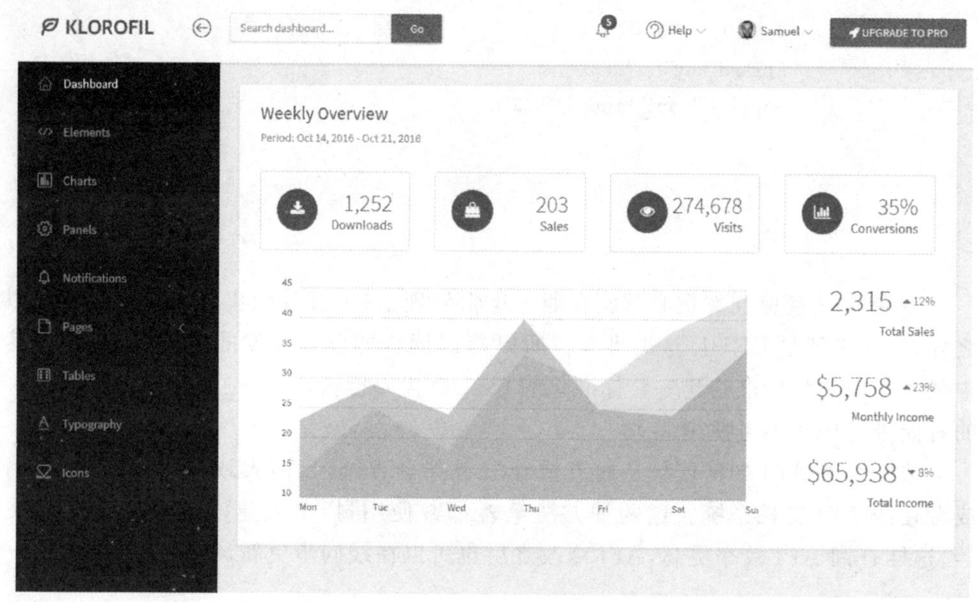

图 6-7　后台管理系统首页示例图

后台页面设计和开发，也是一个不断迭代修正的过程，建议开发者可以在做的过程中，与使用者多沟通即可更好地确定后台管理系统的具体事宜。

6.4.2　后台管理系统功能开发注意要点

后台管理系统在功能开发上需要注意一下几个要点：

- 注册入口建议关闭，改为后台添加和审核的方式，保证后台系统的安全性；
- 登录一定加上安全校验，登录后建议设定好安全登录时长，建议加上登录日志的功能方便记录后台的登录操作情况；
- 针对多用户多角色的后台系统，需要设计好权限分配，不同权限对应不同的页面集合；
- 列表数据量比较大的时候，需要加上分页功能，这样体验更佳；
- 删除操作建议增加提示，避免误操作删除重要数据；
- 如果需要支持手机也能方便查看后台，请选择好自适应的 H5 模板样式，同时页面也需要做响应式设计。

后台系统开发在大型项目中基本都有核心的后台系统研发团队来做，因为它涉及项目的核心业务数据管理，相较于小程序这类偏前端页面开发的工作而言，后台系统更像一个后端数据库管理系统。大多数后台管理系统是采用基于 WEB 的 B/S 结构开发的，所以它也具备有前后端之分，完整开发下来也比较耗时。如果您仅仅只是用来学习小程序开发的话，那么可以不用开发完整的后台系统，仅开发与小程序页面对应的接口程序即可，测试数据可以直接在数据库中进行新增、修改和删除等操作。需要说明的是，本书中的小程序商城案例限于篇幅有限，书中仅为读者提供了小程序商城前端页面以及后端接口的程序代码示例，如需后台管理系统的示例参考，请下载本书附带的电子资料中的项目代码压缩包并具体查看，该压缩包含小程序商城的后台管理系统代码（目录在 backend/App/Admin）供读者参考之用。

小程序开发实战篇

第7章

小程序商城需求分析和数据库设计

本章主要讲解实战小程序商城项目开发中的需求分析和数据库设计，通过项目前期的分析和设计，我们会对项目整体概貌和大致轮廓有一个初步的理解。

7.1 项目需求分析

每一个软件项目基本都会对项目需求进行分析，这样会更有利于开发者对项目需求加深理解，避免在后期实际开发中出现项目需求大方向上的理解偏差。

7.1.1 项目背景概述

电子商城项目一直以来都是一个非常经典且实用的实战开发项目，而借助小程序的社交赋能所构建成的小程序电商，使得移动电商又有了新的玩法和体验。就目前而言，国内电商平台巨头已经有淘宝天猫、京东、拼多多和苏宁等巨无霸电商平台，对于作为一个想涉足电商行业创业的新手而言，做电商平台已经基本没什么大希望了，反而做小而美的垂直电商项目可能会存在一些机会。电商的新趋势有很多：电商从 PC 网站向手机移动端的转变；从当年卖产品到现在可以卖标准化服务；从中心化平台型电商，发展到以社交电商为新潮流的去中心化电商。电商项目总是在随着市场的变化而不断地更新迭代，呈现出一些新的模式，如果能很好的抓住这波新机会，也是有可能分得移动电商红利的一杯羹的。社交电商新势力拼多多在 2018 年已在美国纳斯达克上市，拼多多小程序如图 7－1 所示。

我之前曾参与过大型企业级订单处理系统和一些创业类的移动互联网产品研发，产品中基本都会带有电商这个重要的子模块，因为电商模块是一个能实现导入正向现金流并最终实现交易闭环的重要功能，所以电商项目是非常有潜在商业价值的项目，一个有商业头脑的技术型创业者，非常适合做一个带有电商功能的互联网项目来实现自己的创业梦想。

7.1.2 业务需求分析

讲了这么多电商项目背景和优势分析，那么如何做一个解决真实需求的电商小程序则是接下来的重点部分。首先，进行项目业务需求分析，这个实战项目的需求其

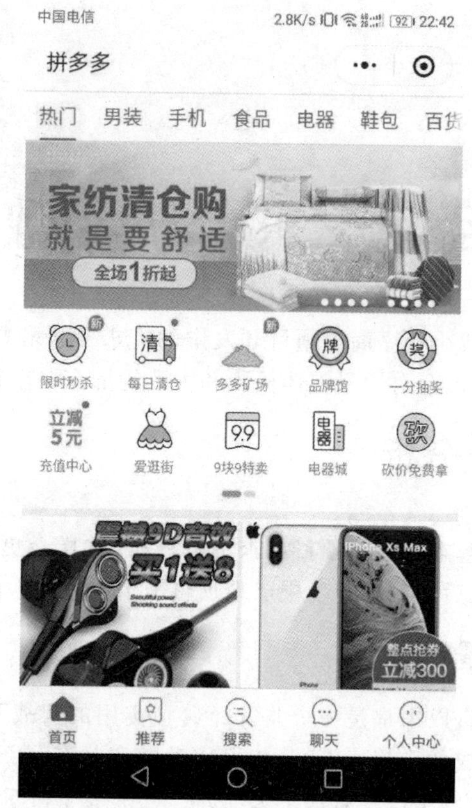

图 7-1　拼多多小程序

实来自于笔者平时在软件项目接单过程中所想到的一个需求:如何将技术服务包装为产品,通过小程序电商这种形式出售给意向客户。业务场景如下:

- 有一个长期合作的甲方客户想付费咨询技术问题;
- 有一个读者想付费购买我的核心代码;
- 有一个客户想单次付费解决一个技术 BUG;
- 有技术爱好者想付费学习某项技术;
- 有客户打算预付项目合作的前期定金款。

我遇到的业务场景其实还有很多,这个项目本质上其实是在出售技术服务或知识类产品。那其实只需要在产品定义上能把技术服务标准化,即可将其转化为标准化的产品来出售。这种业务需求在最近这几年而言非常流行,比如天猫上的产品的安装售后这类技术服务目前已经有专业的技术服务电商平台在做,比如万师傅、户帮户这类平台,万师傅小程序如图 7-2 所示;还有知识型电商也开始被大家所接受和认可,相关应用有:得到 APP,知识星球等。

总之,业务需求是层出不穷的,只需要用合适的技术即可很好地来满足这些需求,从而达到最终的商业变现。

图 7-2 万师傅小程序

7.1.3 产品需求分析

分析完业务需求，接下来进行产品需求分析。产品是用来解决实际业务需求的，这个项目需要做的就是一款电商小程序产品。为了更贴近于真实的实战项目，暂且给本书的实战项目小程序商城取一个产品名：信真小铺，其产品定位为专注于互联网软件技术服务的电商小程序应用。

首先，一款移动端应用在界面布局上需要确定的是首页功能入口。无论是电商类 APP 还是电商小程序，首页功能入口 tab 按钮基本为 5 个：首页、分类、购物车、"我的"和发现，参看小米商城 Lite 小程序，如图 7-3 所示。其中的 4 个按钮：首页、分类、购物车和"我的"，基本是目前应用的"标配"。所以本书中的实战项目就确定加这 4 个功能入口即可。如果读者觉得还可以精简，那么可以只保留：首页、购物车和"我的"这 3 个入口，因为在前期产品数量很少的时候，分类这个入口其实是没必要显示出来的。

图 7-3　小米商城 Lite 小程序

依然以小米商城 Lite 小程序的首页为例，从上到下大致分析一下，有：搜索框、轮播图广告、分类、产品列表和功能入口。其实这就是一个非常典型的移动电商 APP 或小程序的首页布局。其他一些复杂或新颖的首页布局大多是这种经典布局的变种，所以首页布局我们可以参考这个样例来做。

再来看第二个功能入口分类，如图 7-4 所示。分类其实相当于是对产品线的一个划分，我建议，如果有十款产品及其以上，那么是有必要加上分类这个入口的。

购物车功能页面如图 7-5 所示。购物车功能入口是电商类移动应用所必备的功能模块，其功能页面布局已经非常的固化，上半部分为购物车中选购的商品和价格列表，下半部分为结算价格和结算按钮。

“我的”功能页面如图 7-6 所示。“我的”这个功能入口大致等同于个人中心模块，这里主要用来显示用户的订单情况，个人信息等内容。

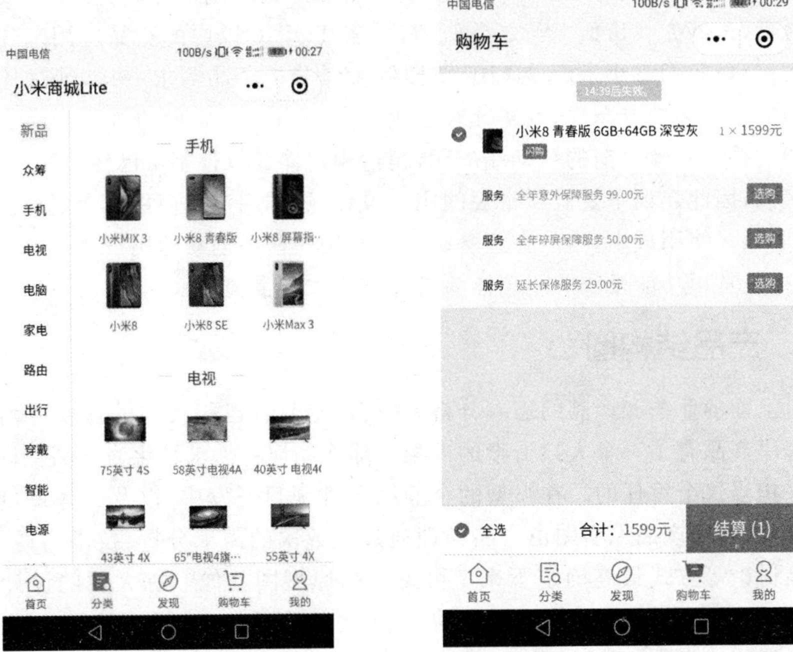

图 7-4　分　类

图 7-5　购物车

图 7-6　我　的

在这里我需要给读者分享一个快速做移动端产品应用的思路：在确定要做某一款应用之后，最好先找市面上与之类似的，排名 TOP10 以内的竞品应用来先分析和体验一番，就像我们拿小米商城 Lite 做样例参考分析产品需求一样，通过体验这些竞品应用的设计和功能布局，其实你在心里已经对自己想要做的这款产品有了一个大致框架的构思。就目前的移动端应用而言，用户体验设计基本已经大同小异，没有太大的区别，因此建议不要做反常规的用户设计，这样用户在使用上会更加的方便和熟悉，少了很多使用成本，当然在此基础之上再做一些小创新的用户体验设计也是可以的，前提是常用功能模块的用户体验设计不要太反常规。

7.1.4 产品结构图

通过上面的业务和产品的需求分析，我们已经对自己想要做的这款“信真小铺”电商小程序产品有了一个大致清晰的框架。那么如何将需求转化为产品？产品原型图就是来担当这个重任的。在大型的企业级软件项目开发中，以及一些重量级的互联网软件应用中，产品原型图由产品经理通过无数次的需求分析、技术沟通、需求确认和产品讨论等方式最终确定下来。可见产品原型图就像一栋大楼的设计图纸一样，一旦确定下来，就不会有太大的变动，因为之后的研发、UI 设计师、测试和运维等人员都会按照这张最初的“设计图纸”来执行、落地整个项目。优秀的产品经理打造的产品原型图可以让研发人员提高开发效率，减少沟通成本，而失败的产品经理则会使得开发成本大大增加、经常变动等，这在大型软件项目中是最可怕的。产品经理常用的设计工具有 Axure PR，简单好用很容易上手；MAC 用户一般用 Sketch 软件来设计原型图。目前市面上也出现了一些可以在线绘制产品原型图的工具，比如 Mockplus，还有一种非常简单直接的方式，便是用纸笔手绘原型图或者用 iPad 绘制。总之工具非常多，读者自行选择。高保真产品原型图示例如图 7-7 所示。

高保真的产品原型设计需要具备一定的交互设计功底，显然这个不是程序员所擅长的，那么有没有适合程序员进行快速开发使用的原型图？有，叫产品结构图（或产品信息架构图）。产品结构图是综合展示产品信息和功能逻辑的图表，简单地说，产品结构图就是产品原型的简化表达。它能够在前期的需求评审中或其他类似场景中作为产品原型的替代，因为产品结构图相较于产品原型的实现成本低，能够快速地对产品功能结构进行增、删、改操作，减少产品经理在这个过程中的实现成本。“信真小铺”电商小程序产品结构图如图 7-8 所示，该图用 XMind 思维导图工具创作。

通过上面的产品结构图，我们可以对所要开发的产品有一个非常清晰的认识，它涵盖了产品的大部分核心功能模块：登录模块、产品模块、购物车模块（含订单）和个人中心模块等。有了这张产品结构图，为接下来的数据建模和数据库设计提供了很直观的展现。

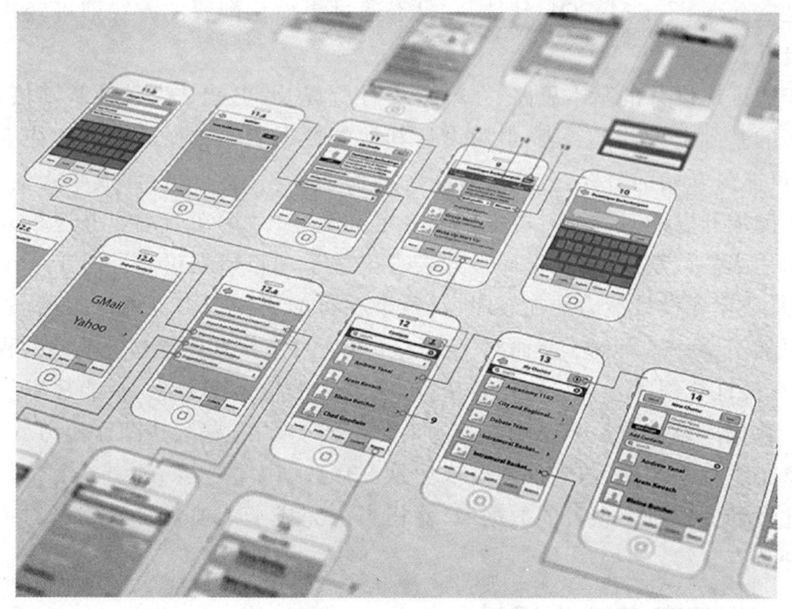

图 7-7　高保真产品原型图

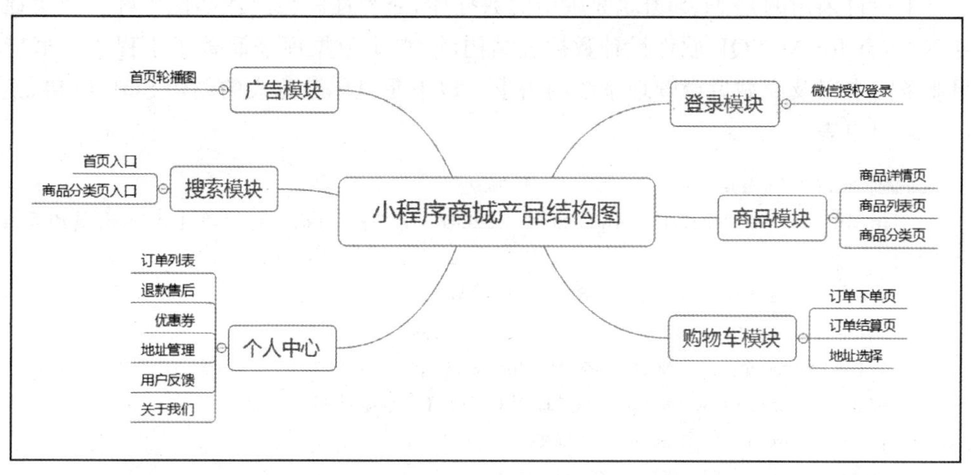

图 7-8　“信真小铺”电商小程序产品结构图

7.2　项目数据库设计

7.2.1　数据库设计概述

经过一番需求分析和产品结构图设计，此时整个项目前期的基础工作已大致完成。接下来需要进行数据建模以及详细的数据库表结构设计。这里采用的数据库设计工具是 Navicat for MySQL 软件，它只需通过界面化配置的方式即可轻松完成表结构的设计。

关于数据库设计的基本思路，在第六章有所提及，也就是先设计基础核心表，再在核心表的基础上设计扩展表，最后设计其他的数据表。数据库设计往往不是一次成型的，可能最开始的时候，先把所有能想到的业务模型对应设计成一张数据表，然后在程序前端开发的时候会对数据表有一定的变动，最后在后端业务逻辑开发的时候也会做适当的调整，因此数据库设计会贯串整个项目研发过程。对于一个开发新手而言，其数据库设计能力往往较弱，对整个项目的架构能力也稍显不足，故而数据库设计更加不会一蹴而就了。而对于熟练开发项目多年且熟悉相关业务的程序员来讲，他们在设计数据库上会显的比较从容，能想到的提前都会想到，能规避的坑能尽量规避。数据库设计是一个熟能生巧的活，经历的项目越多，其设计能力就会越强。其实就整个软件系统而言，它的本质依然是一套数据库信息系统，其核心依然是数据库。因此设计完成一个项目的数据库，就相当于完成了该项目研发工作的一个重要的里程碑节点。

7.2.2 数据库表结构设计

本项目采用的是 MySQL5.6 版本的数据库，兼容性和稳定性都比较好。我是通过 Navicat for MySQL 软件设计数据表结构的，整个数据库表涵盖了小程序商城应用业务需求以及商城后台管理系统的需求。以下是具体数据表的创建 SQL 语句。

➢ 用户表

```
CREATE TABLE 'lr_user' (
   'id' int(11) NOT NULL AUTO_INCREMENT COMMENT '用户表:包括后台管理员、商家会员和普通会员',
   'name' varchar(20) NOT NULL COMMENT '登录账号',
   'uname' varchar(10) DEFAULT NULL COMMENT '昵称',
   'pwd' varchar(50) NOT NULL COMMENT 'MD5 密码',
   'addtime' int(11) NOT NULL DEFAULT '0' COMMENT '创建日期',
   'jifen' float(11,0) DEFAULT '0' COMMENT '积分',
   'photo' varchar(255) DEFAULT NULL COMMENT '头像',
   'tel' char(15) DEFAULT NULL COMMENT '手机',
   'qq_id' varchar(20) NOT NULL DEFAULT '0' COMMENT 'qq',
   'email' varchar(50) DEFAULT NULL COMMENT '邮箱',
   'sex' tinyint(2) NOT NULL DEFAULT '0' COMMENT '性别',
   'del' tinyint(2) NOT NULL DEFAULT '0' COMMENT '状态',
   'openid' varchar(50) NOT NULL COMMENT '授权 ID',
   'source' varchar(10) NOT NULL COMMENT '第三方平台(微信,QQ)',
   PRIMARY KEY ('id')
) ENGINE = InnoDB AUTO_INCREMENT = 1 DEFAULT CHARSET = utf8;
```

➢ 产品表

```
CREATE TABLE 'lr_product' (
  'id' int(11) NOT NULL AUTO_INCREMENT COMMENT '产品表',
  'shop_id' int(11) NOT NULL DEFAULT '0' COMMENT '商铺 id',
  'brand_id' int(11) unsigned DEFAULT NULL COMMENT '品牌 ID',
  'name' varchar(50) NOT NULL COMMENT '产品名称',
  'intro' varchar(100) DEFAULT NULL COMMENT '广告语',
  'pro_number' varchar(100) DEFAULT NULL COMMENT '产品编号',
  'price' decimal(8,2) NOT NULL DEFAULT '0.00' COMMENT '价格',
  'price_yh' decimal(8,2) NOT NULL DEFAULT '0.00' COMMENT '优惠价格',
  'price_jf' int(11) NOT NULL DEFAULT '0' COMMENT '赠送积分',
  'photo_x' varchar(100) DEFAULT NULL COMMENT '小图',
  'photo_d' varchar(100) DEFAULT NULL COMMENT '大图',
  'photo_string' text COMMENT '图片组',
  'content' text COMMENT '商品详情',
  'addtime' int(11) DEFAULT NULL COMMENT '添加时间',
  'updatetime' int(11) NOT NULL COMMENT '修改时间',
  'sort' int(11) DEFAULT '0' COMMENT '排序',
  'renqi' int(11) NOT NULL DEFAULT '0' COMMENT '人气',
  'shiyong' int(11) NOT NULL DEFAULT '0' COMMENT '购买数',
  'num' int(11) NOT NULL DEFAULT '0' COMMENT '数量',
  'type' tinyint(2) NOT NULL DEFAULT '0' COMMENT '是否推荐 1 推荐  0 不推荐',
  'del' tinyint(2) NOT NULL DEFAULT '0' COMMENT '删除状态',
  'del_time' int(10) DEFAULT '0' COMMENT '删除时间',
  'pro_buff' varchar(255) DEFAULT NULL COMMENT '产品属性',
  'cid' int(11) NOT NULL COMMENT '分类 id',
  'company' char(10) DEFAULT NULL COMMENT '单位',
  'is_show' tinyint(1) unsigned NOT NULL DEFAULT '1' COMMENT '是否新品',
  'is_down' tinyint(1) unsigned NOT NULL DEFAULT '0' COMMENT '下架状态',
  'is_hot' tinyint(1) unsigned DEFAULT '0' COMMENT '是否热卖',
  'is_sale' tinyint(1) DEFAULT '0' COMMENT '是否折扣',
  'start_time' int(11) DEFAULT '0' COMMENT '抢购开始时间',
  'end_time' int(11) unsigned DEFAULT '0' COMMENT '抢购结束时间',
  'pro_type' tinyint(1) unsigned NOT NULL DEFAULT '1' COMMENT '产品类型 1 普通 2 抢购产
品',
  PRIMARY KEY ('id')
) ENGINE = InnoDB AUTO_INCREMENT = 1 DEFAULT CHARSET = utf8;
```

➢ 商品品牌表

```
CREATE TABLE 'lr_brand' (
  'id' int(11) unsigned NOT NULL AUTO_INCREMENT COMMENT '产品品牌表',
```

```
  'name' varchar(100) NOT NULL COMMENT '品牌名称',
  'brandprice' float(8,2) NOT NULL DEFAULT '0.00' COMMENT '起始价格',
  'photo' varchar(100) DEFAULT NULL COMMENT '图片',
  'type' tinyint(2) DEFAULT '0' COMMENT '是否推荐',
  'addtime' int(11) DEFAULT NULL COMMENT '添加时间',
  'shop_id' int(11) unsigned DEFAULT '0' COMMENT '店铺 id',
  PRIMARY KEY ('id')
) ENGINE = InnoDB AUTO_INCREMENT = 1 DEFAULT CHARSET = utf8;
```

➤ 商品分类表

```
CREATE TABLE 'lr_category' (
  'id' int(11) NOT NULL AUTO_INCREMENT COMMENT '分类 id',
  'tid' int(11) NOT NULL DEFAULT '0' COMMENT '父级分类 id',
  'name' varchar(50) NOT NULL COMMENT '栏目名称',
  'sort' int(11) NOT NULL DEFAULT '0' COMMENT '排序',
  'addtime' int(11) NOT NULL DEFAULT '0' COMMENT '添加时间',
  'concent' varchar(255) DEFAULT NULL COMMENT '栏目简介',
  'bz_1' varchar(100) DEFAULT NULL COMMENT '缩略图',
  'bz_2' varchar(255) DEFAULT NULL COMMENT '备注字段',
  'bz_3' varchar(100) DEFAULT NULL COMMENT '图标',
  'bz_4' tinyint(2) NOT NULL DEFAULT '0' COMMENT '备用字段',
  'bz_5' varchar(100) DEFAULT NULL COMMENT '推荐略缩图',
  PRIMARY KEY ('id')
) ENGINE = InnoDB AUTO_INCREMENT = 1 DEFAULT CHARSET = utf8;
```

➤ 订单表

```
CREATE TABLE 'lr_order' (
  'id' int(11) NOT NULL AUTO_INCREMENT COMMENT '订单 id',
  'order_sn' varchar(20) NOT NULL COMMENT '订单编号',
  'pay_sn' varchar(20) DEFAULT NULL COMMENT '支付单号',
  'shop_id' int(11) NOT NULL DEFAULT '0' COMMENT '商家 ID',
  'uid' int(11) NOT NULL DEFAULT '0' COMMENT '用户 ID',
  'price' decimal(9,2) NOT NULL DEFAULT '0.00' COMMENT '价格',
  'amount' decimal(9,2) DEFAULT '0.00' COMMENT '优惠后价格',
  'addtime' int(10) NOT NULL DEFAULT '0' COMMENT '购买时间',
  'del' tinyint(2) NOT NULL DEFAULT '0' COMMENT '删除状态',
  'type' enum('weixin','alipay','cash') DEFAULT 'weixin' COMMENT '支付方式',
  'price_h' decimal(9,2) NOT NULL DEFAULT '0.00' COMMENT '真实支付金额',
  'status' tinyint(2) NOT NULL DEFAULT '10' COMMENT '订单状态{0,已取消 10 未付款 20 代
发货 30 待收货 40 待评价 50 交易完成 51 交易关闭',
  'vid' int(11) DEFAULT '0' COMMENT '优惠券 ID',
  'receiver' varchar(15) NOT NULL COMMENT '收货人',
```

```
'tel' char(15) NOT NULL COMMENT '联系方式',
'address_xq' varchar(50) NOT NULL COMMENT '地址详情',
'code' int(11) NOT NULL COMMENT '邮编',
'post' int(11) DEFAULT NULL COMMENT '快递 ID',
'remark' varchar(255) DEFAULT NULL COMMENT '买家留言',
'post_remark' varchar(255) NOT NULL COMMENT '邮费信息',
'product_num' int(11) NOT NULL DEFAULT '1' COMMENT '商品数量',
'trade_no' varchar(50) DEFAULT NULL COMMENT '微信交易单号',
'kuaidi_name' varchar(10) DEFAULT NULL COMMENT '快递名称',
'kuaidi_num' varchar(20) DEFAULT NULL COMMENT '运单号',
'back' enum('1','2','0') DEFAULT '0' COMMENT '标识客户是否有发起退款 1 申请退款 2 已退
款',
'back_remark' varchar(255) DEFAULT NULL COMMENT '退款原因',
'back_addtime' int(11) DEFAULT '0' COMMENT '申请退款时间',
'order_type' tinyint(2) DEFAULT '1' COMMENT '订单类型 1 普通订单 2 抢购订单',
PRIMARY KEY ('id')
) ENGINE = InnoDB AUTO_INCREMENT = 1 DEFAULT CHARSET = utf8;
```

➢ 订单商品信息表

```
CREATE TABLE 'lr_order_product' (
  'id' int(11) NOT NULL AUTO_INCREMENT COMMENT '订单商品信息表',
  'pid' int(11) NOT NULL DEFAULT '0' COMMENT '商品 id',
  'pay_sn' varchar(20) DEFAULT NULL COMMENT '支付单号',
  'order_id' int(11) NOT NULL DEFAULT '0' COMMENT '订单 id',
  'name' varchar(50) NOT NULL COMMENT '产品名称',
  'price' decimal(8,2) NOT NULL DEFAULT '0.00' COMMENT '价格',
  'photo_x' varchar(100) DEFAULT NULL COMMENT '商品图',
  'pro_buff' varchar(255) DEFAULT NULL COMMENT '产品属性',
  'addtime' int(11) NOT NULL COMMENT '添加时间',
  'num' int(11) NOT NULL DEFAULT '1' COMMENT '购买数量',
  'pro_guige' varchar(50) DEFAULT NULL COMMENT '规格 id 和规格名称',
  PRIMARY KEY ('id')
) ENGINE = InnoDB AUTO_INCREMENT = 1 DEFAULT CHARSET = utf8;
```

➢ 购物车表

```
CREATE TABLE 'lr_shopping_char' (
  'id' int(11) NOT NULL AUTO_INCREMENT COMMENT '购物车表',
  'pid' int(11) NOT NULL COMMENT '商品 ID',
  'price' decimal(9,2) NOT NULL DEFAULT '0.00' COMMENT '商品单价',
  'num' int(11) NOT NULL DEFAULT '1' COMMENT '数量',
  'buff' varchar(255) NOT NULL COMMENT '属性(序列化格式)',
```

```
  'addtime' int(10) NOT NULL COMMENT '添加时间',
  'uid' int(11) NOT NULL COMMENT '用户 ID',
  'shop_id' int(11) NOT NULL DEFAULT '0' COMMENT '商家 ID',
  'gid' int(11) DEFAULT '0' COMMENT '规格 id',
  'type' tinyint(2) DEFAULT '2' COMMENT '0 是热卖,1 是团购,2 是普通商品',
  PRIMARY KEY ('id')
) ENGINE = InnoDB AUTO_INCREMENT = 1 DEFAULT CHARSET = utf8;
```

➢ 收货地址表

```
CREATE TABLE 'lr_address' (
  'id' int(11) NOT NULL AUTO_INCREMENT COMMENT '地址 id',
  'name' varchar(10) NOT NULL COMMENT '收货人',
  'tel' char(15) NOT NULL COMMENT '联系方式',
  'sheng' int(11) NOT NULL DEFAULT '0' COMMENT '省 id',
  'city' int(11) NOT NULL DEFAULT '0' COMMENT '市 id',
  'quyu' int(11) NOT NULL DEFAULT '0' COMMENT '区域 id',
  'address' varchar(255) NOT NULL COMMENT '收货地址(不加省市区)',
  'address_xq' varchar(255) NOT NULL COMMENT '省市区+详细地址',
  'code' int(11) NOT NULL DEFAULT '0' COMMENT '邮政编号',
  'uid' int(11) NOT NULL DEFAULT '0' COMMENT '用户 ID',
  'is_default' tinyint(2) NOT NULL DEFAULT '0' COMMENT '是否默认地址 1 默认',
  PRIMARY KEY ('id'),
  UNIQUE KEY 'id' ('id') USING BTREE
) ENGINE = InnoDB AUTO_INCREMENT = 1 DEFAULT CHARSET = utf8 COMMENT = '用户收货地址表';
```

➢ 省市区地址联动表

```
CREATE TABLE 'lr_china_city' (
  'id' int(11) NOT NULL AUTO_INCREMENT COMMENT '主键 id',
  'tid' int(11) DEFAULT '0' COMMENT '父级 id',
  'name' varchar(255) DEFAULT NULL,
  'code' varchar(255) DEFAULT NULL,
  'head' varchar(1) DEFAULT NULL,
  'type' tinyint(2) NOT NULL DEFAULT '0',
  PRIMARY KEY ('id')
) ENGINE = InnoDB DEFAULT CHARSET = utf8 COMMENT = '中国省市区地址联动表';
```

➢ 广告信息表

```
CREATE TABLE 'lr_guanggao' (
  'id' int(11) NOT NULL AUTO_INCREMENT COMMENT '子页广告管理表',
  'name' varchar(20) DEFAULT NULL COMMENT '广告名称',
  'photo' varchar(100) DEFAULT NULL COMMENT '图片',
  'addtime' int(11) NOT NULL DEFAULT '0' COMMENT '添加时间',
```

```
  'sort' int(11) NOT NULL DEFAULT '0',
  'type' enum('product','news','partner','index') DEFAULT 'index' COMMENT '广告类型',
  'action' varchar(255) NOT NULL COMMENT '链接值',
  'position' tinyint(2) unsigned DEFAULT '1' COMMENT '广告位置 1 首页轮播',
  PRIMARY KEY ('id')
) ENGINE = InnoDB AUTO_INCREMENT = 1 DEFAULT CHARSET = utf8;
```

➢ 优惠券表

```
CREATE TABLE 'lr_voucher' (
  'id' int(11) unsigned NOT NULL AUTO_INCREMENT COMMENT '店铺优惠券表',
  'shop_id' int(11) unsigned NOT NULL DEFAULT '0' COMMENT '店铺 ID',
  'title' varchar(100) DEFAULT NULL COMMENT '优惠券名称',
  'full_money' decimal(9,2) NOT NULL DEFAULT '0.00' COMMENT '满多少钱',
  'amount' decimal(9,2) NOT NULL DEFAULT '0.00' COMMENT '减多少钱',
  'start_time' int(11) NOT NULL DEFAULT '0' COMMENT '开始时间',
  'end_time' int(11) NOT NULL DEFAULT '0' COMMENT '结束时间',
  'point' int(11) DEFAULT '0' COMMENT '所需积分',
  'count' int(11) unsigned NOT NULL DEFAULT '1' COMMENT '发行数量',
  'receive_num' int(11) unsigned DEFAULT '0' COMMENT '领取数量',
  'addtime' int(11) NOT NULL DEFAULT '0' COMMENT '添加时间',
  'type' tinyint(1) NOT NULL DEFAULT '1' COMMENT '优惠券类型',
  'del' tinyint(1) unsigned DEFAULT '0' COMMENT '删除状态',
  'proid' text COMMENT '产品 ID',
  PRIMARY KEY ('id')
) ENGINE = InnoDB AUTO_INCREMENT = 1 DEFAULT CHARSET = utf8;
```

➢ 优惠券领取记录表

```
CREATE TABLE 'lr_user_voucher' (
  'id' int(11) unsigned NOT NULL AUTO_INCREMENT COMMENT '会员优惠券领取记录',
  'uid' int(11) unsigned NOT NULL DEFAULT '0' COMMENT '会员 ID',
  'vid' int(11) unsigned NOT NULL DEFAULT '0' COMMENT '优惠券 id',
  'shop_id' int(11) unsigned NOT NULL DEFAULT '0' COMMENT '商铺 ID',
  'full_money' decimal(9,2) NOT NULL DEFAULT '0.00' COMMENT '满多少钱',
  'amount' decimal(9,2) NOT NULL DEFAULT '0.00' COMMENT '减多少钱',
  'start_time' int(11) DEFAULT '0' COMMENT '开始时间',
  'end_time' int(11) DEFAULT '0' COMMENT '结束时间',
  'addtime' int(11) NOT NULL DEFAULT '0' COMMENT '领取时间',
  'status' tinyint(2) unsigned DEFAULT '1' COMMENT '使用状态 1 未使用 2 已使用 3 已失
效',
  PRIMARY KEY ('id')
) ENGINE = InnoDB AUTO_INCREMENT = 1 DEFAULT CHARSET = utf8;
```

➢ 搜索记录表

```
CREATE TABLE 'lr_search_record' (
  'id' int(11) unsigned NOT NULL AUTO_INCREMENT COMMENT '搜索记录表',
  'uid' int(11) NOT NULL DEFAULT '0' COMMENT '会员 ID',
  'keyword' varchar(255) NOT NULL COMMENT '搜索内容',
  'num' int(11) NOT NULL DEFAULT '1' COMMENT '搜索次数',
  'addtime' int(11) DEFAULT '0' COMMENT '搜索时间',
  PRIMARY KEY ('id')
) ENGINE = InnoDB AUTO_INCREMENT = 1 DEFAULT CHARSET = utf8;
```

➢ 意见反馈表

```
CREATE TABLE 'lr_fankui' (
  'id' int(11) NOT NULL AUTO_INCREMENT,
  'uid' int(11) NOT NULL DEFAULT '0' COMMENT '用户 ID',
  'message' varchar(255) NOT NULL COMMENT '反馈内容',
  'addtime' int(11) NOT NULL DEFAULT '0' COMMENT '反馈时间',
  PRIMARY KEY ('id')
) ENGINE = InnoDB AUTO_INCREMENT = 1 DEFAULT CHARSET = utf8;
```

第8章 小程序商城前端程序开发

本章主要讲解小程序商城的前端页面开发内容，包括小程序全局基础模块开发配置和小程序商城具体页面的模板、样式、逻辑和配置代码编写。本章内容基本涵盖了前几章中对小程序核心基础、框架组件和框架 API 的综合理解与运用。

8.1 小程序商城前端页面开发概述

到目前为止，需求分析、产品结构图以及数据库设计均已准备就绪，接下来其实就要进入实际的程序代码 coding 过程了。一般而言，大多是先开发前端页面，后开发后端业务逻辑部分，因为前端页面是所见即所得的直观呈现，在开发的过程中可以实时调整和修改，在页面开发过程中可以预留后端接口，在稍后的后端开发中再来补充完善这些后端接口部分的工作。如果前后端开发不是一个人，是分开的，那么建议提前先约定好接口，然后前端和后端开发可以同时并行开发程序，这样的协作编程方式可以很高效地完成开发工作，尤其对于那种工作量相对大一点的软件项目是非常有帮助的。

由于微信小程序已经约定好了一些规则，所以在小程序前端页面开发上是相对比较省心的。首选需要完成小程序全局入口的代码编写和配置工作，然后再分页面对各个具体的页面进行模板、样式、逻辑和配置代码的编写，这样两大步即可非常轻松地实现小程序前端页面的展现功能。

8.2 项目全局基础模块代码清单

8.2.1 全局入口 app.js

App 全局入口文件为 app.js，主要用于注册一个小程序以及微信登录等，代码如下：

```
App({
  d: {
    hostUrl: 'https://mini.laohuzx.com/index.php',//请填写自己的主机 URL
```

```
    appId:"请填写自己的 APPID",
    appKey:"请填写自己的 APPKEY",
    ceshiUrl:'https://mini.laohuzx.com/index.php',//请填写自己的测试 URL
  },
  //小程序初始化完成时触发,全局只触发一次
  onLaunch: function () {
    //调用 API 从本地缓存中获取数据
    var logs = wx.getStorageSync('logs') || []
    logs.unshift(Date.now())
    wx.setStorageSync('logs', logs);
    //login
    this.getUserInfo();
  },
  getUserInfo:function(cb){
    var that = this
    if(this.globalData.userInfo){
      typeof cb == "function" && cb(this.globalData.userInfo)
    }else{
      //调用登录接口
      wx.login({
        success: function (res) {
          //console.log(res);
          var code = res.code;
          //get wx user simple info
          wx.getUserInfo({
            withCredentials: true,
            success: function (res) {
              //如果已经授权过那么会执行这里的代码,console.log("已授权标记");
              that.globalData.userInfo = res.userInfo;
              typeof cb == "function" && cb(that.globalData.userInfo);
              //get user sessionKey
              that.getUserSessionKey(code);
            },
            fail: function () {
              //获取用户信息失败后,请跳转授权页面
              wx.showModal({
                title: '警告',
                content: '尚未进行授权,请点击确定跳转到授权页面进行授权。',
                success: function (res) {
                  if (res.confirm) {
                    console.log('用户点击确定')
                    wx.navigateTo({
```

```
                    url：'../tologin/tologin',
                  })
                }
              }
            })
          }

        });

      }
    });
  }
},

getUserSessionKey:function(code){

  //用户的订单状态
  var that = this;
  wx.request({
    url：that.d.ceshiUrl + '/Api/Login/getsessionkey',
    method:'post',
    data：{
      code：code
    },
    header：{
      'Content-Type'：  'application/x-www-form-urlencoded'
    },
    success：function (res) {
      //--init data
      var data = res.data;
      if(data.status==0){
        wx.showToast({
          title：data.err,
          duration：2000
        });
        return false;
      }

      that.globalData.userInfo['sessionId'] = data.session_key;
      that.globalData.userInfo['openid'] = data.openid;
      that.onLoginUser();
    },
```

```
        fail:function(e){
          wx.showToast({
            title:'网络异常！err:getsessionkeys',
            duration:2000
          });
        },
      });
    },
    //授权登录
    onLoginUser:function(){
      var that = this;
      var user = that.globalData.userInfo;
      wx.request({
        url:that.d.ceshiUrl + '/Api/Login/authlogin',
        method:'post',
        data:{
          SessionId:user.sessionId,
          gender:user.gender,
          NickName:user.nickName,
          HeadUrl:user.avatarUrl,
          openid:user.openid
        },
        header:{
          'Content-Type': 'application/x-www-form-urlencoded'
        },
        success:function (res) {
          //--init data
          var data = res.data.arr;
          var status = res.data.status;
          if(status! =1){
            wx.showToast({
              title:res.data.err,
              duration:3000
            });
            return false;
          }
          that.globalData.userInfo['id'] = data.ID;
          that.globalData.userInfo['NickName'] = data.NickName;
          that.globalData.userInfo['HeadUrl'] = data.HeadUrl;
          var userId = data.ID;
          if (! userId){
            wx.showToast({
```

```
                title：'登录失败！',
                duration：3000
              });
              return false;
            }
            that.d.userId = userId;
          },
          fail:function(e){
            wx.showToast({
              title：'网络异常！ err:authlogin',
              duration：2000
            });
          },
        });
      },

    globalData:{
        userInfo:null
      },

      onPullDownRefresh：function (){
        wx.stopPullDownRefresh();
      }

    });
```

8.2.2 全局配置 app.json

全局页面配置文件为 app.json，主要用于小程序页面路由配置和窗体参数设置，代码如下：

```
{
  "pages":[
    "pages/index/index",
    "pages/tologin/tologin",
    "pages/product/detail",
    "pages/search/search",
    "pages/user/user",
    "pages/user/dingdan",
    "pages/user/daifahuo",
    "pages/user/daishouhuo",
    "pages/user/success",
```

```
    "pages/user/shouhou",
    "pages/user/tuihuo",
    "pages/user/feedback",
    "pages/order/pay",
    "pages/order/detail",
    "pages/cart/cart",
    "pages/listdetail/listdetail",
    "pages/address/address",
    "pages/address/user-address/user-address",
    "pages/category/index",
    "pages/ritual/ritual",
    "pages/synopsis/synopsis",
    "pages/myritual/myritual"
      ],
  "window":{
    "backgroundTextStyle":"dark",
    "navigationBarBackgroundColor":"#fff",
    "navigationBarTitleText":"信真小铺",
    "navigationBarTextStyle":"black",
    "enablePullDownRefresh":true
  },
  "tabBar":{
    "backgroundColor":"#fafafa",
    "borderStyle":"white",
    "selectedColor":"#b4282d",
    "color":"#666",
    "list":[
      {
        "pagePath":"pages/index/index",
        "iconPath":"static/images/ic_menu_choice_nor.png",
        "selectedIconPath":"static/images/ic_menu_choice_pressed.png",
        "text":"首页"
      },
      {
        "pagePath":"pages/category/index",
        "iconPath":"static/images/ic_menu_sort_nor.png",
        "selectedIconPath":"static/images/ic_menu_sort_pressed.png",
        "text":"分类"
      },
      {
        "pagePath":"pages/cart/cart",
        "iconPath":"static/images/ic_menu_shoping_nor.png",
```

```
        "selectedIconPath": "static/images/ic_menu_shoping_pressed.png",
        "text": "购物车"
      },
      {
        "pagePath": "pages/user/user",
        "iconPath": "static/images/ic_menu_me_nor.png",
        "selectedIconPath": "static/images/ic_menu_me_pressed.png",
        "text": "我的"
      }
    ]
  },
  "networkTimeout": {
    "request": 10000,
    "downloadFile": 10000
  },
  "debug": true
}
```

8.2.3 全局样式 app.wxss

全局样式文件为 app.wxss，主要用于设置全局样式，代码如下：

```
/**app.wxss**/
.container {
  box-sizing: border-box;
  background-color: #f4f4f4;
  font-family: PingFangSC-Light,helvetica,'Heiti SC';
}

view,image,text,navigator{
  box-sizing: border-box;
  padding:0;
  margin:0;

}

view,text{
  font-family: PingFangSC-Light,helvetica,'Heiti SC';
  font-size: 29rpx;
  color: #333;
}
.df{
```

```
    display: -webkit-box;
    display: -webkit-flex;
    display: -ms-flexbox;
    display: flex;
}
.df_1{
    -webkit-box-flex: 1;
    -webkit-flex: 1;
    -ms-flex: 1;
    flex: 1;
    -webkit-tap-highlight-color: transparent;
}
.df_2{
    -webkit-box-flex: 2;
    -webkit-flex: 2;
    -ms-flex: 2;
    flex: 2;
    -webkit-tap-highlight-color: transparent;
}
```

8.2.4 项目配置 project.config.json

项目配置文件为 project.config.json，主要用于对小程序开发者工具进行个性化配置，代码如下：

```
{
    "description": "项目配置文件。",
    "setting": {
        "urlCheck": true,
        "es6": true,
        "postcss": true,
        "minified": true,
        "newFeature": true
    },
    "compileType": "miniprogram",
    "libVersion": "1.6.6",
    "appid": "请填写您自己的 APPID",
    "projectname": "%E4%BF%A1%E7%9C%9F%E5%B0%8F%E9%93%BA",
    "condition": {
        "search": {
            "current": -1,
            "list": []
        },
        "conversation": {
            "current": -1,
```

```
            "list": []
        },
        "plugin": {
            "current": -1,
            "list": []
        },
        "game": {
            "list": []
        },
        "miniprogram": {
            "current": -1,
            "list": []
        }
    }
}
```

8.3 首页代码清单

小程序商城首页效果如图 8-1 和图 8-2 所示。

图 8-1 首 页

图 8-2 首页(续)

代码详情如下所示。

8.3.1 wxml 模板代码

```
<!-- index.wxml -->
<view class="container">
  <view class="search">
    <navigator url="/pages/search/search" class="input">
      <image class="icon"></image>
      <text class="txt">商品搜索，共{{goodsCount}}款好物</text>
    </navigator>
  </view>
  <swiper class="banner" indicator-dots="true" autoplay="true" interval="3000" duration="1000">
    <swiper-item wx:for="{{imgUrls}}" wx:key="{{item.id}}">
      <navigator url="{{item.action}}">
        <image src="{{item.photo}}" background-size="cover"></image>
      </navigator>
    </swiper-item>
  </swiper>

  <view class="m-menu">
    <navigator class="item" url="../listdetail/listdetail?cat_id={{item.id}}" wx:for="{{category}}" wx:key="{{item.id}}">
      <image src="{{item.bz_1}}" background-size="cover"></image>
      <text>{{item.name}}</text>
    </navigator>
<navigator url="../listdetail/listdetail"  class="item">
<image src="../../static/images/more.png" background-size="cover"></image>
      <text>更多</text>
</navigator>
  </view>


  <view class="a-section a-new" wx:if="{{newGoods.length > 0}}">
    <view class="h">
      <view>
        <navigator url="../newGoods/newGoods">
          <text class="txt">新品首发</text>
        </navigator>
      </view>
    </view>
```

```
        < view class = "b" >
          < view class = "item" wx:for = "{{newGoods}}" wx:for - index = "index" wx:for - i-
tem = "item" wx:key = "{{item.id}}" >
            < navigator url = "../product/detail? productId = {{item.id}}" >
              < image class = "img" src = "{{item.photo_x}}" background - size = "cover" >
</image >
              < text class = "name" > {{item.name}} </text >
              < text class = "price" > ¥ {{item.price_yh}} </text >
            </navigator >
          </view >
        </view >
      </view >
      < view class = "a - section a - quan" >
        < view class = "quan" >
          < navigator url = "../ritual/ritual" >
            < image src = "../../static/images/quan.jpg" background - size = "cover" > </
image >
          </navigator >
        </view >
      </view >

    </view >
```

8.3.2 wxss 样式代码

```
.search {
  height: 88rpx;
  width: 100 % ;
  padding: 0 30rpx;
  background: #fff;
  display: flex;
  align - items: center;
}

.search .input {
  width: 690rpx;
  height: 56rpx;
  background: #ededed;
  border - radius: 8rpx;
  display: flex;
  align - items: center;
```

```
    justify - content: center;
  }

  .search .icon {
    background: url(http://cdn.xinzhenkj.com/search2 - 2fb94833aa.png) center no - re-
peat;
    background - size: 100 % ;
    width: 28rpx;
    height: 28rpx;
  }

  .search .txt {
    height: 42rpx;
    line - height: 42rpx;
    color: #666;
    padding - left: 10rpx;
    font - size: 30rpx;
  }
  .banner {
    width: 750rpx;
    height: 417rpx;
  }

  .banner image {
    width: 100 % ;
    height: 417rpx;
  }

  .m - menu {
    display: flex;
    height: 181rpx;
    width: 750rpx;
    flex - flow: row nowrap;
    align - items: center;
    justify - content: space - between;
    background - color: #fff;
  }

  .m - menu .item {
    flex: 1;
    display: block;
    padding: 20rpx 0;
```

```
}

.m - menu image {
  display: block;
  width: 58rpx;
  height: 58rpx;
  margin: 0 auto;
  margin - bottom: 12rpx;
}

.m - menu text {
  display: block;
  font - size: 24rpx;
  text - align: center;
  margin: 0 auto;
  line - height: 1;
  color: #333;
}

.a - section {
  width: 750rpx;
  height: auto;
  overflow: hidden;
  background: #fff;
  color: #333;
  margin - top: 20rpx;
}
.a - section .quan {

}
.quan image {width:100 %;height:251rpx;}

.a - section .h {
  display: flex;
  flex - flow: row nowrap;
  align - items: center;
  justify - content: center;
  height: 130rpx;
}

.a - section .h .txt {
  padding - right: 30rpx;
```

```
  background: url("") right 4rpx no-repeat;
  background-size: 16.656rpx 27rpx;
  display: inline-block;
  height: 36rpx;
  font-size: 33rpx;
  line-height: 36rpx;
}

.a-brand .b {
  width: 750rpx;
  height: auto;
  overflow: hidden;
  position: relative;
}

.a-brand .wrap {
  position: relative;
}

.a-brand .img {
  position: absolute;
  left: 0;
  top: 0;
}

.a-brand .mt {
  position: absolute;
  z-index: 2;
  padding: 27rpx 31rpx;
  left: 0;
  top: 0;
}

.a-brand .mt .brand {
  display: block;
  font-size: 33rpx;
  height: 43rpx;
  color: #333;
}

.a-brand .mt .price, .a-brand .mt .unit {
  font-size: 25rpx;
```

```
  color: #999;
}

.a-brand .item-1 {
  float: left;
  width: 375rpx;
  height: 252rpx;
  overflow: hidden;
  border-top: 1rpx solid #fff;
  margin-left: 1rpx;
}

.a-brand .item-1:nth-child(2n+1){
  margin-left: 0;
  width: 374rpx;
}

.a-brand .item-1 .img {
  width: 375rpx;
  height: 253rpx;
}

.a-new .b {
  width: 750rpx;
  height: auto;
  overflow: hidden;
  padding: 0 31rpx 45rpx 31rpx;
}

.a-new .b .item {
  float: left;
  width: 302rpx;
  margin-top: 10rpx;
  margin-left: 21rpx;
  margin-right: 21rpx;
}

.a-new .b .item-b {
  margin-left: 42rpx;
}

.a-new .b .img {
```

```
    width: 302rpx;
    height: 302rpx;
}

.a-new .b .name {
    text-align: center;
    display: block;
    width: 302rpx;
    height: 35rpx;
    margin-bottom: 14rpx;
    overflow: hidden;
    font-size: 30rpx;
    color: #333;
}

.a-new .b .price {
    display: block;
    text-align: center;
    line-height: 30rpx;
    font-size: 30rpx;
    color: #b4282d;
}

.a-popular {
    width: 750rpx;
    height: auto;
    overflow: hidden;
}

.a-popular .b .item {
    border-top: 1px solid #d9d9d9;
    margin: 0 20rpx;
    height: 264rpx;
    width: 710rpx;
}

.a-popular .b .img {
    margin-top: 12rpx;
    margin-right: 12rpx;
    float: left;
    width: 240rpx;
    height: 240rpx;
```

```
}

.a-popular .b .right {
  float: left;
  height: 264rpx;
  width: 456rpx;
  display: flex;
  flex-flow: row nowrap;
}

.a-popular .b .text {
  display: flex;
  flex-wrap: nowrap;
  flex-direction: column;
  justify-content: center;
  overflow: hidden;
  height: 264rpx;
  width: 456rpx;
}

.a-popular .b .name {
  width: 456rpx;
  display: block;
  color: #333;
  line-height: 50rpx;
  font-size: 30rpx;
}

.a-popular .b .desc {
  width: 456rpx;
  display: block;
  color: #999;
  line-height: 50rpx;
  font-size: 25rpx;
}

.a-popular .b .price {
  width: 456rpx;
  display: block;
  color: #b4282d;
  line-height: 50rpx;
  font-size: 33rpx;
```

```
}

.a-topic .b {
  height: 533rpx;
  width: 750rpx;
  padding: 0 0 48rpx 0;
}

.a-topic .b .list {
  height: 533rpx;
  width: 750rpx;
  white-space: nowrap;
}

.a-topic .b .item {
  display: inline-block;
  height: 533rpx;
  width: 680.5rpx;
  margin-left: 30rpx;
  overflow: hidden;
}

.a-topic .b .item:last-child {
  margin-right: 30rpx;
}

.a-topic .b .img {
  height: 387.5rpx;
  width: 680.5rpx;
  margin-bottom: 30rpx;
}

.a-topic .b .np {
  height: 35rpx;
  margin-bottom: 13.5rpx;
  color: #333;
  font-size: 30rpx;
}

.a-topic .b .np .price {
  margin-left: 20.8rpx;
  color: #b4282d;
```

```
}

.a-topic .b .desc {
  display: block;
  height: 30rpx;
  color: #999;
  font-size: 24rpx;
  white-space: nowrap;
  overflow: hidden;
  text-overflow:ellipsis;
}

.good-grid {
  width: 750rpx;
  height: auto;
  overflow: hidden;
}

.good-grid .h {
  display: flex;
  flex-flow: row nowrap;
  align-items: center;
  justify-content: center;
  height: 130rpx;
  font-size: 33rpx;
  color: #333;
}

.good-grid .b {
  width: 750rpx;
  padding: 0 6.25rpx;
  height: auto;
  overflow: hidden;
}

.good-grid .b .item {
  float: left;
  background: #fff;
  width: 365rpx;
  margin-bottom: 6.25rpx;
  height: 452rpx;
  overflow: hidden;
```

```
  text - align: center;
}

.good - grid .b .item .a{
  height: 452rpx;
  width: 100 % ;
}

.good - grid .b .item - b {
  margin - left: 6.25rpx;
}

.good - grid .item .img {
  margin - top: 20rpx;
  width: 302rpx;
  height: 302rpx;
}

.good - grid .item .name {
  display: block;
  width: 365.625rpx;
  padding: 0 20rpx;
  overflow: hidden;
  height: 35rpx;
  margin: 11.5rpx 0 22rpx 0;
  text - align: center;
  font - size: 30rpx;
  color: #333;
}

.good - grid .item .price {
  display: block;
  width: 365.625rpx;
  height: 30rpx;
  text - align: center;
  font - size: 30rpx;
  color: #b4282d;
}

.good - grid .more - item{
  height: 100 % ;
  width: 100 % ;
```

```
}

.more-a{
  height: 100%;
  width: 100%;
  display: flex;
  flex-direction: column;
  align-items: center;
  justify-content: center;
}

.good-grid .more-a .txt{
  height: 33rpx;
  width: 100%;
  line-height: 33rpx;
  color: #333;
  font-size: 33rpx;
}

.good-grid .more-a .icon{
  margin: 60rpx auto 0 auto;
  width: 70rpx;
  height: 70rpx;
}
```

8.3.3 js 逻辑代码

```
//获取小程序实例
var app = getApp();

Page({
  data: {
    imgUrls: [],
    indicatorDots: true,
    autoplay: true,
    interval: 5000,
    duration: 1000,
    circular: true,
goodsCount: [],
    //productData: [],
    proCat:[],
```

```
      page: 2,
      index: 2,
      // 滑动
      imgUrl: [],
      kbs:[],
      lastcat:[],
      category:[],
  newGoods: [],
      floorGoods: []
    },
  //跳转商品列表页
  listdetail:function(e){
      console.log(e.currentTarget.dataset.title)
      wx.navigateTo({
        url: '../listdetail/listdetail? title='+ e.currentTarget.dataset.title,
        success: function(res){
          // success
        },
        fail: function() {
          // fail
        },
        complete: function() {
          // complete
        }
      })
    },

  //点击加载更多
  getMore:function(e){
    var that = this;
    var page = that.data.page;
    wx.request({
        url: app.d.ceshiUrl + '/Api/Index/getlist',
        method:'post',
        data: {page:page},
        header: {
          'Content-Type':  'application/x-www-form-urlencoded'
        },
        success: function (res) {
          var prolist = res.data.prolist;
          if(prolist == ''){
            wx.showToast({
```

```
          title：'没有更多数据！',
          duration：2000
        })；
        return false；
      }
      //that.initProductData(data)；
      //that.setData({
      //  page：page + 1，
      //  productData：that.data.productData.concat(prolist)
      //})；
      //endInitData
    },
    fail：function(e){
      wx.showToast({
        title：'网络异常！',
        duration：2000
      })；
    }
  })
},

  changeIndicatorDots：function (e) {
    this.setData({
      indicatorDots：! this.data.indicatorDots
    })
  },
  changeAutoplay：function (e) {
    this.setData({
      autoplay：! this.data.autoplay
    })
  },
  intervalChange：function (e) {
    this.setData({
      interval：e.detail.value
    })
  },
  durationChange：function (e) {
    this.setData({
      duration：e.detail.value
    })
  },
```

```
onLoad: function (options) {
  var that = this;
  wx.request({
    url: app.d.ceshiUrl + '/Api/Index/index',
    method:'post',
    data: {},
    header: {
      'Content-Type': 'application/x-www-form-urlencoded'
    },
    success: function (res) {
      //顶部轮播图广告
      var ggtop = res.data.ggtop;
      var prolist = res.data.prolist;
      //商品数量
  var goodsCount = res.data.goodsCount;
      //商品分类
      var category = res.data.category;
      //新品
  var newGoods = res.data.newGoods;
      that.setData({
        imgUrls:ggtop,
        //productData:prolist,
goodsCount: goodsCount,
        category: category,
newGoods: res.data.newGoods,
      });
    },
    fail:function(e){
      wx.showToast({
        title: '网络异常！',
        duration: 2000
      });
    },
  })

},
onShareAppMessage: function () {
  return {
    title: '首页',
    desc: '信真小铺-信守真品的自营电商小程序',
    path: '/pages/index/index',
    success: function(res) {
```

```
            // 分享成功
        },
        fail: function(res) {
            // 分享失败
        }
    }
  }
});
```

8.3.4 json 配置代码

```
{
  "navigationBarTitleText": "信真小铺"
}
```

8.4 商品分类页代码清单

商城商品分类页效果如图 8－3 所示。

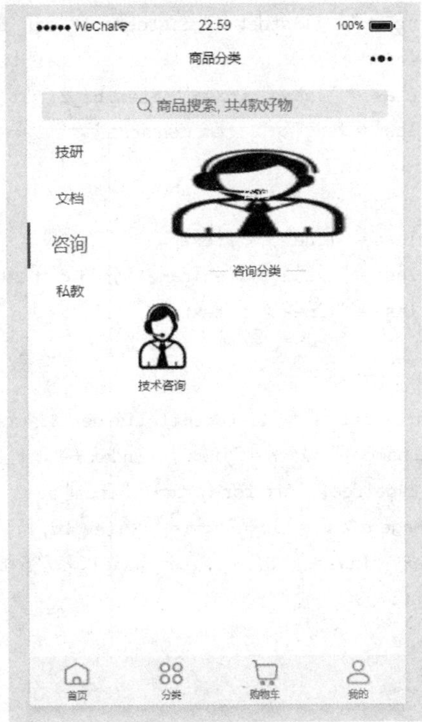

图 8－3 分类页

代码详情如下所示。

8.4.1 wxml 模板代码

```
<view class="container">
  <view class="search">
    <navigator url="/pages/search/search" class="input">
      <image class="icon"></image>
      <text class="txt">商品搜索，共{{goodsCount}}款好物</text>
    </navigator>
  </view>
  <view class="catalog">
    <scroll-view class="nav" scroll-y="true">
      <view class="item {{ item.id == currType ? 'active' : ''}}" wx:for="{{types}}" data-id="{{item.id}}" data-index="{{index}}" wx:key="typeId" data-type-id="{{item.id}}" bindtap="tapType">{{item.name}}</view>
    </scroll-view>
    <scroll-view class="cate {{ item.id == currType ? 'show' : 'hide'}}" scroll-y="true" wx:for="{{types}}" >
      <navigator url="../listdetail/listdetail?cat_id={{item.id}}" class="banner" >
        <image class="image" src="{{item.bz_2}}"></image>
        <view class="txt">{{item.concent}}</view>
      </navigator>
      <view class="hd">
        <text class="line"></text>
        <text class="txt">{{item.name}}分类</text>
        <text class="line"></text>
      </view>
      <view class="bd">
        <navigator url="../listdetail/listdetail?cat_id={{item.id}}&title={{item.name}}" class="item {{(index+1) % 3 == 0 ? 'last' : ''}}" wx:for="{{typeTree}}" wx:for-item="item">
          <image class="icon" src="{{item.bz_1}}"></image>
          <text class="txt">{{item.name}}</text>
        </navigator>
      </view>
    </scroll-view>
  </view>
</view>
```

8.4.2 wxss 样式代码

```
page {
  height: 100%;
}

.container {
  background: #f9f9f9;
  height: 100%;
  width: 100%;
  display: flex;
  flex-direction: column;
}

.search {
  height: 88rpx;
  width: 100%;
  padding: 0 30rpx;
  background: #fff;
  display: flex;
  align-items: center;
}

.search .input {
  width: 690rpx;
  height: 56rpx;
  background: #ededed;
  border-radius: 8rpx;
  display: flex;
  align-items: center;
  justify-content: center;
}

.search .icon {
  background: url(http://cdn.xinzhenkj.com/search2-2fb94833aa.png) center no-repeat;
  background-size: 100%;
  width: 28rpx;
  height: 28rpx;
}
```

```
.search .txt {
  height: 42rpx;
  line-height: 42rpx;
  color: #666;
  padding-left: 10rpx;
  font-size: 30rpx;
}

.catalog {
  flex: 1;
  width: 100%;
  background: #fff;
  display: flex;
  border-top: 1px solid #fafafa;
}

.catalog .nav {
  width: 162rpx;
  height: 100%;
}

.catalog .nav .item {
  text-align: center;
  line-height: 90rpx;
  width: 162rpx;
  height: 90rpx;
  color: #333;
  font-size: 28rpx;
  border-left: 6rpx solid #fff;
}

.catalog .nav .item.active {
  color: #ab2b2b;
  font-size: 36rpx;
  border-left: 6rpx solid #ab2b2b;
}

.catalog .cate {
  border-left: 1px solid #fafafa;
  flex: 1;
  height: 100%;
  padding: 0 30rpx 0 30rpx;
```

```
}

.banner {
  display: block;
  height: 222rpx;
  width: 100%;
  position: relative;
}

.banner .image {
  position: absolute;
  top: 30rpx;
  left: 0;
  border-radius: 4rpx;
  height: 192rpx;
  width: 100%;
}

.banner .txt {
  position: absolute;
  top: 30rpx;
  text-align: center;
  color: #fff;
  font-size: 28rpx;
  left: 0;
  height: 192rpx;
  line-height: 192rpx;
  width: 100%;
}
.show {display:block;}
.hide {display:none;}
.catalog .hd {
  height: 108rpx;
  width: 100%;
  display: flex;
  justify-content: center;
  align-items: center;
}

.catalog .hd .txt {
  font-size: 24rpx;
  text-align: center;
```

```
  color: #333;
  padding: 0 10rpx;
  width: auto;
}

.catalog .hd .line {
  width: 40rpx;
  height: 1px;
  background: #d9d9d9;
}

.catalog .bd {
  height: auto;
  width: 100%;
  overflow: hidden;
}

.catalog .bd .item {
  display: block;
  float: left;
  height: 216rpx;
  width: 144rpx;
  margin-right: 34rpx;
}

.catalog .bd .item.last {
  margin-right: 0;
}

.catalog .bd .item .icon {
  height: 144rpx;
  width: 144rpx;
}

.catalog .bd .item .txt {
  display: block;
  text-align: center;
  font-size: 24rpx;
  color: #333;
  height: 72rpx;
  width: 144rpx;
}
```

8.4.3 js 逻辑代码

```
var app = getApp();
 Page({
     data: {
         typeTree: {}, // 数据缓存
         currType: 0 ,
         // 当前类型
         "types": [],
         typeTree: [],
     goodsCount: []
     },

     onLoad: function (option){
         var that = this;
         wx.request({
             url: app.d.ceshiUrl + '/Api/Category/index',
             method:'post',
             data: {},
             header: {
                 'Content-Type': 'application/x-www-form-urlencoded'
             },
             success: function (res) {
                 var status = res.data.status;
                 if(status==1) {
                     var list = res.data.list;
                     var catList = res.data.catList;
             var goodsCount = res.data.goodsCount;
                     that.setData({
                         types:list,
                         typeTree:catList,
             goodsCount:goodsCount,
                     });
                 } else {
                     wx.showToast({
                         title:res.data.err,
                         duration:2000,
                     });
                 }
      that.setData({
             currType: 2
```

```
        });
      console.log(list)

            },
            error:function(e){
                wx.showToast({
                    title:'网络异常！',
                    duration:2000,
                });
            },

        });
    },
    tapType: function (e){
        var that = this;
        const currType = e.currentTarget.dataset.typeId;

        that.setData({
            currType: currType
        });
        console.log(currType);
        wx.request({
            url: app.d.ceshiUrl + '/Api/Category/getcat',
            method:'post',
            data: {cat_id:currType},
            header: {
                'Content-Type': 'application/x-www-form-urlencoded'
            },
            success: function (res) {
                var status = res.data.status;
                if(status == 1) {
                    var catList = res.data.catList;
                    that.setData({
                        typeTree:catList,
                    });
                } else {
                    wx.showToast({
                        title:res.data.err,
                        duration:2000,
                    });
                }
            },
```

```
                error:function(e){
                    wx.showToast({
                        title:'网络异常!',
                        duration:2000,
                    });
                }
            });
        },
        // 加载品牌、二级类目数据
        getTypeTree (currType) {
            const me = this, _data = me.data;
            if(! _data.typeTree[currType]){
                request({
                    url: ApiList.goodsTypeTree,
                    data: {typeId: + currType},
                    success: function (res) {
                        _data.typeTree[currType] = res.data.data;
                        me.setData({
                            typeTree: _data.typeTree
                        });
                    }
                });
            }
        },
            //右上角分享按钮
        onShareAppMessage: function () {
        return {
            title: '产品分类',
      desc: '信真小铺微信小程序商城',
            path: '/pages/category/index',
      data:{data}
            }
    }
    })
```

8.4.4 json 配置代码

```
{
  "navigationBarTitleText": "产品分类"
}
```

8.5 商品详情页代码清单

商城产品内容页效果如图 8-4 和图 8-5 所示。

图 8-4 商品详情页

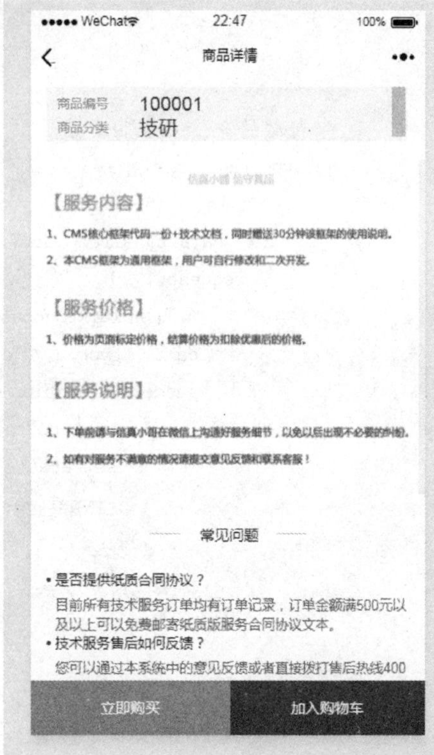

图 8-5 商品详情页(续)

代码详情如下所示。

8.5.1 wxml 模板代码

```
<scroll-view class="container" scroll-y="true">
<view>
<swiper class="goodsimgs" indicator-dots="true" autoplay="true" interval="
3000" duration="1000">
<swiper-item wx:for="{{bannerItem}}" wx:key="{{item.id}}">
<image src="{{item}}" background-size="cover"></image>
</swiper-item>
</swiper>
<view class="service-policy">
```

```
< view class = "item" > 30 分钟极速响应 < /view >
< view class = "item" > 24 小时快速退款 < /view >
< view class = "item" >满 888 元送好礼 < /view >
< /view >
< view class = "goods - info" >
< view class = "c" >
< text class = "name" > {{itemData. name}} < /text >
< text class = "desc" > {{itemData. intro}} < /text >
< text class = "price" > ¥ {{itemData. price_yh}} < /text >
< view class = "brand" >
< navigator url = "../listdetail/listdetail? brandId = {{itemData. brand}}" >
< text > {{itemData. brand}} < /text >
< /navigator >
< /view >
< /view >
< /view >

< view class = "goods - attr" >
< view class = "t" >商品参数 < /view >
< view class = "l" >
< view class = "item" >
< text class = "left" >商品编号 < /text > < text class = "right" > {{itemData. pro_num-
ber}} < /text >
< text class = "left" >商品分类 < /text > < text class = "right" > {{itemData. cat_
name}} < /text >
< /view >
< /view >
< /view >
< view class = "detail" >
< import src = "../../wxParse/wxParse. wxml"/ >
< template is = "wxParse" data = "{{wxParseData:content. nodes}}"/ >
< /view >
< view class = "common - problem" >
< view class = "h" >
< view class = "line" > < /view >
< text class = "title" >常见问题 < /text >
< /view >
< view class = "b" >
< view class = "item" >
< view class = "question - box" > < text class = "spot" > < /text > < text class = "
question" >是否提供纸质合同协议? < /text > < /view >
< view class = "answer" >目前所有技术服务订单均有订单记录,订单金额满 500 元以及以
```

```
上可以免费邮寄纸质版服务合同协议文本。</view>
    <view class="question-box"><text class="spot"></text><text class="
question">技术服务售后如何反馈? </text></view>
    <view class="answer">您可以通过本系统中的意见反馈或者直接拨打售后热线 400-
123-4567 进行反馈。</view>
    </view>
    </view>
    </view>
    <view class="related-goods">
    <view class="h">
    <view class="line"></view>
    <text class="title">大家都在看</text>
    </view>
    <view class="b">
    <view class="item" wx:for="{{newGoods}}" wx:key="{{item.id}}">
    <navigator url="../product/detail? productId={{item.id}}">
    <image class="img" src="{{item.photo_x}}" background-size="cover"></image>
    <text class="name">{{item.name}}</text>
    <text class="price">¥{{item.price_yh}}</text>
    </navigator>
    </view>
    </view>
    </view>
    </view>

    <view class="bottom-btn">
    <view class="c" bindtap="addShopCart" data-status="1" data-type="buynow">立
即购买</view>
    <view class="r" bindtap="addShopCart" data-type="addcart">加入购物车</
view>
    </view>

    </scroll-view>
```

8.5.2 wxss 样式代码

```
.container{
    margin-bottom: 100rpx;
}
.goodsimgs{
  width: 750rpx;
```

```
  height: 750rpx;
}

.goodsimgs image{
  width: 750rpx;
  height: 750rpx;
}

.service-policy{
    width: 750rpx;
    height: 73rpx;
    background: #f4f4f4;
    padding: 0 31.25rpx;
    display: flex;
    flex-flow: row nowrap;
    align-items: center;
    justify-content: space-between;
}

.service-policy .item{
    background: url(http://nos.netease.com/mailpub/hxm/yanxuan-wap/p/20150730/
style/img/icon-normal/servicePolicyRed-518d32d74b.png) 0 center no-repeat;
    background-size: 10rpx;
    padding-left: 15rpx;
    display: flex;
    align-items: center;
    font-size: 25rpx;
    color: #666;
}

.goods-info{
    width: 750rpx;
    height: 306rpx;
    overflow: hidden;
    background: #fff;
}

.goods-info .c{
    display: block;
    width: 718.75rpx;
    height: 100%;
    margin-left: 31.25rpx;
```

```
    padding: 38rpx 31.25rpx 38rpx 0;
    border-bottom: 1px solid #f4f4f4;
}

.goods-info .c text{
    display: block;
    width: 687.5rpx;
    text-align: center;
}

.goods-info .name{
    height: 41rpx;
    margin-bottom: 5.208rpx;
    font-size: 41rpx;
    line-height: 41rpx;
}

.goods-info .desc{
    height: 43rpx;
    margin-bottom: 41rpx;
    font-size: 24rpx;
    line-height: 36rpx;
    color: #999;
}

.goods-info .price{
    height: 35rpx;
    font-size: 35rpx;
    line-height: 35rpx;
    color: #b4282d;
}

.goods-info .brand{
    margin-top: 23rpx;
    min-height: 40rpx;
    text-align: center;
}

.goods-info .brand text{
    display: inline-block;
    width: auto;
```

```
        padding: 2px 30rpx 2px 10.5rpx;
        line-height: 35.5rpx;
        border: 1px solid #f48f18;
        font-size: 25rpx;
        color: #f48f18;
        border-radius: 4px;
        background: url(http://nos.netease.com/mailpub/hxm/yanxuan-wap/p/20150730/style/img/icon-normal/detailTagArrow-18bee52dab.png) 95% center no-repeat;
        background-size: 10.75rpx 18.75rpx;
    }

    .section-nav{
        width: 750rpx;
        height: 108rpx;
        background: #fff;
        margin-bottom: 20rpx;
    }

    .section-nav .t{
        float: left;
        width: 600rpx;
        height: 108rpx;
        line-height: 108rpx;
        font-size: 29rpx;
        color: #333;
        margin-left: 31.25rpx;
    }

    .section-nav .i{
        float: right;
        width: 52rpx;
        height: 52rpx;
        margin-right: 16rpx;
        margin-top: 28rpx;
    }

    .section-act .t{
        float: left;
        display: flex;
        align-items: center;
        width: 600rpx;
        height: 108rpx;
```

```
    overflow: hidden;
    line-height: 108rpx;
    font-size: 29rpx;
    color: #999;
    margin-left: 31.25rpx;
}

.section-act .label{
    color: #999;
}

.section-act .tag{
    display: flex;
    align-items: center;
    padding:0 10rpx;
    border-radius: 3px;
    height: 37rpx;
    width: auto;
    color: #f48f18;
    overflow: hidden;
    border: 1px solid #f48f18;
    font-size: 25rpx;
    margin:0 10rpx;
}

.section-act .text{
    display: flex;
    align-items: center;
    height: 37rpx;
    width: auto;
    overflow: hidden;
    color: #f48f18;
    font-size: 29rpx;
}

.comments{
    width: 100%;
    height: auto;
    padding-left:30rpx;
    background: #fff;
    margin: 20rpx 0;
}
```

```
.comments .h{
    height: 102.5rpx;
    line-height: 100.5rpx;
    width: 718.75rpx;
    padding-right: 16rpx;
    border-bottom: 1px solid #d9d9d9;
}

.comments .h .t{
    display: block;
    float: left;
    width: 50%;
    font-size: 38.5rpx;
    color: #333;
}

.comments .h .i{
    display: block;
    float: right;
    width: 164rpx;
    height: 100.5rpx;
    line-height: 100.5rpx;
    background: url(http://nos.netease.com/mailpub/hxm/yanxuan-wap/p/20150730/
style/img/icon-normal/address-right-990628faa7.png) right center no-repeat;
    background-size: 52rpx;
}

.comments .b{
    height: auto;
    width: 720rpx;
}

.comments .item{
    height: auto;
    width: 720rpx;
    overflow: hidden;
}

.comments .info{
    height: 127rpx;
    width: 100%;
```

```
    padding: 33rpx 0 27rpx 0;
}

.comments .user{
    float: left;
    width: auto;
    height: 67rpx;
    line-height: 67rpx;
    font-size: 0;
}

.comments .user image{
    float: left;
    width: 67rpx;
    height: 67rpx;
    margin-right: 17rpx;
    border-radius: 50%;
}

.comments .user text{
    display: inline-block;
    width: auto;
    height: 66rpx;
    overflow: hidden;
    font-size: 29rpx;
    line-height: 66rpx;
}

.comments .time{
    display: block;
    float: right;
    width: auto;
    height: 67rpx;
    line-height: 67rpx;
    color: #7f7f7f;
    font-size: 25rpx;
    margin-right: 30rpx;
}

.comments .content{
    width: 720rpx;
    padding-right: 30rpx;
```

```
    line-height: 45.8rpx;
    font-size: 29rpx;
    margin-bottom: 24rpx;
}

.comments .imgs{
    width: 720rpx;
    height: auto;
    margin-bottom: 25rpx;
}

.comments .imgs .img{
    height: 150rpx;
    width: 150rpx;
    margin-right: 28rpx;
}


.comments .spec{
    width: 720rpx;
    padding-right: 30rpx;
    line-height: 30rpx;
    font-size: 24rpx;
    color: #999;
    margin-bottom: 30rpx;
}


.goods-attr{
    width: 750rpx;
    height: auto;
    overflow: hidden;
    padding: 0 31.25rpx 25rpx 31.25rpx;
    background: #fff;
}

.goods-attr .t{
    width: 687.5rpx;
    height: 104rpx;
    line-height: 104rpx;
    font-size: 38.5rpx;
}
```

```
.goods - attr .item{
    width: 687.5rpx;
    height: 100rpx;
    padding: 11rpx 20rpx;
    margin - bottom: 11rpx;
    background: #f7f7f7;
    font - size: 38.5rpx;
}

.goods - attr .left{
    float: left;
    font - size: 25rpx;
    width: 134rpx;
    height: 45rpx;
    line - height: 45rpx;
    overflow: hidden;
    color: #999;
}

.goods - attr .right{
    float: left;
    font - size: 36.5rpx;
    margin - left: 20rpx;
    width: 480rpx;
    height: 45rpx;
    line - height: 45rpx;
    overflow: hidden;
    color: #333;
}

.detail{
    width: 750rpx;
    height: auto;
    overflow: hidden;
}

.detail image{
    width: 750rpx;
    display: block;
}
```

```
.common-problem{
    width: 750rpx;
    height: auto;
    overflow: hidden;
}

.common-problem .h{
    position: relative;
    height: 145.5rpx;
    width: 750rpx;
    padding: 56.25rpx 0;
    background: #fff;
    text-align: center;
}

.common-problem .h .line{
    display: inline-block;
    position: absolute;
    top: 72rpx;
    left: 0;
    z-index: 2;
    height: 1px;
    margin-left: 225rpx;
    width: 300rpx;
    background: #ccc;
}

.common-problem .h .title{
    display: inline-block;
    position: absolute;
    top: 56.125rpx;
    left: 0;
    z-index: 3;
    height: 33rpx;
    margin-left: 285rpx;
    width: 180rpx;
    background: #fff;
}

.common-problem .b{
  width: 750rpx;
```

```
  height: auto;
  overflow: hidden;
  padding: 0rpx 30rpx;
  background: #fff;

}

.common-problem .item{
  height: auto;
  overflow: hidden;
  padding-bottom: 25rpx;
}

.common-problem .question-box .spot{
  float: left;
  display: block;
  height: 8rpx;
  width: 8rpx;
  background: #b4282d;
  border-radius: 50%;
  margin-top: 11rpx;
}

.common-problem .question-box .question{
  float: left;
  line-height: 30rpx;
  padding-left: 8rpx;
  display: block;
  font-size: 26rpx;
  padding-bottom: 15rpx;
  color: #303030;
}

.common-problem .answer{
  line-height: 36rpx;
  padding-left: 16rpx;
  font-size: 26rpx;
  color: #787878;
}


.related-goods{
```

```
    width: 750rpx;
    height: auto;
    overflow: hidden;
}

.related-goods .h{
    position: relative;
    height: 145.5rpx;
    width: 750rpx;
    padding: 56.25rpx 0;
    background: #fff;
    text-align: center;
    border-bottom: 1px solid #f4f4f4;
}

.related-goods .h .line{
    display: inline-block;
    position: absolute;
    top: 72rpx;
    left: 0;
    z-index: 2;
    height: 1px;
    margin-left: 225rpx;
    width: 300rpx;
    background: #ccc;
}

.related-goods .h .title{
    display: inline-block;
    position: absolute;
    top: 56.125rpx;
    left: 0;
    z-index: 3;
    height: 33rpx;
    margin-left: 285rpx;
    width: 180rpx;
    background: #fff;
}

.related-goods .b{
  width: 750rpx;
  height: auto;
```

```
  overflow: hidden;
}

.related - goods .b .item{
  float: left;
  background: #fff;
  width: 375rpx;
  height: auto;
  overflow: hidden;
  text - align: center;
  padding: 15rpx 31.25rpx;
  border - right: 1px solid #f4f4f4;
  border - bottom: 1px solid #f4f4f4;
}

.related - goods .item .img{
  width: 311.45rpx;
  height: 311.45rpx;
}

.related - goods .item .name{
  display: block;
  width: 311.45rpx;
  height: 35rpx;
  margin: 11.5rpx 0 15rpx 0;
  text - align: center;
  overflow: hidden;
  font - size: 30rpx;
  color: #333;
}

.related - goods .item .price{
  display: block;
  width: 311.45rpx;
  height: 30rpx;
  text - align: center;
  font - size: 30rpx;
  color: #b4282d;
}

.bottom - btn{
    position: fixed;
```

```
    left:0;
    bottom:0;
    z-index: 10;
    width: 750rpx;
    height: 100rpx;
    display: flex;
    background: #fff;
}

.bottom-btn .l{
    float: left;
    height: 100rpx;
    width: 162rpx;
    border: 1px solid #f4f4f4;
    display: flex;
    align-items: center;
    justify-content: center;

}

.bottom-btn .l.l-collect{
    border-right: none;
    border-left: none;
    text-align: center;
}


.bottom-btn .l.l-cart .box{
    height: 62rpx;
    width: 52rpx;
}

.bottom-btn .l.l-cart .cart-count{
    height: 28rpx;
    width: 28rpx;
    z-index: 10;
    position: absolute;
    top: 0;
    right:0;
    background: #b4282d;
    text-align: center;
    font-size: 18rpx;
```

```
    color: #fff;
    line-height: 28rpx;
    border-radius: 50%;

}

.bottom-btn .l.l-cart .icon{

    position: absolute;
    top: 20rpx;
    left:0;
    margin-left:200rpx;
}


.bottom-btn .l .icon{
    display: block;
    height: 62rpx;
    width: 52rpx;
margin-left:30rpx;
}


.bottom-btn .c{
    float: left;
    height: 100rpx;
    line-height: 96rpx;
    flex: 1;
    text-align: center;
    color: #fff;
    border-top: 1px solid #f4f4f4;
    border-bottom: 1px solid #f4f4f4;
border:1px solid #f48f18;
background: #f48f18;

}

.bottom-btn .r{
    border:1px solid #b4282d;
    background: #b4282d;
    float: left;
    height: 100rpx;
```

```
    line-height: 96rpx;
    flex: 1;
    text-align: center;
    color: #fff;
}
@import "../../wxParse/wxParse.wxss";

.attr-pop{
    width: 100%;
    height: 100%;
    padding: 31.25rpx;
    background: #fff;
}

.attr-pop .img-info{
    width: 687.5rpx;
    height: 177rpx;
    overflow: hidden;
    margin-bottom: 41.5rpx;
}

.attr-pop .img{
    float: left;
    height: 177rpx;
    width: 177rpx;
    background: #f4f4f4;
    margin-right: 31.25rpx;
}

.attr-pop .info{
    float: left;
    height: 177rpx;
    display: flex;
    align-items: center;
}

.attr-pop .p{
    font-size: 33rpx;
    color: #333;
    height: 33rpx;
    line-height: 33rpx;
    margin-bottom: 10rpx;
```

```
}

.attr-pop .a{
    font-size: 29rpx;
    color: #333;
    height: 40rpx;
    line-height: 40rpx;
}

.spec-con{
    width: 100%;
    height: auto;
    overflow: hidden;
}

.spec-con .name{
    height: 32rpx;
    margin-bottom: 22rpx;
    font-size: 29rpx;
    color: #333;
}

.spec-con .values{
    height: auto;
    margin-bottom: 31.25rpx;
    font-size: 0;
}

.spec-con .value{
    display: inline-block;
    height: 62rpx;
    padding: 0 35rpx;
    line-height: 56rpx;
    text-align: center;
    margin-right: 25rpx;
    margin-bottom: 16.5rpx;
    border: 1px solid #333;
    font-size: 25rpx;
    color: #333;
}
```

```
.spec-con .value.disable{
    border: 1px solid #ccc;
    color: #ccc;
}

.spec-con .value.selected{
    border: 1px solid #b4282d;
    color: #b4282d;
}

.number-item .selnum{
    width: 322rpx;
    height: 71rpx;
    border: 1px solid #ccc;
    display: flex;
}

.number-item .cut{
    width: 93.75rpx;
    height: 100%;
    text-align: center;
    line-height: 65rpx;
}

.number-item .number{
    flex: 1;
    height: 100%;
    text-align: center;
    line-height: 68.75rpx;
    border-left: 1px solid #ccc;
    border-right: 1px solid #ccc;
    float: left;
}

.number-item .add{
    width: 93.75rpx;
    height: 100%;
    text-align: center;
    line-height: 65rpx;
}
```

8.5.3 js 逻辑代码

```
//获取应用实例
var app = getApp();
//引入这个插件,使 html 内容自动转换成 wxml 内容
var WxParse = require('../../wxParse/wxParse.js');
Page({
  firstIndex: -1,
  data:{
    bannerApp:true,
    winWidth: 0,
    winHeight: 0,
    currentTab: 0, //tab 切换
    productId:0,
    itemData:{},
    bannerItem:[],
  newGoods: [],
    buynum:1,
    // 产品图片轮播
    indicatorDots: true,
    autoplay: true,
    interval: 5000,
    duration: 1000,
    // 属性选择
    firstIndex: -1,
    //准备数据
    //数据结构:以一组一组来进行设定
     commodityAttr:[],
     attrValueList: []
  },
  toCar() {
    wx.switchTab({
      url: '/pages/cart/cart'
    })
  },
  // 弹窗
  setModalStatus: function (e) {
    var animation = wx.createAnimation({
      duration: 200,
      timingFunction: "linear",
      delay: 0
```

```
  })

  this.animation = animation
  animation.translateY(300).step();

  this.setData({
    animationData: animation.export()
  })

  if (e.currentTarget.dataset.status == 1) {

    this.setData(
      {
        showModalStatus: true
      }
    );
  }
  setTimeout(function () {
    animation.translateY(0).step()
    this.setData({
      animationData: animation
    })
    if (e.currentTarget.dataset.status == 0) {
      this.setData(
        {
          showModalStatus: false
        }
      );
    }
  }.bind(this), 200)
},
// 加减
changeNum:function  (e) {
  var that = this;
  if (e.target.dataset.alphaBeta == 0) {
      if (this.data.buynum < = 1) {
          buynum:1
      }else{
          this.setData({
              buynum:this.data.buynum - 1
          })
      };
```

```
    }else{
        this.setData({
            buynum:this.data.buynum + 1
        })
    };
  },
  // 传值
  onLoad: function (option) {
    var that = this;
    that.setData({
      productId: option.productId,
    });
    that.loadProductDetail();

  },
// 商品详情数据获取
  loadProductDetail:function(){
    var that = this;
    wx.request({
      url: app.d.ceshiUrl + '/Api/Product/index',
      method:'post',
      data: {
        pro_id: that.data.productId,
      },
      header: {
        'Content-Type':  'application/x-www-form-urlencoded'
      },
      success: function (res) {
        // -- init data
        var status = res.data.status;
        if(status == 1) {
          var pro = res.data.pro;
          var content = pro.content;
        var newGoods = res.data.newGoods;
          //that.initProductData(data);
          WxParse.wxParse('content', 'html', content, that, 3);
          that.setData({
            itemData:pro,
            bannerItem:pro.img_arr,
            commodityAttr:res.data.commodityAttr,
            attrValueList:res.data.attrValueList,
          newGoods: res.data.newGoods,
```

```
        });
      } else {
        wx.showToast({
          title:res.data.err,
          duration:2000,
        });
      }
    },
    error:function(e){
      wx.showToast({
        title:'网络异常！',
        duration:2000,
      });
    },
  });
},
// 属性选择
onShow: function () {
  this.setData({
    includeGroup: this.data.commodityAttr
  });
  this.distachAttrValue(this.data.commodityAttr);
  // 只有一个属性组合的时候默认选中
  // console.log(this.data.attrValueList);
  if (this.data.commodityAttr.length == 1) {
    for (var i = 0; i < this.data.commodityAttr[0].attrValueList.length; i++) {
      this.data.attrValueList[i].selectedValue = this.data.commodityAttr[0].attrValueList[i].attrValue;
    }
    this.setData({
      attrValueList: this.data.attrValueList
    });
  }
},
/* 获取数据 */
distachAttrValue: function (commodityAttr) {
  /**
    将后台返回的数据组合成类似
    {
      attrKey:'型号',
      attrValueList:['1','2','3']
    }
```

```
*/
// 把数据对象的数据(视图使用)写到局部内
var attrValueList = this.data.attrValueList;
// 遍历获取的数据
for (var i = 0; i < commodityAttr.length; i++) {
  for (var j = 0; j < commodityAttr[i].attrValueList.length; j++) {
    var attrIndex = this.getAttrIndex(commodityAttr[i].attrValueList[j].attrK-
ey, attrValueList);
      // console.log('属性索引', attrIndex)
      // 如果还没有属性索引为"-1",此时新增属性并设置属性值数组的第一个值;索
         引大于等于0,表示已存在的属性名的位置
      if (attrIndex >= 0) {
        // 如果属性值数组中没有该值,push新值;否则不处理
        if (! this.isValueExist(commodityAttr[i].attrValueList[j].attrValue, at-
trValueList[attrIndex].attrValues)) {
          attrValueList[attrIndex].attrValues.push(commodityAttr[i].attrValueL-
ist[j].attrValue);
        }
      } else {
        attrValueList.push({
          attrKey: commodityAttr[i].attrValueList[j].attrKey,
          attrValues: [commodityAttr[i].attrValueList[j].attrValue]
        });
      }
    }
  }
// console.log('result', attrValueList)
for (var i = 0; i < attrValueList.length; i++) {
  for (var j = 0; j < attrValueList[i].attrValues.length; j++) {
    if (attrValueList[i].attrValueStatus) {
      attrValueList[i].attrValueStatus[j] = true;
    } else {
      attrValueList[i].attrValueStatus = [];
      attrValueList[i].attrValueStatus[j] = true;
    }
  }
}
this.setData({
  attrValueList: attrValueList
});
},
getAttrIndex: function (attrName, attrValueList) {
```

```
    // 判断数组中的 attrKey 是否有该属性值
    for (var i = 0; i < attrValueList.length; i++) {
      if (attrName == attrValueList[i].attrKey) {
        break;
      }
    }
    return i < attrValueList.length ? i : -1;
  },
  isValueExist: function (value, valueArr) {
    // 判断是否已有属性值
    for (var i = 0; i < valueArr.length; i++) {
      if (valueArr[i] == value) {
        break;
      }
    }
    return i < valueArr.length;
  },
  /* 选择属性值事件 */
  selectAttrValue: function (e) {
    /*
    点选属性值,联动判断其他属性值是否可选
    {
      attrKey:'型号',
      attrValueList:['1','2','3'],
      selectedValue:'1',
      attrValueStatus:[true,true,true]
    }
    console.log(e.currentTarget.dataset);
    */
    var attrValueList = this.data.attrValueList;
    var index = e.currentTarget.dataset.index;//属性索引
    var key = e.currentTarget.dataset.key;
    var value = e.currentTarget.dataset.value;
    if (e.currentTarget.dataset.status || index == this.data.firstIndex) {
        if (e.currentTarget.dataset.selectedvalue == e.currentTarget.dataset.
value) {
          // 取消选中
          this.disSelectValue(attrValueList, index, key, value);
        } else {
          // 选中
          this.selectValue(attrValueList, index, key, value);
        }
```

```
    }
  },
  /* 选中 */
  selectValue: function (attrValueList, index, key, value, unselectStatus) {
    // console.log('firstIndex', this.data.firstIndex);
    var includeGroup = [];
    if (index == this.data.firstIndex && ! unselectStatus) {
      // 如果是第一个选中的属性值,则该属性所有值可选
      var commodityAttr = this.data.commodityAttr;
      // 其他选中的属性值全都置空
      // console.log('其他选中的属性值全都置空', index, this.data.firstIndex, !
        unselectStatus);
      for (var i = 0; i < attrValueList.length; i++) {
        for (var j = 0; j < attrValueList[i].attrValues.length; j++) {
          attrValueList[i].selectedValue = '';
        }
      }
    } else {
      var commodityAttr = this.data.includeGroup;
    }

    // console.log('选中', commodityAttr, index, key, value);
    for (var i = 0; i < commodityAttr.length; i++) {
      for (var j = 0; j < commodityAttr[i].attrValueList.length; j++) {
        if (commodityAttr[i].attrValueList[j].attrKey == key && commodityAttr[i].
            attrValueList[j].attrValue == value) {
          includeGroup.push(commodityAttr[i]);
        }
      }
    }
    attrValueList[index].selectedValue = value;

    this.setData({
      attrValueList: attrValueList,
      includeGroup: includeGroup
    });

    var count = 0;
    for (var i = 0; i < attrValueList.length; i++) {
      for (var j = 0; j < attrValueList[i].attrValues.length; j++) {
        if (attrValueList[i].selectedValue) {
```

```
            count++;
            break;
          }
        }
      }
      if (count < 2) {// 第一次选中,同属性的值都可选
        this.setData({
          firstIndex: index
        });
      } else {
        this.setData({
          firstIndex: -1
        });
      }
    },
    /* 取消选中 */
    disSelectValue: function (attrValueList, index, key, value) {
      var commodityAttr = this.data.commodityAttr;
      attrValueList[index].selectedValue = '';

      // 判断属性是否可选
      for (var i = 0; i < attrValueList.length; i++) {
        for (var j = 0; j < attrValueList[i].attrValues.length; j++) {
          attrValueList[i].attrValueStatus[j] = true;
        }
      }
      this.setData({
        includeGroup: commodityAttr,
        attrValueList: attrValueList
      });

      for (var i = 0; i < attrValueList.length; i++) {
        if (attrValueList[i].selectedValue) {
          this.selectValue(attrValueList, i, attrValueList[i].attrKey, attrValueList
[i].selectedValue, true);
        }
      }
    },

    addShopCart:function(e){ //添加到购物车
      var that = this;
      wx.request({
```

```
        url: app.d.ceshiUrl + '/Api/Shopping/add',
        method:'post',
        data: {
          uid: app.d.userId,
          pid: that.data.productId,
          num: that.data.buynum,
        },
        header: {
          'Content-Type':  'application/x-www-form-urlencoded'
        },
        success: function (res) {
          // //-- init data
          var data = res.data;
          if(data.status == 1){
            var ptype = e.currentTarget.dataset.type;
            if(ptype == 'buynow'){
              wx.redirectTo({
                url: '../order/pay? cartId ='+ data.cart_id
              });
              return;
            }else{
              wx.showToast({
                  title:'加入购物车成功',
                  icon:'success',
                  duration: 2000
              });
            }
          }else{
            wx.showToast({
                  title: data.err,
                  duration: 2000
              });
          }
        },
        fail: function() {
          // fail
          wx.showToast({
            title:'网络异常!',
            duration: 2000
          });
        }
      });
```

```
  },
  bindChange: function (e) {//滑动切换 tab
    var that = this;
    that.setData({ currentTab: e.detail.current });
  },
  initNavHeight:function(){////获取系统信息
    var that = this;
    wx.getSystemInfo({
      success: function (res) {
        that.setData({
          winWidth: res.windowWidth,
          winHeight: res.windowHeight
        });
      }
    });
  },
  bannerClosed:function(){
    this.setData({
      bannerApp:false,
    })
  },
  swichNav: function (e) {//点击 tab 切换
    var that = this;
    if (that.data.currentTab === e.target.dataset.current) {
      return false;
    } else {
      that.setData({
        currentTab: e.target.dataset.current
      })
    }
  },

  //右上角分享按钮
    onShareAppMessage: function () {
    return {
      title: '商品详情',
  desc: '信真小铺 - 信守真品的软件技术服务商',
      path: '/pages/product/detail? productId = ' + id,
  data:{data}
    }
}
});
```

8.5.4 json 配置代码

```
{
    "navigationBarTitleText":"商品详情"
}
```

8.6 购物车页代码清单

购物车页效果如图 8-6 所示。

图 8-6 购物车

代码详情如下所示。

8.6.1 wxml 模板代码

```
< view class = "container" >
  < view class = "service - policy" >
    < view class = "item" > 30 分钟极速响应 < /view >
    < view class = "item" > 24 小时快速退款 < /view >
    < view class = "item" > 满 888 元送好礼 < /view >
  < /view >
  < view class = "no - cart" wx:if = "{{carts == ''}}" >
    < view class = "c" >
      < image src = "http://nos.netease.com/mailpub/hxm/yanxuan - wap/p/20150730/
style/img/icon - normal/noCart - a8fe3f12e5.png" / >
      < text >去添加点什么吧 < /text >
```

```
        </view>
      </view>
    <!-- 购物车空样式 -->

      <view class="cart-view">
        <view class="list">
          <view class="group-item">
            <view class="goods">
              <view class="item edit" wx:for="{{carts}}" data-title="{{item.pro_
name}}" id="{{item.id}}">
          <view class="checkbox checked" wx:if="{{item.selected}}" type="success_
circle" size="20" bindtap="bindCheckbox" data-index="{{index}}"></view>
              <view class="checkbox" wx:else type="circle" size="20" bindtap="
bindCheckbox" data-index="{{index}}"></view>


                <view class="cart-goods">
                  <image class="img" src="{{item.photo_x}}" mode="aspectFill"></
                  image>
                  <view class="info">
                    <view class="t">
                      <text class="name">{{item.pro_name}}</text>
                    <!--    <text class="num">x{{item.BuyCount}}</text>  -->
                    </view>
                    <view class="attr">已选择:{{item.goods_specifition_name_val-
                    ue}}</view>
                    <view class="b">
                      <text class="price">¥{{item.price}}</text>
                      <view class="selnum"><text bindtap="removeShopCard" data-
cartid="{{item.id}}" class="modal-close">x</text>
                        <view class="cut" data-index="{{index}}" bindtap="bindMi-
nus" data-cartid="{{item.id}}">-</view>
                        <input class="number" type="number" bindchange="bindManu-
al" value="{{item.num}}"/>
                        <view class="add" data-index="{{index}}" bindtap="bind-
Plus" data-cartid="{{item.id}}">+</view>
                      </view>
                    </view>
                  </view>
                </view>
              </view>
            </view>
```

```
        </view>

      </view>
      <view class="cart-bottom" bindtap="bindSelectAll">
        <view class="checkbox checked" wx:if="{{selectedAllStatus}}" type="success_circle" size="24">全选</view>
    <view class="checkbox" wx:else type="circle" size="24">全选</view>

        <view class="total">{{total}}</view>

        <view class="checkout" bindtap='bindCheckout'>下单</view>
      </view>
    </view>
  </view>
```

8.6.2 wxss 样式代码

```
page{
    height: 100%;
    min-height: 100%;
    background: #f4f4f4;
}
.container{
    background: #f4f4f4;
    width: 100%;
    height: auto;
    min-height: 100%;
    overflow: hidden;
}
.service-policy{
    width: 750rpx;
    height: 73rpx;
    background: #f4f4f4;
    padding: 0 31.25rpx;
    display: flex;
    flex-flow: row nowrap;
    align-items: center;
    justify-content: space-between;
}

.service-policy .item{
```

```
    background: url(http://nos.netease.com/mailpub/hxm/yanxuan-wap/p/20150730/style/img/icon-normal/servicePolicyRed-518d32d74b.png) 0 center no-repeat;
    background-size: 10rpx;
    padding-left: 15rpx;
    display: flex;
    align-items: center;
    font-size: 25rpx;
    color: #666;
}

.no-cart{
    width: 100%;
    height: auto;
    margin: 0 auto;
}

.no-cart .c{
    width: 100%;
    height: auto;
    margin-top: 200rpx;
}

.no-cart .c image{
    margin: 0 auto;
    display: block;
    text-align: center;
    width: 258rpx;
    height: 258rpx;
}

.no-cart .c text{
    margin: 0 auto;
    display: block;
    width: 258rpx;
    height: 29rpx;
    line-height: 29rpx;
    text-align: center;
    font-size: 29rpx;
    color: #999;
}
```

```
.cart-view{
    width: 100%;
    height: auto;
    overflow: hidden;

}

.cart-view .list{
    height: auto;
    width: 100%;
    overflow: hidden;
    margin-bottom: 120rpx;
}

.cart-view .group-item{
    height: auto;
    width: 100%;
    background: #fff;
    margin-bottom: 18rpx;
}

.cart-view .item{
    height: 164rpx;
    width: 100%;
    overflow: hidden;
}
.cart-view .item .checkbox{
    float: left;
    height: 34rpx;
    width: 34rpx;
    margin: 65rpx 18rpx 65rpx 26rpx;
    background: url(http://nos.netease.com/mailpub/hxm/yanxuan-wap/p/20150730/style/img/icon-normal/checkbox-0e09baa37e.png) no-repeat;
    background-size: 34rpx;
}

.cart-view .item .checkbox.checked{
    background: url(http://nos.netease.com/mailpub/hxm/yanxuan-wap/p/20150730/style/img/icon-normal/checkbox-checked-822e54472a.png) no-repeat;
    background-size: 34rpx;
}
```

```
.cart-view .item .cart-goods{
    float: left;
    height: 164rpx;
    width: 672rpx;
    border-bottom: 1px solid #f4f4f4;
}

.cart-view .item .img{
    float: left;
    height:125rpx;
    width: 125rpx;
    background: #f4f4f4;
    margin: 19.5rpx 18rpx 19.5rpx 0;
}

.cart-view .item .info{
    float: left;
    height:125rpx;
    width: 503rpx;
    margin: 19.5rpx 26rpx 19.5rpx 0;
}

.cart-view .item .t{
    margin: 8rpx 0;
    height: 28rpx;
    font-size: 25rpx;
    color: #333;
    overflow: hidden;
}


.cart-view .item .name{
    height: 28rpx;
    max-width: 310rpx;
    line-height: 28rpx;
    font-size: 25rpx;
    color: #333;
    overflow: hidden;
}

.cart-view .item .num{
    height: 28rpx;
```

```
        line-height: 28rpx;
        float: right;
    }

    .cart-view .item .attr{
        margin-bottom: 17rpx;
        height: 24rpx;
        line-height: 24rpx;
        font-size: 22rpx;
        color: #666;
        overflow: hidden;
    }

    .cart-view .item .b{
        height: 28rpx;
        line-height: 28rpx;
        font-size: 25rpx;
        color: #333;
        overflow: hidden;
    }

    .cart-view .item .price{
        float: left;
    }

    .cart-view .item .open{
        height: 28rpx;
        width: 150rpx;
        display: block;
        float: right;
        background: url(http://nos.netease.com/mailpub/hxm/yanxuan-wap/p/20150730/
style/img/icon-normal/arrowDown-d48093db25.png) right center no-repeat;
        background-size: 25rpx;
        font-size: 25rpx;
        color: #333;
    }

    .cart-view .item.edit .t{

    }

    .cart-view .item.edit .attr{
```

```
    text - align: left;

    padding - right: 25rpx;
    background - size: 12rpx 20rpx;
    margin - bottom: 0rpx;
    height: 39rpx;
    line - height: 39rpx;
    font - size: 24rpx;
    color: #999;
    overflow: hidden;
}

.cart - view .item.edit .b{
    display: flex;
    height: 52rpx;
    overflow: hidden;
}

.cart - view .item.edit .price{
    line - height: 52rpx;
    height: 52rpx;
    flex: 1;
}

.cart - view .item .selnum{
    display: block;
}

.cart - view .item.edit .selnum{
    width: 235rpx;
    height: 52rpx;
    border: 1rpx solid #ccc;
    display: flex;
}

.selnum .cut{
    width: 70rpx;
    height: 100%;
    text - align: center;
    line - height: 50rpx;
}
.modal - close{
```

```
margin - left:228rpx;
position:absolute;
margin - top: - 67rpx;


}
.selnum .number{
    flex: 1;
    height: 100 % ;
    text - align: center;
    line - height: 68.75rpx;
    border - left: 1px solid #ccc;
    border - right: 1px solid #ccc;
    float: left;
}

.selnum .add{
    width: 80rpx;
    height: 100 % ;
    text - align: center;
    line - height: 50rpx;
}


.cart - view .group - item .header{
    width: 100 % ;
    height: 94rpx;
    line - height: 94rpx;
    padding: 0 26rpx;
    border - bottom: 1px solid #f4f4f4;
}

.cart - view .promotion .icon{
    display: inline - block;
    height: 24rpx;
    width: 15rpx;
}

.cart - view .promotion{
    margin - top: 25.5rpx;
    float: left;
    height: 43rpx;
```

```
    width: 480rpx;
    /* margin-right: 84rpx; */
    line-height: 43rpx;
    font-size: 0;
}

.cart-view .promotion .tag{
    border: 1px solid #f48f18;
    height: 37rpx;
    line-height: 31rpx;
    padding: 0 9rpx;
    margin-right: 10rpx;
    color: #f48f18;
    font-size: 24.5rpx;
}

.cart-view .promotion .txt{
    height: 43rpx;
    line-height: 43rpx;
    padding-right: 10rpx;
    color: #333;
    font-size: 29rpx;
    overflow: hidden;
}

.cart-view .get{
    margin-top: 18rpx;
    float: right;
    height: 58rpx;
    padding-left: 14rpx;
    border-left: 1px solid #d9d9d9;
    line-height: 58rpx;
    font-size: 29rpx;
    color: #333;
}

.cart-bottom{
    position: fixed;
    bottom:0;
    left:0;
    height: 100rpx;
    width: 100%;
```

```
    background: #fff;
    display: flex;
}

.cart-bottom .checkbox{
    height: 34rpx;

    padding-left: 60rpx;
    line-height: 34rpx;
    margin: 33rpx 18rpx 33rpx 26rpx;
    background: url(http://nos.netease.com/mailpub/hxm/yanxuan-wap/p/20150730/style/img/icon-normal/checkbox-0e09baa37e.png) no-repeat;
    background-size: 34rpx;
    font-size: 29rpx;
}

.cart-bottom .checkbox.checked{
    background: url(http://nos.netease.com/mailpub/hxm/yanxuan-wap/p/20150730/style/img/icon-normal/checkbox-checked-822e54472a.png) no-repeat;
    background-size: 34rpx;
}

.cart-bottom .total{
    height: 34rpx;
    flex: 1;
    margin: 33rpx 10rpx;
    font-size: 29rpx;
}


.cart-bottom .delete{
    height: 34rpx;
    width: auto;
    margin: 33rpx 18rpx;
    font-size: 29rpx;
}

.cart-bottom .checkout{
    height: 100rpx;
    width: 210rpx;
    text-align: center;
    line-height: 100rpx;
```

```
    font - size: 29rpx;
    background: #b4282d;
    color: #fff;
}
```

8.6.3 js 逻辑代码

```
var app = getApp();
// pages/cart/cart.js
Page({
  data:{
    page:1,
    minusStatuses: ['disabled', 'disabled', 'normal', 'normal', 'disabled'],
    total: 0,
    carts: []
  },

bindMinus: function(e) {
    var that = this;
    var index = parseInt(e.currentTarget.dataset.index);
    var num = that.data.carts[index].num;
    // 如果只有 1 件了,就不允许再减少
    if (num > 1) {
      num --;
    }
    console.log(num);
    var cart_id = e.currentTarget.dataset.cartid;
    wx.request({
      url: app.d.ceshiUrl + '/Api/Shopping/up_cart',
      method:'post',
      data: {
        user_id: app.d.userId,
        num:num,
        cart_id:cart_id
      },
      header: {
        'Content - Type': 'application/x - www - form - urlencoded'
      },
      success: function (res) {
        var status = res.data.status;
        if(status == 1){
```

```
            // 只有大于一件的时候，才是 normal 状态，否则是 disable 状态
            var minusStatus = num < = ? 'disabled' : 'normal';
            // 购物车数据
            var carts = that.data.carts;
            carts[index].num = num;
            // 按钮可用状态
            var minusStatuses = that.data.minusStatuses;
            minusStatuses[index] = minusStatus;
            // 将数值与状态写回
            that.setData({
              minusStatuses: minusStatuses
            });
            that.sum();
          }else{
            wx.showToast({
              title: '操作过快！',
              duration: 2000
            });
          }
        },
        fail: function() {
          // fail
          wx.showToast({
            title: '网络异常！',
            duration: 2000
          });
        }
      });
},

bindPlus: function(e) {
    var that = this;
    var index = parseInt(e.currentTarget.dataset.index);
    var num = that.data.carts[index].num;
    // 自增
    num + + ;
    console.log(num);
    var cart_id = e.currentTarget.dataset.cartid;
    wx.request({
      url: app.d.ceshiUrl + '/Api/Shopping/up_cart',
      method:'post',
      data: {
```

```
        user_id: app.d.userId,
        num:num,
        cart_id:cart_id
      },
      header: {
        'Content-Type': 'application/x-www-form-urlencoded'
      },
      success: function (res) {
        var status = res.data.status;
        if(status == 1){
          // 只有大于一件的时候,才是 normal 状态,否则是 disable 状态
          var minusStatus = num <= 1 ? 'disabled' : 'normal';
          // 购物车数据
          var carts = that.data.carts;
          carts[index].num = num;
          // 按钮可用状态
          var minusStatuses = that.data.minusStatuses;
          minusStatuses[index] = minusStatus;
          // 将数值与状态写回
          that.setData({
            minusStatuses: minusStatuses
          });
          that.sum();
        }else{
          wx.showToast({
            title:'操作过快!',
            duration: 2000
          });
        }
      },
      fail: function() {
        // fail
        wx.showToast({
          title:'网络异常!',
          duration: 2000
        });
      }
    });
},

bindCheckbox: function(e) {
  /* 绑定点击事件,将 checkbox 样式改变为选中与非选中 */
```

```
    //拿到下标值,以在 carts 作遍历指示用
    var index = parseInt(e.currentTarget.dataset.index);
    //原始的 icon 状态
    var selected = this.data.carts[index].selected;
    var carts = this.data.carts;
    // 对勾选状态取反
    carts[index].selected = ! selected;
    // 写回经点击修改后的数组
    this.setData({
      carts: carts
    });
    this.sum()
  },

  bindSelectAll: function() {
    // 环境中目前已选状态
    var selectedAllStatus = this.data.selectedAllStatus;
    // 取反操作
    selectedAllStatus = ! selectedAllStatus;
    // 购物车数据,关键是处理 selected 值
    var carts = this.data.carts;
    // 遍历
    for (var i = 0; i < carts.length; i++) {
      carts[i].selected = selectedAllStatus;
    }
    this.setData({
      selectedAllStatus: selectedAllStatus,
      carts: carts
    });
    this.sum()
  },

  bindCheckout: function() {
    // 初始化 toastStr 字符串
    var toastStr = '';
    // 遍历取出已勾选的 cid
    for (var i = 0; i < this.data.carts.length; i++) {
      if (this.data.carts[i].selected) {
        toastStr += this.data.carts[i].id;
        toastStr += ',';
      }
    }
```

```
    if (toastStr == ''){
      wx.showToast({
        title: '请选择要结算的商品！',
        duration: 2000
      });
      return false;
    }
    //存回 data
    wx.navigateTo({
      url: '../order/pay? cartId=' + toastStr,
    })
  },

  bindToastChange: function() {
    this.setData({
      toastHidden: true
    });
  },

  sum: function() {
      var carts = this.data.carts;
      // 计算总金额
      var total = 0;
      for (var i = 0; i < carts.length; i++) {
        if (carts[i].selected) {
          total += carts[i].num * carts[i].price;
        }
      }
      // 写回经点击修改后的数组
      this.setData({
        carts: carts,
        total: '¥ ' + total
      });
    },

  onLoad:function(options){
      this.loadProductData();
      this.sum();
  },

  onShow:function(){
    this.loadProductData();
```

```
},

removeShopCard:function(e){
    var that = this;
    var cardId = e.currentTarget.dataset.cartid;
    wx.showModal({
      title:'提示',
      content:'你确认移除吗',
      success:function(res) {
        res.confirm && wx.request({
          url:app.d.ceshiUrl + '/Api/Shopping/delete',
          method:'post',
          data:{
            cart_id:cardId,
          },
          header:{
            'Content-Type':  'application/x-www-form-urlencoded'
          },
          success:function (res) {
            //-- init data
            var data = res.data;
            if(data.status == 1){
              //that.data.productData.length = 0;
              that.loadProductData();
            }else{
              wx.showToast({
                title:'操作失败!',
                duration:2000
              });
            }
          },
        });
      },
      fail:function() {
        // fail
        wx.showToast({
          title:'网络异常!',
          duration:2000
        });
      }
    });
  },
```

```
// 数据案例
  loadProductData:function(){
    var that = this;
    wx.request({
      url: app.d.ceshiUrl + '/Api/Shopping/index',
      method:'post',
      data: {
        user_id: app.d.userId
      },
      header: {
        'Content-Type': 'application/x-www-form-urlencoded'
      },
      success: function (res) {
        //-- init data
        var cart = res.data.cart;
        that.setData({
          carts:cart,
        });
        //endInitData
      },
    });
  },

})
```

8.6.4 json 配置代码

```
{
"navigationBarTitleText": "购物车"
}
```

8.7 订单确认页代码清单

订单确认页如图 8-7 和图 8-8 所示。需要注意的是,当没有收货地址的时候需要先添加收货地址,有地址的情况下即可直接微信支付。

图 8-7　订单确认页

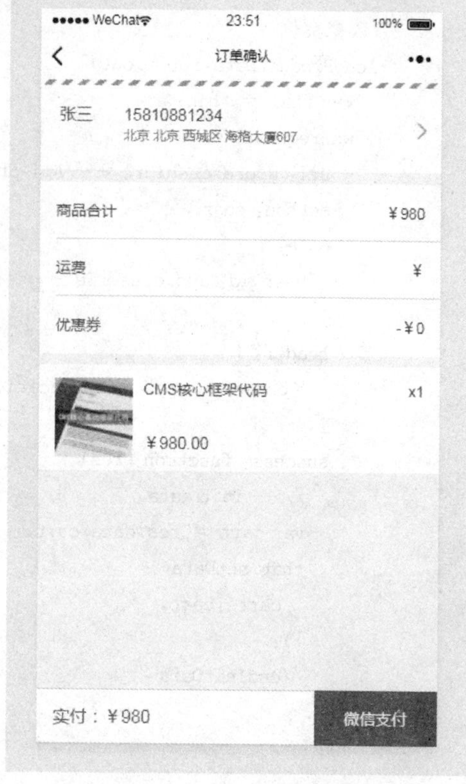

图 8-8　订单确认页(续)

代码详情如下所示。

8.7.1　wxml 模板代码

```
< view class = "container" >
< form type = "submit" report - submit = "true" bindsubmit = "submitForm" >
< view class = "address - box" >
< view class = "address - item" bindtap = "selectAddress" wx:if = "{{addemt == 0}}" >
< view class = "l" >
    < text class = "name" > {{address.name}} < /text >
    < text class = "default" wx:if = "{{checkedAddress.is_default  === 1}}" > 默认 < /
    text >
< /view >
< view class = "m" >
    < text class = "mobile" > {{address.tel}} < /text >
    < text class = "address" > {{address.address_xq}} < /text >
< /view >
```

```
< view class = "r" >
    < image src = "/static/images/address_right.png" > < /image >
< /view >
< /view >
< view class = "address - item address - empty" bindtap = "addAddress" wx:else >
< navigator url = "../address/user - address/user - address? cartId = {{cartId}}" hov-
er - class = "none" >
< view class = "m" >
            还没有收货地址,去添加 < /view >
< /navigator >
< view class = "r" >
    < image src = "/static/images/address_right.png" > < /image >
< /view >
< /view >
< /view >

< view class = "coupon - box" wx:if = "{{vou!  = ''}}" >
< view class = "coupon - item" >
< view class = "l" >
    < text class = "name" > 请选择优惠券 < /text >
    < text class = "txt" > 0 张 < /text >
< /view >
< view class = "r" >
    < image src = "/static/images/address_right.png" > < /image >
< /view >
< /view >
< /view >

< view class = "order - box" >
< view class = "order - item" >
< view class = "l" >
    < text class = "name" > 商品合计 < /text >
< /view >
< view class = "r" >
    < text class = "txt" > ¥ {{total}} < /text >
< /view >
< /view >
< view class = "order - item" >
< view class = "l" > < text class = "name" > 运费 < /text > < /view >
< view class = "r" > < text class = "txt" > ¥ {{freightPrice}} < /text > < /view >
< /view >
< view class = "order - item no - border" >
```

```
<view class="l"><text class="name">优惠券</text></view>
<view class="r">
<text class="txt">-¥0</text>
</view>
</view>
</view>

<view class="goods-items">
<view class="item" wx:for="{{productData}}" wx:key="{{item.id}}">
    <view class="img">
        <image src="{{item.photo_x}}"></image>
    </view>
    <view class="info">
        <view class="t">
            <text class="name">{{item.name}}</text>
            <text class="number">x{{item.num}}</text>
        </view>
        <view class="m">{{item.goods_specifition_name_value}}</view>
        <view class="b">¥{{item.price}}</view>
    </view>
</view>
</view>

<view class="order-total" wx:if="{{addemt==0}}">
<view class="l">实付:¥{{total}}</view>
<view class="r" type="button" id="wxPay" disabled="{{btnDisabled}}" formType
="submit" bindtap="createProductOrderByWX">微信支付</view>
</view>
<view class="order-total" wx:else>
<view class="l">实付:¥{{total}}</view>
<view class="r" disabled="{{btnDisabled}}" style="background-color:#1AAD16;
color:#FFFFFF;">请先添加收货地址</view>
</view>
</form>

</view>
```

8.7.2 wxss 样式代码

```
page{
    height:100%;
    background:#f4f4f4;
```

```
}

.address - box{
    width: 100 % ;
    height: 166.55rpx;
    background: url ('http://yanxuan. nosdn. 127. net/hxm/yanxuan - wap/p/20161201/
style/img/icon - normal/address - bg - bd30f2bfeb. png') 0 0 repeat - x;
    background - size: 62.5rpx 10.5rpx;
    margin - bottom: 20rpx;
    padding - top: 10.5rpx;
}

.address - item{
    display: flex;
    height: 155.55rpx;
    background: #fff;
    padding: 41.6rpx 0 41.6rpx 31.25rpx;
}

.address - item.address - empty{
  line - height: 75rpx;
  text - align: center;
}

.address - box .l{
    width: 125rpx;
    height: 100 % ;
}

.address - box .l .name{
    margin - left: 6.25rpx;
    margin - top: - 7.25rpx;
    display: block;
    width: 125rpx;
    height: 43rpx;
    line - height: 43rpx;
    font - size: 30rpx;
    color: #333;
    margin - bottom: 5rpx;

}
```

```
.address - box .l .default{
    margin - left: 6.25rpx;
    display: block;
    width: 62rpx;
    height: 33rpx;
    border - radius: 5rpx;
    border: 1px solid #b4282d;
    font - size: 20.5rpx;
    text - align: center;
    line - height: 29rpx;
    color: #b4282d;
}

.address - box .m{
    flex: 1;
    height: 72.25rpx;
    color: #999;
}

.address - box .mobile{
    display: block;
    height: 29rpx;
    line - height: 29rpx;
    margin - bottom: 6.25rpx;
    font - size: 30rpx;
    color: #333;
}

.address - box .address{
    display: block;
    height: 37.5rpx;
    line - height: 37.5rpx;
    font - size: 25rpx;
    color: #666;
}

.address - box .r{
    width: 77rpx;
    height: 77rpx;
    display: flex;
    justify - content: center;
    align - items: center;
```

```
}

.address - box .r image{
    width: 52.078rpx;
    height: 52.078rpx;
}

.coupon - box{
    width: 100 % ;
    height: auto;
    overflow: hidden;
    background: #fff;
}

.coupon - box .coupon - item{
    width: 100 % ;
    height: 108.3rpx;
    overflow: hidden;
    background: #fff;
    display: flex;
    padding - left: 31.25rpx;
}

.coupon - box .l{
    flex: 1;
    height: 43rpx;
    line - height: 43rpx;
    padding - top: 35rpx;
}

.coupon - box .l .name{
    float: left;
    font - size: 30rpx;
    color: #666;
}

.coupon - box .l .txt{
    float: right;
    font - size: 30rpx;
    color: #666;
}
```

```
.coupon - box .r{
    margin - top: 15.5rpx;
    width: 77rpx;
    height: 77rpx;
    display: flex;
    justify - content: center;
    align - items: center;
}

.coupon - box .r image{
    width: 52.078rpx;
    height: 52.078rpx;
}

.order - box{
    margin - top: 20rpx;
    width: 100 % ;
    height: auto;
    overflow: hidden;
    background: #fff;
}

.order - box .order - item{
    height: 104.3rpx;
    overflow: hidden;
    background: #fff;
    display: flex;
    margin - left: 31.25rpx;
    padding - right: 31.25rpx;
    padding - top: 26rpx;
    border - bottom: 1px solid #d9d9d9;
}

.order - box .order - item .l{
    float: left;
    height: 52rpx;
    width: 50 % ;
    line - height: 52rpx;
    overflow: hidden;
}

.order - box .order - item .r{
```

```
    float: right;
    text-align: right;
    width: 50%;
    height: 52rpx;
    line-height: 52rpx;
    overflow: hidden;
}

.order-box .order-item.no-border{
    border-bottom: none;
}

.goods-items{
    margin-top: 20rpx;
    width: 100%;
    height: auto;
    overflow: hidden;
    background: #fff;
    padding-left: 31.25rpx;
    margin-bottom: 120rpx;
}

.goods-items .item{
    height: 192rpx;
    padding-right: 31.25rpx;
    display: flex;
    align-items: center;
    border-bottom: 1px solid rgba(0,0,0,0.15);
}

.goods-items .item.no-border{
    border-bottom: none;
}


.goods-items .item:last-child{
    border-bottom: none;
}

.goods-items .img{
    height: 145.83rpx;
    width: 145.83rpx;
```

```
    background - color: #f4f4f4;
    margin - right: 20rpx;
}

.goods - items .img image{
    height: 145.83rpx;
    width: 145.83rpx;
}

.goods - items .info{
    flex: 1;
    height: 145.83rpx;
    padding - top: 5rpx;
}

.goods - items .t{
    height:  33rpx;
    line - height: 33rpx;
    margin - bottom: 10rpx;
    overflow: hidden;
    font - size: 30rpx;
    color: #333;
}

.goods - items .t .name{
    display: block;
    float: left;
}

.goods - items .t .number{
    display: block;
    float: right;
    text - align: right;
}

.goods - items .m {
    height:  29rpx;
    overflow: hidden;
    line - height: 29rpx;
    margin - bottom: 25rpx;
    font - size: 25rpx;
    color: #666;
```

```
}

.goods - items .b {
    height:  41rpx;
    overflow: hidden;
    line - height: 41rpx;
    font - size: 30rpx;
    color: #333;
}

.order - total{
    position: fixed;
    left:0;
    bottom: 0;
    height: 100rpx;
    width: 100% ;
    display: flex;
}

.order - total .l{
    flex: 1;
    height: 100rpx;
    line - height: 100rpx;
    color: #b4282d;
    background: #fff;
    font - size: 33rpx;
    padding - left: 31.25rpx;
    border - top: 1rpx solid rgba(0,0,0,0.2);
    border - bottom: 1rpx solid rgba(0,0,0,0.2);
}

.order - total .r{
    width: 233rpx;
    height: 100rpx;
    background: #b4282d;
    border: 1px solid #b4282d;
    line - height: 100rpx;
    text - align: center;
    color: #fff;
    font - size: 30rpx;
}
```

8.7.3 js 逻辑代码

```
var app = getApp();
// pages/order/pay.js
Page({
  data:{
    itemData:{},
    userId:0,
    paytype:'weixin',//微信支付
    remark:'',
    cartId:0,
    addrId:0,//收货地址//测试
    btnDisabled:false,
    productData:[],
    address:{},
    total:0,
    vprice:0,
    vid:0,
    addemt:1,
    vou:[]
  },
  onLoad:function(options){
    var uid = app.d.userId;
    this.setData({
      cartId: options.cartId,
      userId: uid
    });
    this.loadProductDetail();
  },
  loadProductDetail:function(){
    var that = this;
    wx.request({
      url: app.d.ceshiUrl + '/Api/Payment/buy_cart',
      method:'post',
      data: {
        cart_id: that.data.cartId,
        uid: that.data.userId,
      },
      header: {
        'Content-Type': 'application/x-www-form-urlencoded'
      },
```

```
        success: function (res) {
          //that.initProductData(res.data);
          var adds = res.data.adds;
          if (adds){
            var addrId = adds.id;
            that.setData({
              address: adds,
              addrId: addrId
            });
          }
          that.setData({
            addemt: res.data.addemt,
            productData:res.data.pro,
            total: res.data.price,
            vprice: res.data.price,
            vou: res.data.vou,
          });
          //endInitData
        },
      });
    },

    remarkInput:function(e){
      this.setData({
        remark: e.detail.value,
      })
    },

  //选择优惠券
    getvou:function(e){
      var vid = e.currentTarget.dataset.id;
      var price = e.currentTarget.dataset.price;
      var zprice = this.data.vprice;
      var cprice = parseFloat(zprice) - parseFloat(price);
      this.setData({
        total: cprice,
        vid: vid
      })
    },

  //微信支付
    createProductOrderByWX:function(e){
```

```
    this.setData({
      paytype: 'weixin',
    });

    this.createProductOrder();
  },


  //确认订单
  createProductOrder:function(){
    this.setData({
      btnDisabled:false,
    })

    //创建订单
    var that = this;
    wx.request({
      url: app.d.ceshiUrl + '/Api/Payment/payment',
      method:'post',
      data: {
        uid: that.data.userId,
        cart_id: that.data.cartId,
        type:that.data.paytype,
        aid: that.data.addrId,//地址的 id
        remark: that.data.remark,//用户备注
        price: that.data.total,//总价
        vid: that.data.vid,//优惠券 id
      },
      header: {
        'Content-Type':  'application/x-www-form-urlencoded'
      },
      success: function (res) {
        //--init data
        var data = res.data;
        if(data.status == 1){
          //创建订单成功
          if(data.arr.pay_type == 'cash'){
              wx.showToast({
                title:"请自行联系商家进行发货!",
                duration:3000
              });
              return false;
```

```
            }
            if(data.arr.pay_type == 'weixin'){
              //微信支付
              that.wxpay(data.arr);
            }
          }else{
            wx.showToast({
              title:"下单失败!",
              duration:2500
            });
          }
        },
        fail: function (e) {
          wx.showToast({
            title: '网络异常! err:createProductOrder',
            duration: 2000
          });
        }
      });
    },

    //调起微信支付
    wxpay: function(order){
        wx.request({
          url: app.d.ceshiUrl + '/Api/Wxpay/wxpay',
          data: {
            order_id:order.order_id,
            order_sn:order.order_sn,
            uid:this.data.userId,
          },
          method: 'POST', // OPTIONS, GET, HEAD, POST, PUT, DELETE, TRACE, CONNECT
          header: {
            'Content - Type':  'application/x - www - form - urlencoded'
          }, // 设置请求的 header
          success: function(res){

            if(res.data.status == 1){
              var order = res.data.arr;
              console.log(order);
              wx.requestPayment({
                timeStamp: order.timeStamp,
                nonceStr: order.nonceStr,
```

```
            package: order.package,
            signType: 'MD5',
            paySign: order.paySign,
            success: function(res){
              wx.showToast({
                title:"支付成功!",
                duration:2000,
              });
              setTimeout(function(){
                wx.navigateTo({
                  url: '../user/dingdan? currentTab = 1&otype = deliver',
                });
              },2500);
            },
            fail: function(res) {
              wx.showToast({
                title:res,
                duration:3000
              })
            }
          })
        }else{
          wx.showToast({
            title: res.data.err,
            duration: 2000
          });
        }
      },
      fail: function() {
        // fail
        wx.showToast({
          title: '网络异常! err:wxpay',
          duration: 2000
        });
      }
    })
  },

});
```

8.7.4 json 配置代码

```
{
    "navigationBarTitleText": "订单确认"
}
```

8.8 订单详情页代码清单

订单详情页效果如图 8-9 所示。

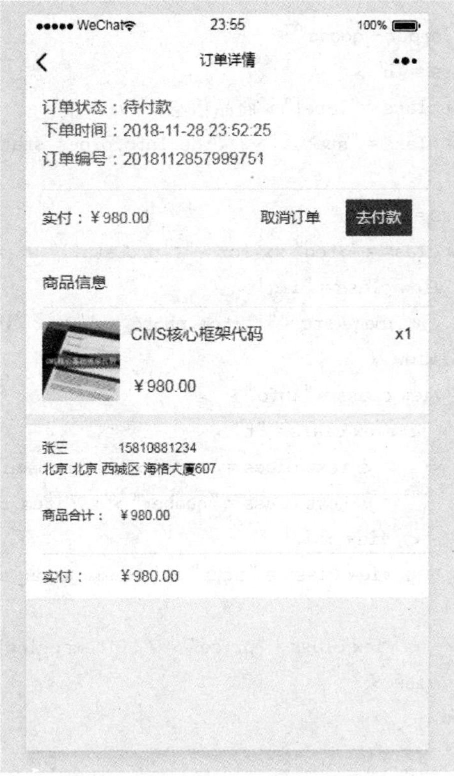

图 8-9 订单详情页

程序代码如下所示。

8.8.1 wxml 模板代码

```
< view class = "container" >
    < view class = "order - info" >
    < view class = "item - a" > 订单状态:{{orderData.order_status}} < /view >
```

```
        <view class="item-a">下单时间:{{orderData.addtime}}</view>
        <view class="item-b">订单编号:{{orderData.order_sn}}</view>
        <view class="item-c">
            <view class="l">实付:<text class="cost">¥{{orderData.price}}
</text></view>
            <view class="r">
                <view class="btn" bindtap="cancelOrder">取消订单</view>
                <view class="btn active" bindtap="payOrder">去付款</view>
            </view>
        </view>
    </view>

    <view class="order-goods">
        <view class="h">
            <view class="label">商品信息</view>
            <view class="status">{{orderInfo.order_status_text}}</view>
        </view>
        <view class="goods">
            <view class="item" wx:for="{{proData}}" wx:key="key">
                <view class="img">
                    <image src="{{item.photo_x}}"></image>
                </view>
                <view class="info">
                    <view class="t">
                        <text class="name">{{item.name}}</text>
                        <text class="number">x{{item.num}}</text>
                    </view>
                    <view class="attr">{{item.goods_specifition_name_value}}
</view>
                    <view class="price">¥{{item.price}}</view>
                </view>
            </view>
        </view>
    </view>

    <view class="order-bottom">
        <view class="address">
            <view class="t">
                <text class="name">{{orderData.receiver}}</text>
                <text class="mobile">{{orderData.tel}}</text>
            </view>
            <view class="b">{{orderData.address_xq}}</view>
```

```
        </view>
        <view class="total">
            <view class="t">
                <text class="label">商品合计：</text>
                <text class="txt">¥{{orderData.price}}</text>
            </view>
            <view class="t" wx:if="{{orderData.order_status == '待收货'}}">
                <text class="label">运费：</text>
                <text class="txt">¥{{orderInfo.freight_price}}</text>
            </view>
        </view>
        <view class="pay-fee">
            <text class="label">实付：</text>
            <text class="txt">¥{{orderData.price}}</text>
        </view>
    </view>
</view>
```

8.8.2 wxss 样式代码

```
page{
    height: 100%;
    width: 100%;
    background: #f4f4f4;
}

.order-info{
    padding-top: 25rpx;
    background: #fff;
    height: auto;
    overflow: hidden;
}

.item-a{
    padding-left: 31.25rpx;
    height: 42.5rpx;
    padding-bottom: 12.5rpx;
    line-height: 30rpx;
    font-size: 30rpx;
    color: #666;
}
```

```
.item - b{
    padding - left: 31.25rpx;
    height: 29rpx;
    line - height: 29rpx;
    margin - top: 12.5rpx;
    margin - bottom: 41.5rpx;
    font - size: 30rpx;
    color: #666;
}

.item - c{
    margin - left: 31.25rpx;
    border - top: 1px solid #f4f4f4;
    height: 103rpx;
    line - height: 103rpx;
}

.item - c .l{
    float: left;
}

.item - c .r{
    height: 103rpx;
    float: right;
    display: flex;
    align - items: center;
    padding - right: 16rpx;
}

.item - c .r .btn{
    float: right;
}

.item - c .cost{
    color: #b4282d;
}

.item - c .btn{
    line - height: 66rpx;
    border - radius: 5rpx;
    text - align: center;
    margin: 0 15rpx;
```

```
    padding: 0 20rpx;
    height: 66rpx;
}

.item-c .btn.active{
    background: #b4282d;
    color: #fff;
}

.order-goods{
    margin-top: 20rpx;
    background: #fff;
}

.order-goods .h{
    height: 93.75rpx;
    line-height: 93.75rpx;
    margin-left: 31.25rpx;
    border-bottom: 1px solid #f4f4f4;
    padding-right: 31.25rpx;
}

.order-goods .h .label{
    float: left;
    font-size: 30rpx;
    color: #333;
}

.order-goods .h .status{
    float: right;
    font-size: 30rpx;
    color: #b4282d;
}

.order-goods .item{
    display: flex;
    align-items: center;
    height: 192rpx;
    margin-left: 31.25rpx;
    padding-right: 31.25rpx;
    border-bottom: 1px solid #f4f4f4;
}
```

```
.order－goods .item:last－child{
    border－bottom: none;
}

.order－goods .item .img{
    height: 145.83rpx;
    width: 145.83rpx;
    background: ＃f4f4f4;
}

.order－goods .item .img image{
    height: 145.83rpx;
    width: 145.83rpx;
}

.order－goods .item .info{
    flex: 1;
    height: 145.83rpx;
    margin－left: 20rpx;
}

.order－goods .item .t{
    margin－top: 8rpx;
    height: 33rpx;
    line－height: 33rpx;
    margin－bottom: 10.5rpx;
}

.order－goods .item .t .name{
    display: block;
    float: left;
    height: 33rpx;
    line－height: 33rpx;
    color: ＃333;
    font－size: 30rpx;
}

.order－goods .item .t .number{
    display: block;
    float: right;
    height: 33rpx;
```

```
    text - align: right;
    line - height: 33rpx;
    color: #333;
    font - size: 30rpx;
}

.order - goods .item .attr{
    height: 29rpx;
    line - height: 29rpx;
    color: #666;
    margin - bottom: 25rpx;
    font - size: 25rpx;
}

.order - goods .item .price{
    height: 30rpx;
    line - height: 30rpx;
    color: #333;
    font - size: 30rpx;
}

.order - bottom{
    margin - top: 20rpx;
    padding - left: 31.25rpx;
    height: auto;
    overflow: hidden;
    background: #fff;
}

.order - bottom .address{
    height: 128rpx;
    padding - top: 25rpx;
    border - bottom: 1px solid #f4f4f4;
}

.order - bottom .address .t{
    height: 35rpx;
    line - height: 35rpx;
    margin - bottom: 7.5rpx;
}

.order - bottom .address .name{
```

```
    display: inline-block;
    height: 35rpx;
    width: 140rpx;
    line-height: 35rpx;
    font-size: 25rpx;
}

.order-bottom .address .mobile{
    display: inline-block;
    height: 35rpx;
    line-height: 35rpx;
    font-size: 25rpx;
}

.order-bottom .address .b{
    height: 35rpx;
    line-height: 35rpx;
    font-size: 25rpx;
}

.order-bottom .total{
    height: 106rpx;
    padding-top: 20rpx;
    border-bottom: 1px solid #f4f4f4;
}

.order-bottom .total .t{
    height: 25rpx;
    line-height: 25rpx;
    margin-bottom: 7.5rpx;
    display: flex;
}

.order-bottom .total .label{
    width: 140rpx;
    display: inline-block;
    height: 35rpx;
    line-height: 35rpx;
    font-size: 25rpx;
}

.order-bottom .total .txt{
```

```
        flex: 1;
        display: inline-block;
        height: 35rpx;
        line-height: 35rpx;
        font-size: 25rpx;
    }

    .order-bottom .pay-fee{
        height: 81rpx;
        line-height: 81rpx;
    }

    .order-bottom .pay-fee .label{
        display: inline-block;
        width: 140rpx;
        color: #b4282d;
    }

    .order-bottom .pay-fee .txt{
        display: inline-block;
        width: 140rpx;
        color: #b4282d;
    }
```

8.8.3 js 逻辑代码

```
var app = getApp();
// pages/order/detail.js
Page({
  data:{
    orderId:0,
    orderData:{},
    proData:[],
  },
  onLoad:function(options){
    this.setData({
      orderId: options.orderId,
    })
    this.loadProductDetail();
  },
  loadProductDetail:function(){
```

```
    var that = this;
    wx.request({
      url: app.d.ceshiUrl + '/Api/Order/order_details',
      method:'post',
      data: {
        order_id: that.data.orderId,
      },
      header: {
        'Content-Type':  'application/x-www-form-urlencoded'
      },
      success: function (res) {

        var status = res.data.status;
        if(status==1){
          var pro = res.data.pro;
          var ord = res.data.ord;
          that.setData({
            orderData: ord,
            proData:pro
          });
        }else{
          wx.showToast({
            title: res.data.err,
            duration: 2000
          });
        }
      },
      fail: function () {
          // fail
          wx.showToast({
            title: '网络异常！',
            duration: 2000
          });
      }
    });
  },

})
```

8.8.4 json 配置代码

```
{
    "navigationBarTitleText":"订单详情"
}
```

8.9 地址管理页代码清单

地址管理页效果如图 8 - 10 所示。

图 8 - 10 地址管理

代码详情如下所示。

8.9.1 wxml 模板代码

```
<!-- 地址管理 -->
<radio-group class="radio-group" wx:for="{{address}}" wx:key="key">
<view hidden="{{hiddenAddress}}" class="address" >
  <view class="address-icon" bindtap="setDefault" data-id="{{item.id}}">
    <radio checked="{{item.is_default==1? true:false}}" value="{{index}}" />
  </view>

  <view class="address-detail">
    <view class="address-name-phone">
      <text class="address-name">{{item.name}}</text>
      <text class="address-phone">{{item.tel}}</text>
```

```
        </view>
        <view class="address-info">{{item.address_xq}}</view>
        <view class="address-edit">
          <view>
            <icon></icon>
            <text hidden="{{item.is_default==0? false:true}}"></text>
          </view>
          <view>
            <text hidden="{{item.is_default==0? false:true}}" bindtap="setDefault"
data-id="{{item.id}}">设置默认</text>
            <text hidden="{{item.is_default==0? false:true}}">|</text>
            <text data-id="{{item.id}}" bindtap="delAddress">删除</text>
          </view>
        </view>
      </view>
    </view>
    </radio-group>
    <navigator url="../address? cartId={{cartId}}">
      <view hidden="{{hiddenAddress}}" class="add-address">
        <image class="add-address-icon"></image>
        <view>新增地址</view>
      </view>
    </navigator>
```

8.9.2 wxss 样式代码

```
/*地址管理*/
page {
  background-color: #efeff4;
  font-size: 10pt;
  -webkit-user-select: none;
  user-select: none;
  width: 100%;
  overflow-x: hidden;
}

.address {
  display: flex;
  flex-wrap: wrap;
  background-color: #fff;
  margin-top: 1px;
```

```
  padding: 30rpx;
  margin: 30rpx 0;
}

.address-icon {
  width: 100rpx;
}

.address-detail {
  width: 590rpx;
}

.address-name-phone {
  margin-bottom: 20rpx;
  font-size: 11pt;
  font-weight: 900;
}

.address-name{
  margin-right: 20rpx;
}

.address-info {
  margin-bottom: 30rpx;
}

.address-edit {
  display: flex;
  justify-content: space-between;
  flex-wrap: wrap;
  border-top: 1px #efeff4 solid;
  padding-top: 30rpx;
}

.add-address {
  display: flex;
  align-items: center;
  margin-top: 20px;
  padding: 18rpx;
  background-color: #fff;
  font-size: 11pt;
  font-weight: 900;
```

```
}

.add-address-icon {
  height: 20px;
  width: 20px;
}
```

8.9.3 js 逻辑代码

```
// pages/address/user-address/user-address.js
var app = getApp()
Page({
  data: {
    address: [],
    radioindex: '',
    pro_id:0,
    num:0,
    cartId:0
  },
  onLoad: function (options) {
    var that = this;
    // 页面初始化 options 为页面跳转所带来的参数
    var cartId = options.cartId;
    console.log(app.d.userId);
    wx.request({
      url: app.d.ceshiUrl + '/Api/Address/index',
      data: {
        user_id:app.d.userId,
      },
      method: 'POST', // OPTIONS, GET, HEAD, POST, PUT, DELETE, TRACE, CONNECT
      header: {// 设置请求的 header
        'Content-Type':  'application/x-www-form-urlencoded'
      },

      success: function (res) {
        // success
        var address = res.data.adds;
        console.log(address);
        if (address == '') {
          var address = []
        }
```

```
        that.setData({
          address: address,
          cartId: cartId,
        })
      },
      fail: function () {
        // fail
        wx.showToast({
          title: '网络异常！',
          duration: 2000
        });
      }
    })

  },

  onReady: function () {
    // 页面渲染完成
  },
  setDefault: function(e) {
    var that = this;
    var addrId = e.currentTarget.dataset.id;
    wx.request({
      url: app.d.ceshiUrl + '/Api/Address/set_default',
      data: {
        uid:app.d.userId,
        addr_id:addrId
      },
      method: 'POST', // OPTIONS, GET, HEAD, POST, PUT, DELETE, TRACE, CONNECT
      header: {// 设置请求的 header
        'Content-Type':  'application/x-www-form-urlencoded'
      },

      success: function (res) {
        // success
        var status = res.data.status;
        var cartId = that.data.cartId;
        if(status == 1){
          if (cartId) {
            wx.redirectTo({
              url: '../../order/pay?cartId=' + cartId,
            });
```

```
            return false;
          }

          wx.showToast({
            title: '操作成功!',
            duration: 2000
          });

          that.DataonLoad();
        }else{
          wx.showToast({
            title: res.data.err,
            duration: 2000
          });
        }
      },
      fail: function () {
        // fail
        wx.showToast({
          title: '网络异常!',
          duration: 2000
        });
      }
    })
  },
  delAddress: function (e) {
    var that = this;
    var addrId = e.currentTarget.dataset.id;
    wx.showModal({
      title: '提示',
      content: '你确认移除吗',
      success: function(res) {
        res.confirm && wx.request({
          url: app.d.ceshiUrl + '/Api/Address/del_adds',
          data: {
            user_id:app.d.userId,
            id_arr:addrId
          },
          method: 'POST', // OPTIONS, GET, HEAD, POST, PUT, DELETE, TRACE, CONNECT
          header: {// 设置请求的 header
            'Content-Type':  'application/x-www-form-urlencoded'
          },
```

```
        success: function (res) {
          // success
          var status = res.data.status;
          if(status == 1){
            that.DataonLoad();
          }else{
            wx.showToast({
              title: res.data.err,
              duration: 2000
            });
          }
        },
        fail: function () {
          // fail
          wx.showToast({
            title: '网络异常! ',
            duration: 2000
          });
        }
      });
    }
  });

},
DataonLoad: function () {
  var that = this;
  // 页面初始化 options 为页面跳转所带来的参数
  wx.request({
    url: app.d.ceshiUrl + '/Api/Address/index',
    data: {
      user_id:app.d.userId,
    },
    method: 'POST', // OPTIONS, GET, HEAD, POST, PUT, DELETE, TRACE, CONNECT
    header: {// 设置请求的 header
      'Content-Type': 'application/x-www-form-urlencoded'
    },

    success: function (res) {
      // success
      var address = res.data.adds;
      if (address == '') {
        var address = []
      }
      that.setData({
```

```
          address: address,
        })
      },
      fail: function () {
        // fail
        wx.showToast({
          title: '网络异常! ',
          duration: 2000
        });
      }
    })

  },
})
```

8.9.4 json 配置代码

```
{
    "navigationBarTitleText": "地址管理"
}
```

8.10 保存地址页代码清单

保存地址页效果图如图 8-11 所示。

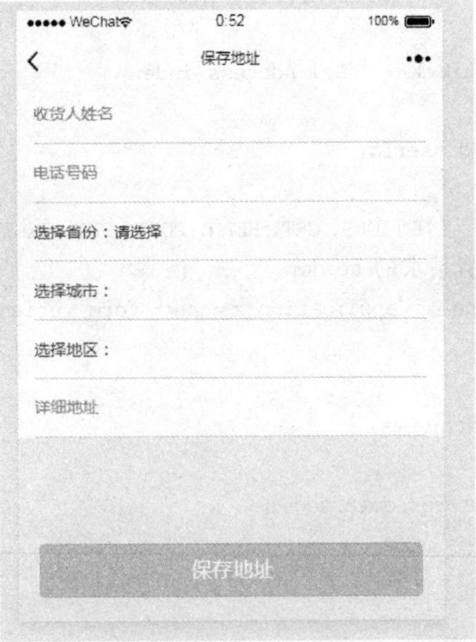

图 8-11 保存地址

程序代码如下所示。

8.10.1 wxml 模板代码

```
<view class="container">
  <form bindsubmit="formSubmit" bindreset="formReset">
    <view class="section">
      <input type="text" hidden="true" value="{{mid}}" name="user_id"/>
      <input type="text" name="name" placeholder="收货人姓名" />
    </view>
    <view class="section">
      <input type="text" name="phone" placeholder="电话号码" />
    </view>
    <view class="section">
      <picker bindchange="bindPickerChangeshengArr" value="{{shengIndex}}" range
="{{shengArr}}" data-id="{{shengId[shengIndex]}}">
        <view class="picker">
          选择省份:{{shengArr[shengIndex]}}
          <input hidden="true" name="province" value="{{shengArr[shengIn-
dex]}}"/>
        </view>
      </picker>
    </view>
    <view class="section">
      <picker bindchange="bindPickerChangeshiArr" value="{{shiIndex}}" range="
{{shiArr}}">
        <view class="picker">
          选择城市:{{shiArr[shiIndex]}}
          <input hidden="true" name="city" value="{{shiArr[shiIndex]}}"/>
        </view>
      </picker>
    </view>
    <view class="section">
      <picker bindchange="bindPickerChangequArr" value="{{quIndex}}" range="
{{quArr}}">
        <view class="picker">
          选择地区:{{quArr[quIndex]}}
          <input hidden="true" name="town" value="{{quArr[quIndex]}}"/>
        </view>
      </picker>
    </view>
    <view class="section">
```

```
        < input type = "text" class = "ww" name = "address" placeholder = "详细地址" / >
      < /view >
      < view >
        < label > < /label >
      < /view >
      < view class = "btn - area" >
        < button formType = "submit" > 保存地址 < /button >
      < /view >
    < /form >
  < /view >
```

8.10.2 wxss 样式代码

```
page {
  background: #efeff4;
}

/* pages/address/address.wxss */

.container {
  font - size: 12px;
  background - color: white;
}

.section {
  height: 100rpx;
  border - bottom: 1px lightgrey solid;
  display: flex;
  align - items: center;
  margin: 0 30rpx;
}

.btn - area {
  display: flex;
  justify - content: center;
}

button {
  width: 90 % ;
  position: fixed;
  bottom: 40rpx;
```

```
    color: #fff;
    background-color: #fc0;
  }

  .picker {
    width: 600rpx;
    height: 100%;
  }
  .ww{
    width: 250px;
  }
```

8.10.3 js 逻辑代码

```
var app = getApp();
 Page({
 data: {
     shengArr: [],//省级数组
     shengId: [],//省级 id 数组
     shiArr: [],//城市数组
     shiId: [],//城市 id 数组
     quArr: [],//区数组
     shengIndex: 0,
     shiIndex: 0,
     quIndex: 0,
     mid: 0,
     sheng:0,
     city:0,
     area:0,
     code:0,
     cartId:0
   },
   formSubmit: function (e) {
     var adds = e.detail.value;
     var cartId = this.data.cartId;
     wx.request({
       url: app.d.ceshiUrl + '/Api/Address/add_adds',
       data: {
         user_id:app.d.userId,
         receiver: adds.name,
         tel: adds.phone,
```

```
        sheng: this.data.sheng,
        city: this.data.city,
        quyu: this.data.area,
        adds: adds.address,
        code: this.data.code,
      },
      method: 'POST', // OPTIONS, GET, HEAD, POST, PUT, DELETE, TRACE, CONNECT
      header: {// 设置请求的 header
        'Content-Type': 'application/x-www-form-urlencoded'
      },
      success: function (res) {
        // success
        var status = res.data.status;
        if(status == 1){
          wx.showToast({
            title: '保存成功！',
            duration: 2000
          });
        }else{
          wx.showToast({
            title: res.data.err,
            duration: 2000
          });
        }
        wx.redirectTo({
          url: 'user-address/user-address? cartId=' + cartId
        });
      },
      fail: function () {
        // fail
        wx.showToast({
          title: '网络异常！',
          duration: 2000
        });
      }
    })


  },
  onLoad: function (options) {
    // 生命周期函数——监听页面加载
    var that = this;
```

```
    that.setData({
      cartId: options.cartId
    })
    //获取省级城市
    wx.request({
      url: app.d.ceshiUrl + '/Api/Address/get_province',
      data: {},
      method: 'POST',
      success: function (res) {
        var status = res.data.status;
        var province = res.data.list;
        var sArr = [];
        var sId = [];
        sArr.push('请选择');
        sId.push('0');
        for (var i = 0; i < province.length; i++) {
          sArr.push(province[i].name);
          sId.push(province[i].id);
        }
        that.setData({
          shengArr: sArr,
          shengId: sId
        })
      },
      fail: function () {
        // fail
        wx.showToast({
          title: '网络异常!',
          duration: 2000
        });
      },
    })

},

bindPickerChangeshengArr: function (e) {
  this.setData({
    shengIndex: e.detail.value,
    shiArr: [],
    shiId: [],
    quArr:[],
    quiId: []
```

```
    });
    var that = this;
    wx.request({
      url: app.d.ceshiUrl + '/Api/Address/get_city',
      data: {sheng:e.detail.value},
      method: 'POST', // OPTIONS, GET, HEAD, POST, PUT, DELETE, TRACE, CONNECT
      header: {// 设置请求的 header
        'Content-Type': 'application/x-www-form-urlencoded'
      },
      success: function (res) {
        // success
        var status = res.data.status;
        var city = res.data.city_list;

        var hArr = [];
        var hId = [];
        hArr.push('请选择');
        hId.push('0');
        for (var i = 0; i < city.length; i++) {
          hArr.push(city[i].name);
          hId.push(city[i].id);
        }
        that.setData({
          sheng:res.data.sheng,
          shiArr: hArr,
          shiId: hId
        })
      },
      fail: function () {
        // fail
        wx.showToast({
          title: '网络异常!',
          duration: 2000
        });
      },

    })
  },
  bindPickerChangeshiArr: function (e) {
    this.setData({
      shiIndex: e.detail.value,
      quArr:[],
```

```
      quiId: []
    })
    var that = this;
    wx.request({
      url: app.d.ceshiUrl + '/Api/Address/get_area',
      data: {
        city:e.detail.value,
        sheng:this.data.sheng
      },
      method: 'POST', // OPTIONS, GET, HEAD, POST, PUT, DELETE, TRACE, CONNECT
      header: {// 设置请求的 header
        'Content-Type': 'application/x-www-form-urlencoded'
      },
      success: function (res) {
        var status = res.data.status;
        var area = res.data.area_list;

        var qArr = [];
        var qId = [];
        qArr.push('请选择');
        qId.push('0');
        for (var i = 0; i < area.length; i++) {
          qArr.push(area[i].name)
          qId.push(area[i].id)
        }
        that.setData({
          city:res.data.city,
          quArr: qArr,
          quiId: qId
        })
      },
      fail: function () {
        // fail
        wx.showToast({
          title: '网络异常!',
          duration: 2000
        });
      }
    })
  },
  bindPickerChangequArr: function (e) {
    console.log(this.data.city)
```

```
        this.setData({
          quIndex: e.detail.value
        });
        var that = this;
        wx.request({
          url: app.d.ceshiUrl + '/Api/Address/get_code',
          data: {
            quyu:e.detail.value,
            city:this.data.city
          },
          method: 'POST', // OPTIONS, GET, HEAD, POST, PUT, DELETE, TRACE, CONNECT
          header: {// 设置请求的 header
            'Content - Type': 'application/x - www - form - urlencoded'
          },
          success: function (res) {
            that.setData({
              area:res.data.area,
              code:res.data.code
            })
          },
          fail: function () {
            // fail
            wx.showToast({
              title: '网络异常！',
              duration: 2000
            });
          }
        })
      }

})
```

8.10.4 json 配置代码

```
{
    "navigationBarTitleText": "保存地址"
}
```

8.11 优惠券页代码清单

优惠券页效果图如图 8－12 所示。

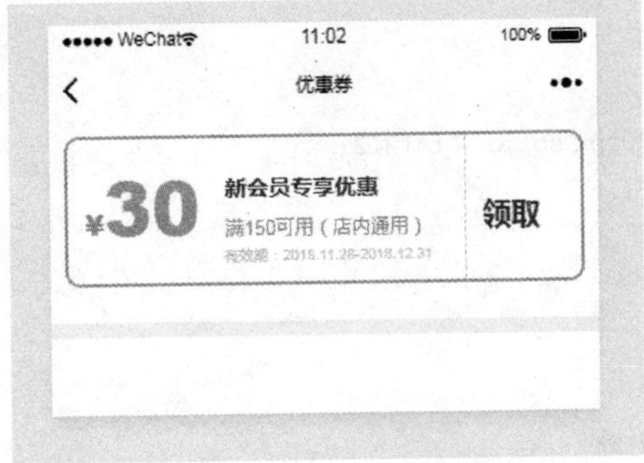

图 8-12 优惠券

程序代码如下所示。

8.11.1 wxml 模板代码

```
<view class="orderDetails" wx:for="{{vou}}">
 <view class="orderListDetails">
 <view class="productInfo">
            <text class="x">¥ </text>
            <text class="y">{{item.amount}}</text>

    </view>

 <view class="pr">
            <text class="n">{{item.title}}</text>
            <text class="b">满{{item.full_money}}可用({{item.desc}})</text>
    </view>
    <view class="hh" bindtap="getvou" data-vid="{{item.id}}">
      <view class="hb" wx:if="{{item.point > 0}}">积分兑换</view>
      <view class="hb" wx:else>领取</view>
    </view>
 </view>
           <view class="c">
              有效期:{{item.start_time}}-{{item.end_time}}</view>
 </view>
```

8.11.2 wxss 样式代码

```
page {
   background: white;
```

```
}

.orderDetails {
  border-bottom: 20rpx solid #f3f3f3;
  background: #fff;
}

.orderListTitle {
  height: 100rpx;
  line-height: 100rpx;
  border-bottom: 1rpx solid #f3f3f3;
}

.orderListTitle .userImg {
  float: left;
  width: 70rpx;
  height: 70rpx;
  border-radius: 35rpx;
  margin: 15rpx;
}

.orderListTitle .userImg image {
  border-radius: 35rpx;
  width: 100%;
  height: 100%;
}

.orderListTitle .userName {
  padding-right: 50rpx;
  background: url(../../../images/member_list_arrow.png) no-repeat 90% center;
  background-size: 35rpx 35rpx;
}

.orderListTitle .orderStatus {
  float: right;
  margin-right: 20rpx;
  color: #2f7b27;
  font-size: 34rpx;
}

.orderListDetails {
  margin: 10px;
```

```
  display: flex;
  height: 200rpx;
  border-bottom: 1rpx solid #f3f3f3;
  border:2px solid #b4a078;
  border-radius: 20rpx;
}

.orderListDetails .productImg {
  flex: 1.3;
  height: 217rpx;
  margin-top: 20rpx;
  margin-left: 20rpx;
}

.orderListDetails .productImg image {
  width: 100%;
  height: 100%;
}

.orderListDetails .productInfo {
  flex: 1.5;
  padding: 20rpx 10rpx;
  text-align: center;
}

.pr {
  flex: 2.6;
  padding: 0 10rpx;
  line-height: 44rpx;
}

.n {
  margin-top: 47rpx;
  font-size: 30rpx;
  display: inline-block;
  line-height: 40rpx;
  font-weight: 900;
  overflow: hidden;
}

.b {
  margin-top: 8rpx;
```

```
  font - size: 27rpx;
  display: inline - block;
  line - height: 30rpx;
  color: #999;
}

.y {
  font - size: 98rpx;
  font - weight: 900;
  color: #b4a078;
}

.x {
  color: #b4a078;
  font - size: 38rpx;
  font - weight: 900;
}

.hh {
  font - weight: 800;
  color: #b4a078;
  flex: 1.3;
  padding: 20rpx 0rpx;
  text - align: center;
  border - left: 1px dashed   #ccc;
}

.hb {
  width: 80 %;
  margin - top: 38rpx;
  font - size: 40rpx;
  padding: 18rpx;
  border - radius: 4rpx;
}

.c {
  font - size: 21rpx;
  position: relative;
  top: - 73rpx;
  left: 237rpx;
  width: 400rpx;
  color: #ccc;
```

```
}

.classname {
  font-size: 25rpx;
  color: #ccc;
}
```

8.11.3 js 逻辑代码

```
// pages/ritual/ritual.js
var app = getApp();
Page({
  data:{
    vou:[],
  },
  getvou:function(e){
    var vid = e.currentTarget.dataset.vid;
    var uid = app.d.userId;
    wx.request({
      url: app.d.ceshiUrl + '/Api/Voucher/get_voucher',
      method:'post',
      data: {vid:vid,uid:uid},
      header: {
        'Content-Type':  'application/x-www-form-urlencoded'
      },
      success: function (res) {
        var status = res.data.status;
        if(status == 1){
          wx.showToast({
            title: '领取成功！',
            duration: 2000
          });
        }else{
          wx.showToast({
            title: res.data.err,
            duration: 2000
          });
        }
        //endInitData
      },
      fail:function(e){
```

```
        wx.showToast({
          title：'网络异常！',
          duration：2000
        });
      },
    });
  },
  onLoad:function(options){
    // 页面初始化 options 为页面跳转所带来的参数
    var that = this;
    wx.request({
      url：app.d.ceshiUrl + '/Api/Voucher/index',
      method:'post',
      data：{},
      header：{
        'Content - Type'： 'application/x - www - form - urlencoded'
      },
      success：function (res) {
        var vou = res.data.vou;
        that.setData({
          vou:vou,
        });
        //endInitData
      },
      error:function(e){
        wx.showToast({
          title：'网络异常！',
          duration：2000
        });
      },
    });
  },
  onReady:function(){
    // 页面渲染完成
  },
  onShow:function(){
    // 页面显示
  },
  onHide:function(){
    // 页面隐藏
  },
  onUnload:function(){
    // 页面关闭
```

```
},
    //右上角分享按钮
  onShareAppMessage: function () {
  return {
    title: '优惠券',
    desc: '信真小铺 - 信守真品的软件技术服务商',
    path: '/pages/ritual/ritual',
 data:{data}
    }
}
})
```

8.11.4 json 配置代码

```
{
  "backgroundTextStyle":"light",
  "navigationBarTitleText": "优惠券"
}
```

8.12 个人中心页代码清单

个人中心页效果图如图 8 - 13 所示。

图 8 - 13 个人中心

程序代码如下所示。

8.12.1 wxml 模板代码

```
< view class = "container" >
  < view class = "profile - info" bindtap = "goLogin" >
    < image class = "avatar" src = "{{userInfo.avatarUrl}}" > < /image >
    < view class = "info" >
      < text class = "name" > {{userInfo.nickName}} < /text >
    < /view >
  < /view >

  < view class = "user - menu" >
    < view class = "item" >
      < navigator url = "../user/dingdan? currentTab = 0" class = "a" >
        < text class = "icon order" > < /text >
        < text class = "txt" > 待付款订单 < /text >
      < /navigator >
    < /view >
 < view class = "item" >
      < navigator url = "../user/daifahuo? currentTab = 0" class = "a" >
        < text class = "icon order" > < /text >
        < text class = "txt" > 待发货订单 < /text >
      < /navigator >
    < /view >
 < view class = "item" >
      < navigator url = "../user/daishouhuo? currentTab = 0" class = "a" >
        < text class = "icon order" > < /text >
        < text class = "txt" > 待收货订单 < /text >
      < /navigator >
    < /view >
 < view class = "item" >
      < navigator url = "../user/success? currentTab = 0" class = "a" >
        < text class = "icon order" > < /text >
        < text class = "txt" > 已完成订单 < /text >
      < /navigator >
    < /view >

    < view class = "item" >
      < navigator url = "../user/shouhou" class = "a" >
        < text class = "icon kefu" > < /text >
        < text class = "txt" > 退款售后 < /text >
```

```
      </navigator>
    </view>

    <view class="item">
      <navigator url="../myritual/myritual" class="a">
        <text class="icon coupon"></text>
        <text class="txt">优惠券</text>
      </navigator>
    </view>

    <view class="item">
      <navigator url="../address/user-address/user-address" class="a">
        <text class="icon address"></text>
        <text class="txt">地址管理</text>
      </navigator>
    </view>

    <view class="item item-bottom">
      <navigator url="../user/feedback" class="a">
        <text class="icon feedback"></text>
        <text class="txt">用户反馈</text>
      </navigator>
    </view>

    <view class="item">
      <navigator url="../synopsis/synopsis" class="a">
        <text class="icon security"></text>
        <text class="txt">关于我们</text>
      </navigator>
    </view>

  </view>

<view class="logout" bindtap="exitLogin">信真小铺 匠心出品</view>
</view>
```

8.12.2 wxss 样式代码

```
page{
    height: 100%;
    width: 100%;
```

```
    background: #f4f4f4;
}
.container{
    background: #f4f4f4;
    height: auto;
    overflow: hidden;
    width: 100%;
}
.profile-info{
    width: 100%;
    height: 280rpx;
    display: flex;
    flex-wrap: wrap;
    align-items: center;
    justify-content: flex-start;
    padding: 0 30.25rpx;
    background: #b4a078;
}

.profile-info .avatar{
    height: 148rpx;
    width: 148rpx;
    border-radius: 50%;
}

.profile-info .info{
    flex: 1;
    height: 85rpx;
    padding-left: 31.25rpx;
}

.profile-info .name{
    display: block;
    height: 45rpx;
    line-height: 45rpx;
    color: #fff;
    font-size: 37.5rpx;
    margin-bottom: 10rpx;
}

.profile-info .level{
    display: block;
```

```
    height: 30rpx;
    line-height: 30rpx;
    margin-bottom: 10rpx;
    color: #7f7f7f;
    font-size: 30rpx;
}

.user-menu{
    width: 100%;
    height: auto;
    overflow: hidden;
    background: #fff;
}

.user-menu .item{
    float: left;
    width: 33.33333%;
    height: 187.5rpx;
    border-right: 1px solid rgba(0,0,0,.15);
    border-bottom: 1px solid rgba(0,0,0,.15);
    text-align: center;
}

.user-menu .item .a{
  display: flex;
  width: 100%;
  height: 100%;
  flex-direction: column;
  align-items: center;
  justify-content: center;
}

.user-menu .item.no-border {
    border-right: 0;
}

.user-menu .item.item-bottom {
    border-bottom: none;
}

.user-menu .icon{
    margin: 0 auto;
```

```
    display: block;
    height: 52.803rpx;
    width: 52.803rpx;
    margin-bottom: 16rpx;
}

.user-menu .icon.order{
    background: url(http://cdn.xinzhenkj.com/ucenter-sdf6a55ee56-f2c2b9c2f0.png)
0 -437.5rpx no-repeat;
    background-size: 52.803rpx;
}

.user-menu .icon.coupon{
    background: url(http://cdn.xinzhenkj.com/ucenter-sdf6a55ee56-f2c2b9c2f0.png)
0 -62.4997rpx no-repeat;
    background-size: 52.803rpx;
}

.user-menu .icon.gift{
    background: url(http://cdn.xinzhenkj.com/ucenter-sdf6a55ee56-f2c2b9c2f0.png)
0 -187.5rpx no-repeat;
    background-size: 52.803rpx;
}

.user-menu .icon.address{
    background: url(http://cdn.xinzhenkj.com/ucenter-sdf6a55ee56-f2c2b9c2f0.png)
0 0 no-repeat;
    background-size: 52.803rpx;
}

.user-menu .icon.security{
    background: url(http://cdn.xinzhenkj.com/ucenter-sdf6a55ee56-f2c2b9c2f0.png)
0 -500rpx no-repeat;
    background-size: 52.803rpx;
}

.user-menu .icon.kefu{
    background: url(http://cdn.xinzhenkj.com/ucenter-sdf6a55ee56-f2c2b9c2f0.png)
0 -312.5rpx no-repeat;
    background-size: 52.803rpx;
}
```

```
.user - menu .icon.help{
    background: url(http://cdn.xinzhenkj.com/ucenter - sdf6a55ee56 - f2c2b9c2f0.png)
0 -250rpx no - repeat;
    background - size: 52.803rpx;
}

.user - menu .icon.feedback{
    background: url(http://cdn.xinzhenkj.com/ucenter - sdf6a55ee56 - f2c2b9c2f0.png)
0 -125rpx no - repeat;
    background - size: 52.803rpx;
}


.user - menu .txt{
    display: block;
    height: 24rpx;
    width: 100 % ;
    font - size: 24rpx;
    color: #333;
}

.logout{
    margin - top: 50rpx;
    height: 101rpx;
    width: 100 % ;
    line - height: 101rpx;
    text - align: center;
    background: #fff;
    color: #333;
    font - size: 30rpx;
}
```

8.12.3 js 逻辑代码

```
// pages/user/user.js
var app = getApp()
Page( {
  data: {
    userInfo: {},
    orderInfo:{},
    userListInfo: [ {
```

```
          icon: '../../images/iconfont-dingdan.png',
          text: '我的订单',
          isunread: true,
          unreadNum: 2
        }, {
          icon: '../../images/iconfont-shouhuodizhi.png',
          text: '收货地址管理'
        }, {
          icon: '../../images/iconfont-kefu.png',
          text: '联系客服'
        }, {
          icon: '../../images/iconfont-help.png',
          text: '常见问题'
        }],
        loadingText: '加载中...',
        loadingHidden: false,
    },
    onLoad: function () {
        var that = this
        //调用应用实例的方法获取全局数据
        app.getUserInfo(function(userInfo){
          //更新数据
          that.setData({
            userInfo:userInfo,
            loadingHidden: true
          })
        });

        console.log("个人中心:--" + app.d.userId);

        this.loadOrderStatus();
    },
    onShow:function(){
      this.loadOrderStatus();
    },
    loadOrderStatus:function(){
      //获取用户订单数据
      var that = this;
      console.log("个人中心2:--" + app.d.userId);
      wx.request({
        url: app.d.ceshiUrl + '/Api/User/getorder',
```

```
        method:'post',
        data: {
          userId:app.d.userId,
        },
        header: {
          'Content - Type':  'application/x - www - form - urlencoded'
        },
        success: function (res) {
          // -- init data
          var status = res.data.status;
          if(status == 1){
            var orderInfo = res.data.orderInfo;
            that.setData({
              orderInfo: orderInfo
            });
          }else{
            wx.showToast({
              title: '非法操作.',
              duration: 2000
            });
          }
        },
        error:function(e){
          wx.showToast({
            title: '网络异常! ',
            duration: 2000
          });
        }
      });
    },
    getPhoneNumber: function(e) {
      console.log(e.detail.errMsg)
      console.log(e.detail.iv)
      console.log(e.detail.encryptedData)
      if (e.detail.errMsg == 'getPhoneNumber:fail user deny'){
        wx.showModal({
          title: '提示',
          showCancel: false,
```

```
                content：'未授权',
                success：function (res) { }
            })
        } else {
            wx.showModal({
                title：'提示',
                showCancel：false,
                content：'同意授权',
                success：function (res) { }
            })
        }
    },
    onShareAppMessage：function () {
        return {
            title：'用户信息',
            path：'/pages/index/index',
            success：function (res) {
                // 分享成功
            },
            fail：function (res) {
                // 分享失败
            }
        }
    }
})
```

8.12.4 json 配置代码

```
{
    "navigationBarTitleText"："个人中心"
}
```

8.13 我的订单列表页代码清单

我的订单列表页包含待付款、待发货、待收货和已完成这几个状态，由于订单列表页布局基本类似，故我以“我的待付款订单页”作为样例，如图 8-14 所示。

程序代码如下所示。

图 8-14　我的待付款订单

8.13.1　wxml 模板代码

```
<!-- 待付款 -->
<view class="container">
<view class="orders">
<navigator class="order" wx:for="{{orderList0}}" wx:key="key">
<view class="h">
<view class="l">订单编号:{{item.order_sn}}</view>
<view class="r">{{item.order_status_text}}</view>
</view>
<view class="goods">
<view class="img"><image src="{{item.photo_x}}"></image></view>
<view class="info">
<text class="name">{{item.name}}</text>

<text class="number">共{{item.pro_count}}件商品 X 单价:¥ {{item.price_yh}} 数量:X{{item.product_num}}</text>
</view>
<view class="status"></view>
</view>
<view class="b">
<view class="l">实付:¥{{item.price}}</view>
<view class="r">
<navigator url="../order/detail? orderId={{item.id}}" class="btn">订单详情</navigator>
<button class="btn" bindtap="removeOrder" data-order-id="{{item.id}}">取消订单</button>
```

```
< button class = "btn" bindtap = "payOrderByWechat" data - orderId = "{{item.id}}" data - ordersn = "{{item.order_sn}}" wx:if = "{{item.type == 'weixin'}}" > 微信付款 < /button >
</view >
</view >
</navigator >
</view >
</view >
```

8.13.2 wxss 样式代码

```
page{
    height: 100 % ;
    width: 100 % ;
    background: #f4f4f4;
}

.orders{
    height: auto;
    width: 100 % ;
    overflow: hidden;
}

.order{
    margin - top: 20rpx;
    background: #fff;
}

.order .h{
    height: 83.3rpx;
    line - height: 83.3rpx;
    margin - left: 31.25rpx;
    padding - right: 31.25rpx;
    border - bottom: 1px solid #f4f4f4;
    font - size: 30rpx;
    color: #333;
}

.order .h .l{
    float: left;
}
```

```
.order .h .r{
    float: right;
    color: #b4282d;
    font-size: 24rpx;
}

.order .goods{
    display: flex;
    align-items: center;
    height: 199rpx;
    margin-left: 31.25rpx;
}

.order .goods .img{
    height:145.83rpx;
    width:145.83rpx;
    background: #f4f4f4;
}

.order .goods .img image{
    height:145.83rpx;
    width:145.83rpx;
}

.order .goods .info{
    height: 145.83rpx;
    flex: 1;
    padding-left: 20rpx;
}

.order .goods .name{
    margin-top: 0rpx;
    display: block;
    height: 44rpx;
    line-height: 44rpx;
    color: #333;
    font-size: 30rpx;
}

.order .goods .number{
    display: block;
    height: 37rpx;
```

```
    line-height: 37rpx;
    color: #666;
    font-size: 26rpx;
    margin-top:63rpx;

}

.order .goods .status{
    width:105rpx;
    color: #b4282d;
    font-size: 25rpx;
}

.order .b{
    height: 103rpx;
    line-height: 103rpx;
    margin-left: 31.25rpx;
    padding-right: 31.25rpx;
    border-top: 1px solid #f4f4f4;
    font-size: 30rpx;
    color: #333;
}

.order .b .l{
    float: left;
}

.order .b .r{
    float: right;
}

.order .b .btn{
    margin-top: 19rpx;
    height: 64.5rpx;
    line-height: 64.5rpx;
    text-align: center;
    padding: 0 20rpx;
    border-radius: 5rpx;
    font-size: 28rpx;
    color: #fff;
    background: #b4282d;
    float:left;
```

```
margin-right:10rpx;

}
```

8.13.3 js 逻辑代码

```
// pages/user/dingdan.js
//获取应用实例
var app = getApp();
var common = require("../../utils/common.js");
Page({
  data: {
    winWidth: 0,
    winHeight: 0,
    // tab 切换
    currentTab: 0,
    isStatus:'pay',//订单状态码:10,待付款;20,待发货;30,待收货;40,待评价;50,交易
完成
    page:0,
    refundpage:0,
    orderList0:[],
    orderList1:[],
    orderList2:[],
    orderList3:[],
    orderList4:[],
  },
  onLoad: function(options) {
    this.initSystemInfo();
    this.setData({
      currentTab: parseInt(options.currentTab),
      isStatus:options.otype
    });

    if(this.data.currentTab == 4){
      this.loadReturnOrderList();
    }else{
      this.loadOrderList();
    }
  },
  getOrderStatus:function(){
    return this.data.currentTab == 0 ? 1 : this.data.currentTab == 2 ? 2 :this.data.
```

```
currentTab == 3 ? 3:0;
    },

    //取消订单
    removeOrder:function(e){
        var that = this;
        var orderId = e.currentTarget.dataset.orderId;
        wx.showModal({
          title: '提示',
          content: '你确定要取消订单吗？',
          success: function(res) {
            res.confirm && wx.request({
              url: app.d.ceshiUrl + '/Api/Order/orders_edit',
              method:'post',
              data: {
                id: orderId,
                type:'cancel',
              },
              header: {
                'Content-Type': 'application/x-www-form-urlencoded'
              },
              success: function (res) {
                // -- init data
                var status = res.data.status;
                if(status == 1){
                  wx.showToast({
                    title: '操作成功！',
                    duration: 2000
                  });
                  that.loadOrderList();
                }else{
                  wx.showToast({
                    title: res.data.err,
                    duration: 2000
                  });
                }
              },
              fail: function () {
                // fail
                wx.showToast({
                  title: '网络异常！',
                  duration: 2000
```

```
                })；
              }
            })；

          }
        })；
      },

      //确认收货
    recOrder:function(e){
        var that = this;
        var orderId = e.currentTarget.dataset.orderId;
        wx.showModal({
          title：'提示'，
          content：'你确定已收到宝贝吗？'，
          success：function(res) {
            res.confirm && wx.request({
              url：app.d.ceshiUrl + '/Api/Order/orders_edit'，
              method:'post'，
              data：{
                id：orderId，
                type:'receive'，
              }，
              header：{
                'Content - Type'：  'application/x - www - form - urlencoded'
              }，
              success：function (res) {
                // -- init data
                var status = res.data.status;
                if(status == 1){
                  wx.showToast({
                    title：'操作成功！'，
                    duration：2000
                  })；
                  that.loadOrderList()；
                }else{
                  wx.showToast({
                    title：res.data.err，
                    duration：2000
                  })；
                }
              }，
```

```
            fail: function () {
              // fail
              wx.showToast({
                title: '网络异常！',
                duration: 2000
              });
            }
          });

        }
      });
    },

    loadOrderList: function(){
      var that = this;
      wx.request({
        url: app.d.ceshiUrl + '/Api/Order/index',
        method:'post',
        data: {
          uid:app.d.userId,
          order_type:that.data.isStatus,
          page:that.data.page,
        },
        header: {
          'Content-Type': 'application/x-www-form-urlencoded'
        },
        success: function (res) {
          //-- init data
          var status = res.data.status;
          var list = res.data.ord;
          switch(that.data.currentTab){
            case 0:
              that.setData({
                orderList0: list,
              });
              break;
            case 1:
              that.setData({
                orderList1: list,
              });
              break;
            case 2:
```

```
            that.setData({
              orderList2: list,
            });
            break;
          case 3:
            that.setData({
              orderList3: list,
            });
            break;
          case 4:
            that.setData({
              orderList4: list,
            });
            break;
        }
      },
      fail: function () {
        // fail
        wx.showToast({
          title: '网络异常！',
          duration: 2000
        });
      }
    });
  },

loadReturnOrderList:function(){
    var that = this;
    wx.request({
      url: app.d.ceshiUrl + '/Api/Order/order_refund',
      method:'post',
      data: {
        uid:app.d.userId,
        page:that.data.refundpage,
      },
      header: {
        'Content-Type':  'application/x-www-form-urlencoded'
      },
      success: function (res) {
        //--init data
        var data = res.data.ord;
        var status = res.data.status;
```

```
          if(status == 1){
            that.setData({
              orderList4: that.data.orderList4.concat(data),
            });
          }else{
            wx.showToast({
              title: res.data.err,
              duration: 2000
            });
          }
        },
        fail: function () {
          // fail
          wx.showToast({
            title:'网络异常!',
            duration: 2000
          });
        }
      });
    },

    // returnProduct:function(){
    // },
    initSystemInfo:function(){
      var that = this;

      wx.getSystemInfo( {
        success: function( res ) {
          that.setData( {
            winWidth: res.windowWidth,
            winHeight: res.windowHeight
          });
        }
      });
    },
    bindChange: function(e) {
      var that = this;
      that.setData( { currentTab: e.detail.current });
    },
    swichNav: function(e) {
      var that = this;
      if( that.data.currentTab === e.target.dataset.current ) {
```

```
      return false;
    } else {
      var current = e.target.dataset.current;
      that.setData({
        currentTab: parseInt(current),
        isStatus: e.target.dataset.otype,
      });

      //没有数据就进行加载
      switch(that.data.currentTab){
          case 0:
            ! that.data.orderList0.length && that.loadOrderList();
            break;
          case 1:
            ! that.data.orderList1.length && that.loadOrderList();
            break;
          case 2:
            ! that.data.orderList2.length && that.loadOrderList();
            break;
          case 3:
            ! that.data.orderList3.length && that.loadOrderList();
            break;
          case 4:
            that.data.orderList4.length = 0;
            that.loadReturnOrderList();
            break;
        }
    };
  },


  payOrderByWechat: function (e) {
    var order_id = e.currentTarget.dataset.orderId;
    var order_sn = e.currentTarget.dataset.ordersn;
    if(! order_sn){
      wx.showToast({
        title: "订单异常!",
        duration: 2000,
      });
      return false;
    }
    wx.request({
```

```
url: app.d.ceshiUrl + '/Api/Wxpay/wxpay',
data: {
  order_id: order_id,
  order_sn: order_sn,
  uid: app.d.userId,
},
method: 'POST', // OPTIONS, GET, HEAD, POST, PUT, DELETE, TRACE, CONNECT
header: {
  'Content - Type': 'application/x - www - form - urlencoded'
}, // 设置请求的 header
success: function (res) {
  if (res.data.status == 1) {
    var order = res.data.arr;
    wx.requestPayment({
      timeStamp: order.timeStamp,
      nonceStr: order.nonceStr,
      package: order.package,
      signType: 'MD5',
      paySign: order.paySign,
      success: function (res) {
        wx.showToast({
          title: "支付成功!",
          duration: 2000,
        });
        setTimeout(function () {
          wx.navigateTo({
            url: '../user/dingdan? currentTab = 1&otype = deliver',
          });
        }, 3000);
      },
    //fail: function (res) {
    //wx.showToast({
    //title: res,
    //duration: 3000
    //})
    //}
    })
  } else {
    wx.showToast({
      title: res.data.err,
      duration: 2000
    });
```

```
        }
      },
      fail：function (e) {
        // fail
        wx.showToast({
          title：'网络异常！',
          duration：2000
        });
      }
    })
  },

})
```

8.13.4 json 配置代码

```
{
    "navigationBarTitleText"："我的待付款订单"
}
```

8.14 搜索页代码清单

搜索页效果图如图 8－15 所示。

图 8－15 搜索

程序代码如下所示。

8.14.1 wxml 模板代码

```
<!-- pages/search/search.wxml -->
<view class="container">
<view class="search df">
    <input class="df_1" placeholder="请输入你有搜索的内容" auto-focus focus="{{focus}}" value="{{searchValue}}" bindinput="searchValueInput"/>
    <button bindtap="doSearch"> <image class="searchcion" src="/images/search.png"></image></button>
</view>
<view class="cont" wx:if="{{hotKeyShow}}">
    <text class="font_14">热门搜索</text>
    <view class="w100">
        <button wx:for="{{hotKeyList}}" bindtap="doKeySearch" data-key="{{item.keyword}}">{{item.keyword}}</button>
    </view>
    <text class="font_14 mt10">历史搜索</text>
    <view class="w100">
        <button wx:for="{{historyKeyList}}" bindtap="doKeySearch" data-key="{{item.keyword}}">{{item.keyword}}</button>
    </view>
</view>
<view class="search_no" wx:if="{{!!searchValue && !productData.length}}">
    <view class="font_14"><image class="scimg" src="/images/search_no.png"></image></view>
    <text>没有找到您要的宝贝/(ToT)/~~</text>
</view>

<view class="shop" wx:for="{{productData}}">
    <navigator url="../product/detail?productId={{item.id}}" hover-class="changestyle">
        <image class="sh_slt" src="{{item.photo_x}}"></image>
        <view class="sp_text">
            <view class="sp_tit ovh1">{{item.name}}</view>
            <view class="sp_jg">¥ {{item.price_yh}}</view>
        </view>
    </navigator>
</view></view>
```

8.14.2 wxss 样式代码

```
/* pages/search/search.wxss */
.search{
    padding: 2% 2%;
    background: #b4a078;
}
.search input{
    width: 75%;
    border-radius: 13px;
    background: #fff;
    border: none;
    font-size: 12px;
    padding:1% 2.5%;
    margin-right: 5px;
    }
.search button{
    line-height:0px;
    background: none;
    text-align: center;
    border: none;
    padding: 0px;
}
.search button::after{
    content: none;
}
.searchcion{
    width: 24px;
    height: 24px;
    text-align: center
}
.cont{
     width: 94%;
     padding: 3%;
}
.w100{
    width: 100%;
    padding-bottom: 10px;
}
.w100 button{
    text-align: center;
```

```
    line-height: 20px;
    margin: 3% 2% 0 0;
    display: inline-table;
    padding:5px 10px;
    font-size: 12px;
}
.shop{
    width: 49%;
    background: #fff;
    float: left;
    margin-bottom: 4%;
}
.shop:nth-child(even){
    margin-right: 2%;
}
.sh_slt{
    width: 100%;
    height: 370rpx;
    overflow: hidden;
}
.sp_text{
    float: left;
    line-height: 25px;
    width: 92%;
    padding: 4%;
    }
.sp_tit{
    width: 100%;
    overflow: hidden;
    font-size: 14px;
}
.sp_jg{
    width: 100%;
    overflow: hidden;
    font-size: 14px;
    color: #fc0628;
    line-height: 30px;
}
.search_no{
    width: 100%;
    display: block;
    color: #666;
```

```
    text - align: center;
    font - size: 14px;
}
.scimg{
    width: 100px;
    height: 100px;
    display: inline - block;
    background - size:100px;
}
.font_14{  font - size: 13px;}
```

8.14.3 js 逻辑代码

```
var app = getApp();
// pages/search/search.js
Page({
  data:{
    focus:true,
    hotKeyShow:true,
    historyKeyShow:true,
    searchValue:'',
    page:0,
    productData:[],
    historyKeyList:[],
    hotKeyList:[]
  },
  onLoad:function(options){
    var that = this;
    wx.request({
      url: app.d.ceshiUrl + '/Api/Search/index',
      method:'post',
      data: {uid:app.d.userId},
      header: {
        'Content - Type':  'application/x - www - form - urlencoded'
      },
      success: function (res) {
        var remen = res.data.remen;
        var history = res.data.history;

        that.setData({
          historyKeyList:history,
```

```
            hotKeyList:remen,
          });
        },
        fail:function(e){
          wx.showToast({
            title:'网络异常!',
            duration:2000
          });
        },
      })
    },
    onReachBottom:function(){
        //下拉加载更多
        this.setData({
          page:(this.data.page + 10)
        })

        this.searchProductData();
    },
    doKeySearch:function(e){
      var key = e.currentTarget.dataset.key;
      this.setData({
        searchValue:key,
         hotKeyShow:false,
         historyKeyShow:false,
      });

      this.data.productData.length = 0;
      this.searchProductData();
    },
    doSearch:function(){
      var searchKey = this.data.searchValue;
      if (!searchKey) {
          this.setData({
              focus:true,
              hotKeyShow:true,
              historyKeyShow:true,
          });
          return;
      };

      this.setData({
```

```
        hotKeyShow:false,
        historyKeyShow:false,
      })

      this.data.productData.length = 0;
      this.searchProductData();

      this.getOrSetSearchHistory(searchKey);
    },
    getOrSetSearchHistory:function(key){
      var that = this;
      wx.getStorage({
        key: 'historyKeyList',
        success: function(res) {
            console.log(res.data);

            //console.log(res.data.indexOf(key))
            if(res.data.indexOf(key) >= 0){
              return;
            }

            res.data.push(key);
            wx.setStorage({
              key:"historyKeyList",
              data:res.data,
            });

            that.setData({
              historyKeyList:res.data
            });
        }
      });
    },
    searchValueInput:function(e){
      var value = e.detail.value;
      this.setData({
        searchValue:value,
      });
      if(!value && this.data.productData.length == 0){
        this.setData({
          hotKeyShow:true,
          historyKeyShow:true,
```

```
        });
      }
    },
    searchProductData:function(){
      var that = this;
      wx.request({
        url: app.d.ceshiUrl + '/Api/Search/searches',
        method:'post',
        data: {
          keyword:that.data.searchValue,
          uid: app.d.userId,
          page:that.data.page,
        },
        header: {
          'Content-Type':  'application/x-www-form-urlencoded'
        },
        success: function (res) {
          var data = res.data.pro;
          that.setData({
            productData:that.data.productData.concat(data),
          });
        },
        fail:function(e){
          wx.showToast({
            title:'网络异常!',
            duration: 2000
          });
        },
      });
    },

});
```

8.14.4 json 配置代码

```
{
    "navigationBarTitleText":"搜索"
}
```

限于篇幅有限,文中仅罗列了小程序商城基础核心模块和常用前端页面的代码示例,其他未涉及到的页面代码内容请下载本书附带的电子资料中的项目代码压缩包并具体查看。

第9章

小程序商城后端程序开发

本章主要讲解小程序商城项目的后端接口开发内容，通过本章的学习，希望读者可以更好地理解小程序应用前后端是如何通过接口进行数据交互的，并理解接口开发的思路。

9.1 小程序商城后端接口开发概述

小程序前端页面开发完成之后，接下来需要实现的是与之对应的后端接口开发任务。后端接口的功能主要实现系统核心业务数据处理和前后端数据交互。整个后端系统采用基于 ThinkPHP 的 PHP 框架系统来搭建，该框架成熟稳定且国内开发者众多，具体的使用不再过多赘述，如果读者是其他开发语言或者其他框架的使用者，那么请使用自己最熟悉的开发方式即可，本章的业务逻辑代码可供开发思路参考之用。

限于篇幅有限，此处仅摘录了小程序商城核心功能页面所对应的接口程序代码，其已能涵盖商城的主要核心业务逻辑处理的需要，供学习之用。全部完整的接口代码请下载本书附带的电子资料中的项目代码压缩包并具体查看。还有需要提醒的是，小程序商城的后台管理系统也在该代码包中，请注意查看 backend/App/Admin 目录中的代码，供读者参考之用，本书不再赘述。

9.2 小程序后端接口开发

后端接口的功能很单一，主要负责前后端数据交互和业务逻辑数据处理，无论是采用何种框架，其流程不外乎三步：①接收请求参数；②处理请求并进行业务逻辑数据处理；③返回结果数据给前端。本项目中的接口核心代码主要放在 API 控制器层进行处理，代码示例如下所示。

9.2.1 接口公共父类代码清单

所有接口的公共父类控制器程序文件为 PublicController. class. php，代码如下：

```
<?php
```

```
namespace Api\Controller;
use Think\Controller;

/**
  * 公共父类接口
  *
  * 公共父类 API
  * Class PublicController
  */
class PublicController extends Controller {


    /**
  * 构造函数
  */
    public function _initialize(){
        //php 判断 http 还是 https
        $http_type = ((isset($_SERVER['HTTPS']) && $_SERVER['HTTPS'] == 'on') || (isset($_SERVER['HTTP_X_FORWARDED_PROTO']) && $_SERVER['HTTP_X_FORWARDED_PROTO'] == 'https')) ? 'https://' : 'http://';
  //所有图片路径
    define(__DATAURL__, $http_type.$_SERVER['SERVER_NAME'].__DATA__.'/');
    define(__PUBLICURL__, $http_type.$_SERVER['SERVER_NAME'].__PUBLIC__.'/');
    define(__HTTP__, $http_type);
  }

    /**
     * 查找二级分类下的所有子分类 id,用逗号拼接
     * @param int $id
     * @return string
     */
    public function catid_tree($id=2){
        $Category = M('category');
        $list=$Category->where("tid=".$id)->order('sort desc,id asc')->select();
        //dump($list);exit;
        $cidstr='';
        foreach($list as $v){
            $json[]=$v['id'];
            $num=$Category->where("tid=".$v['id'])->field('id')->count();
```

```
            if( $ num > 0){
                $ json[] = $ this -> catid_tree( $ v['id']);
            }
        }
        $ cidstr. = implode(',', $ json);
        return $ cidstr;
    }


    /**
     * 一次性查出产品分类的所有分类
     * @param int $ id
     * @return array
     */
public function cat_tree( $ id = 2){
    $ Category = M('category');
    $ list = $ Category -> where( "tid = ". $ id) -> field('id,tid,name') -> order
('sort desc,id asc') -> select();
    //echo ' < pre > ';print_r( $ list);exit;
    foreach( $ list as $ v){
        $ num = $ Category -> where("tid = ". $ v['id']) -> count();
        $ subclass = array();
        if( $ num > 0)
        {
            $ subclass = $ this -> cat_tree( $ v['id']);
        }
        $ json[] = array(
            'id' = > $ v['id'] ,
            'name' = > $ v['name'] ,
            'num' = > $ num ,
            'subclass' = > $ subclass,
        );
    }
    return $ json;
}


    /**
     * 导航部分　查找父级分类
     * @param $ categoryID
     * @return array
     */
```

```
function getAllFcateIds( $categoryID)
{
    //初始化 ID 数组
    $array[] = $categoryID;

    do
    {
        $ids = '';
        $where['id'] = array('in', $categoryID);
        $cate = M('category') -> where( $where) -> field('id,tid,name') -> se-
lect();
        // echo M('aaa_cpy_category') -> _sql();
        foreach ( $cate as $v)
        {
            $array[] = $v['tid'];
            $ids .= ',' . $v['tid'];
        }
        $ids = substr( $ids, 1, strlen( $ids));
        $categoryID = $ids;
    }
    while (! empty( $cate));
    // $cates = array();
    foreach ( $array as $key => $va){
        $cates[] = M('category') -> where('id ='. $va) -> field('id,tid,name') -
> find();
        // echo M('aaa_cpy_category') -> _sql();
        //echo $cates[ $key]['name'];
    $cates[ $key]['name'] = str_replace('(系统分类,不要删除)','', $cates[ $key]
['name']);
    }
    array_pop( $cates);
    $ca = array_reverse( $cates);
    //echo "<pre>";
    // print_r( $ca);
    return $ca; //返回数组
}

/**
 * 是否开通 PC
 * @param $val
 * @return int|mixed
 */
```

```
public function ispc( $ val){
//$ val = 1850;//这个为 admin_app 的 id
 $ app = M('admin_app');
 $ val = $ app-> getField('id');
//$ url = $ app-> db(2,DB) -> where('id = '. $ val) -> field('ispcshop,end_time,
name,pcnav_color,ahover_color') -> find();
 $ url = $ app-> where('id = '. $ val) -> field('ispcshop,end_time,name,pcnav_color,
ahover_color') -> find();
//print_r( $ url);exit;
//return $ url;

if( $ url['end_time'] > time()){
    return $ url;
}else{
    return 0;
}
    }


}
```

9.2.2 登录接口代码清单

登录接口控制器文件为 LoginController. class. php，主要处理微信授权登录登出等功能，注册和账号登录代码为冗余扩展设计，代码如下：

```
<? php
namespace Api\Controller;
use Think\Controller;

/**
 * 登录接口控制器
 *
 * 处理前台登入和登出 API
 * Class LoginController
 */
class LoginController extends PublicController {


    /**
     * 前台登录接口
     */
```

```
    public function dologin(){
        session_destroy();
        $name = trim($_POST['username']);//接受“会员账号”
        $pwd  = md5(md5($_POST['pwd']));//接受“会员密码”
        if (!$name || !$pwd) {
            echo json_encode(array('status' => 0,'err' => '请输入账号或密码！'));
        exit();
    }

    $user = M('user');
    $where['name'] = $name;
    $where['pwd'] = $pwd;
    $usrNum = $user->where($where)->find();
    ///echo $user->_sql();exit;
    if($usrNum){
        @session_start();
        $_COOKIE['sessionid'] = session_id();
        //$_SESSION['sessionid'] = session_id();
        $_SESSION['LoginCheck'] = md5($name);
        $_SESSION['LoginName'] = $name;
        $_SESSION['ID'] = $usrNum['id'];
        $_SESSION['photo'] = $usrNum['photo'];

        echo json_encode(array('status' => 1,'session' => $_SESSION));
        exit();
    }else{
        echo json_encode(array('status' => 0,'err' => '账号密码错误！'));
        exit();
    }
}


    /**
     * 授权登录接口
     */
public function authlogin(){
    $openid = $_POST['openid'];
    if (!$openid) {
        echo json_encode(array('status' => 0,'err' => '授权失败！'.__LINE__));
        exit();
    }
    $con = array();
```

```
    $con['openid'] = trim($openid);
    $uid = M('user')->where($con)->getField('id');
    if ($uid) {
        $userinfo = M('user')->where('id='.intval($uid))->find();
        if (intval($userinfo['del']) == 1) {
            echo json_encode(array('status'=>0,'err'=>'账号状态异常!'));
            exit();
        }
        $err = array();
        $err['ID'] = intval($uid);
        $err['NickName'] = $_POST['NickName'];
        $err['HeadUrl'] = $_POST['HeadUrl'];
        echo json_encode(array('status'=>1,'arr'=>$err));
        exit();
    }else{
    $data = array();
    $data['name'] = $_POST['NickName'];
        $data['uname'] = $_POST['NickName'];
        $data['photo'] = $_POST['HeadUrl'];
        $data['sex'] = $_POST['gender'];
        $data['openid'] = $openid;
        $data['source'] = 'wx';
        $data['addtime'] = time();
        if (!$data['openid']) {
            echo json_encode(array('status'=>0,'err'=>'授权失败!'.__LINE__));
            exit();
        }
    $res = M('user')->add($data);
        if ($res) {
            $err = array();
            $err['ID'] = intval($res);
            $err['NickName'] = $data['name'];
            $err['HeadUrl'] = $data['photo'];
            echo json_encode(array('status'=>1,'arr'=>$err));
            exit();
        }else{
            echo json_encode(array('status'=>0,'err'=>'授权失败!'.__LINE__));
            exit();
        }
    }
}
```

```
    /**
     * 前台注册接口
     */
public function register(){
    $name = trim($_POST['user']);
    $pwd  = md5(md5($_POST['pwd']));
    $pwds = md5(md5($_POST['pwds']));
    if($pwd! = $pwds) {
        echo json_encode(array('status'=> 0,'err'=> '两次输入密码不同！'));
        exit();
    }

    $user = M('user');
    $where = array();
    $where['name'] = $name;
    $count = $user-> where($where)-> count();
    if($count) {
        echo json_encode(array('status'=> 0,'err'=> '用户名已被注册了！'));
        exit();
    }

    $check_mob = $user-> where('tel = '.trim($_POST['tel']))-> count();
    if($check_mob) {
        echo json_encode(array('status'=> 0,'err'=> '手机号已存在！'));
        exit();
        }
        $data = array();
        $data['name'] = $name;
        $data['qx'] = 6;

        $data['pwd']    = $pwd;
        $data['tel']    = trim($_POST['tel']);
        $data['addtime'] = time();
        $res = $user-> add($data);
        if ($res) {
            $_SESSION['LoginName'] = $name;
            $_SESSION['ID'] = $res;
            $arr = array();
            $arr['status'] = 1;
            $arr['uid'] = $res;
```

```
        $arr['LoginName'] = $name;
        echo json_encode($arr);
        exit();
    }else{
        echo json_encode(array('status'=> 0,'err'=> '注册失败！'));
        exit();
    }
}


/**
 * 获取 sessionkey 接口
 */
public function getsessionkey(){
    $wx_config = C('weixin');
 $appid = $wx_config['appid'];
 $secret = $wx_config['secret'];

    $code = trim($_POST['code']);

    //记录日志:要写入文件的文件名(可以是任意文件名),如果文件不存在,将会创建
      一个.log.txt 位置在项目的根目录下
    $file  = 'log.txt';
    $content = "内容:获取 sessionkey\n";
    $content .= "appid 内容".$appid."\n";
    $content .= "secret 内容".$secret."\n";
    $content .= "code 内容".$code."\n";
    // 这个函数支持版本(PHP 5)
    if($f  = file_put_contents($file, $content,FILE_APPEND)){
        //echo "成功。<br />";
    }


    if (!$code) {
        echo json_encode(array('status'=> 0,'err'=> '非法操作！'));
        exit();
    }

    if (!$appid || !$secret) {
        echo json_encode(array('status'=> 0,'err'=> '非法操作！'.__LINE__));
        exit();
    }
```

```
        $ get_token_url = 'https://api.weixin.qq.com/sns/jscode2session? appid = '.
$ appid.'&secret = '. $ secret.'&js_code = '. $ code.'&grant_type = authorization_code';
        $ ch = curl_init();
        curl_setopt( $ ch,CURLOPT_URL, $ get_token_url);
        curl_setopt( $ ch,CURLOPT_HEADER,0);
        curl_setopt( $ ch, CURLOPT_RETURNTRANSFER, 1 );
        curl_setopt( $ ch, CURLOPT_CONNECTTIMEOUT, 10);
        $ res = curl_exec( $ ch);
        curl_close( $ ch);

        //记录日志:要写入文件的文件名(可以是任意文件名),如果文件不存在,将会创建
          一个.log.txt 位置在项目的根目录下
        $ file  = 'log.txt';
        $ content = "内容:获取 sessionkey 的 URL 和结果\n";
        $ content . = "get_token_url 内容". $ get_token_url."\n";
        $ content . = "res 内容".json_encode( $ res)."\n";
        // 这个函数支持版本(PHP 5)
        if( $ f  = file_put_contents( $ file, $ content,FILE_APPEND)){
            //echo "成功。 < br / > ";
        }


        echo $ res;
        exit();
    }


    / **
     * 获取 sessionkey 接口
     * 这里的 appid 和 secret 是固化的
     * /
    public function getsessionkeys(){
        $ wx_config = C('weixin');
     $ appid = 'wx5e3dd81af8bf352a';
     $ secret = '30421d00aaee201bc0400994defbeed7';

        $ code = trim( $ _POST['code']);
        if (! $ code) {
            echo json_encode(array('status' = > 0,'err' = > '非法操作! '));
            exit();
        }
```

```
        $ get_token_url = 'https://api.weixin.qq.com/sns/jscode2session? appid = '.
$ appid.'&secret = '. $ secret.'&js_code = '. $ code.'&grant_type = authorization_code';
        $ ch = curl_init();
        curl_setopt( $ ch,CURLOPT_URL, $ get_token_url);
        curl_setopt( $ ch,CURLOPT_HEADER,0);
        curl_setopt( $ ch, CURLOPT_RETURNTRANSFER, 1 );
        curl_setopt( $ ch, CURLOPT_CONNECTTIMEOUT, 10);
        $ res = curl_exec( $ ch);
        curl_close( $ ch);
        echo $ res;
        exit();
    }

    /**
     * 前台退出登录接口
     */
    public function logout(){
        unset( $ _SESSION['uid']);
        unset( $ _SESSION['LoginName']);
        session_destroy();
        echo json_encode(array('status' = > 1));
        exit();
    }

}
```

9.2.3 首页接口代码清单

首页接口控制器文件为 IndexController.class.php，主要处理首页信息的展现，代码如下：

```
<? php
namespace Api\Controller;
use Think\Controller;

/**
 * 首页接口
 *
 * 处理首页信息 API
 * Class IndexController
```

```
 */
class IndexController extends PublicController {

    /**
     * 首页信息接口
     */
    public function index(){

        //顶部轮播图广告
    $ggtop = M('guanggao') -> order('sort desc,id asc') -> field('id,name,photo,action') -
> limit(10) -> select();
        foreach ($ggtop as $k => $v) {
            $ggtop[$k]['photo'] = __DATAURL__.$v['photo'];
            $ggtop[$k]['name'] = urlencode($v['name']);
            $ggtop[$k]['action'] = $v['action'];
        }

        //产品分类
            $category = M('category') -> where('tid = 1') -> field('id,name,bz_1') ->
limit(8) -> select();
        foreach ($category as $k => $v) {
            $category[$k]['bz_1'] = __DATAURL__.$v['bz_1'];
        }


        $product = M('product');
        $goodsCount = $product -> count();
        $this -> assign('goodsCount', $goodsCount);
        //商品列表,目前冗余
        $pro_list = M('product') -> where('del = 0 AND pro_type = 1 AND is_down = 0 AND
                        type = 1') -> order('sort desc,id desc') -> field('id,name,in-
                        tro,photo_x,price_yh,price,shiyong') -> limit(8) -> select();
    foreach ($pro_list as $k => $v) {
            $pro_list[$k]['photo_x'] = __DATAURL__.$v['photo_x'];
    }

        //新品首发
        $newGoods = M('product') -> where('del = 0 AND pro_type = 1 AND is_down = 0 AND
                        type = 1') -> order('sort desc,id desc') -> field('id,name,in-
                        tro,photo_x,price_yh,price,shiyong') -> limit(8) -> select();
        foreach ($newGoods as $k => $v) {
        $newGoods[$k]['photo_x'] = __DATAURL__.$v['photo_x'];
        }
```

```
        echo json_encode(array('ggtop' => $ggtop,'prolist' => $pro_list,'newGoods' =>
$newGoods,'category' => $category,'goodsCount' => $goodsCount));
        exit();
        }

        /**
         * 产品列表
         */
        public function getlist(){
            $page = intval($_REQUEST['page']);
            $limit = intval($page*8) - 8;

            $pro_list = M('product') -> where('del = 0 AND pro_type = 1 AND is_down = 0 AND
                        type = 1') -> order('sort desc,id desc') -> field('id,name,photo
                        _x,price_yh,shiyong') -> limit($limit.',8') -> select();
            foreach ($pro_list as $k => $v) {
                $pro_list[$k]['photo_x'] = __DATAURL__.$v['photo_x'];
            }

            echo json_encode(array('prolist' => $pro_list));
            exit();
        }


    }
```

9.2.4 商品分类接口代码清单

商品分类接口控制器文件为 CategoryController.class.php，主要处理分类信息展现和获取，代码如下：

```
<?php
namespace Api\Controller;
use Think\Controller;

/**
 * 产品分类接口控制器
 *
 * 处理产品分类 API
 * Class CategoryController
 */
class CategoryController extends PublicController {
```

```
/**
 * 产品分类
 */
public function index(){
    $list = M('category')->where('tid=1')->field('id,tid,name,concent,bz_1,bz_2')
->select();
        foreach ($list as $k => $v) {
            $list[$k]['bz_2'] = __DATAURL__.$v['bz_2'];
        }
        $catList = M('category')->where('tid='.intval($list[0]['id']))->field
('id,name,bz_1')->select();
        foreach ($catList as $k => $v) {
            $catList[$k]['bz_1'] = __DATAURL__.$v['bz_1'];
        }

    $product = M('product');
        $goodsCount = $product->count();
        $this->assign('goodsCount', $goodsCount);
        echo json_encode(array('status' => 1,'list' => $list,'catList' =>
$catList,'goodsCount' => $goodsCount));
        exit();
}


/**
 * 获取产品分类信息
 */
public function getcat(){
        $catid = intval($_REQUEST['cat_id']);
        if (!$catid) {
            echo json_encode(array('status' => 0,'err' => '没有找到产品数据.'));
            exit();
        }

        $catList = M('category')->where('tid='.intval($catid))->field('id,
name,bz_1')->select();
        foreach ($catList as $k => $v) {
            $catList[$k]['bz_1'] = __DATAURL__.$v['bz_1'];
        }
        echo json_encode(array('status' => 1,'catList' => $catList));
        exit();
}
```

```
}
```

9.2.5 商品接口代码清单

商品接口控制器文件为 ProductController. class. php，主要用于商品详情、列表信息的获取等，代码如下：

```
<? php
namespace Api\Controller;
use Think\Controller;

/**
 * 产品(商品)接口
 *
 * 处理产品(商品)信息 API
 * Class ProductController
 */
class ProductController extends PublicController {

    /**
     * 获取商品详情信息接口
     */
    public function index(){
        $product = M("product");

        $pro_id = intval( $_REQUEST['pro_id']);
        if (! $pro_id) {
            echo json_encode(array('status' => 0,'err' => '商品不存在或已下架!
'));

            exit();
        }

    $pro = $product -> where('id = '. intval( $pro_id). ' AND del = 0 AND is_down = 0') ->
        find();
        if(! $pro){
            echo json_encode(array('status' => 0,'err' => '商品不存在或已下架! '._
            _LINE__));
            exit();
        }

        $pro['photo_x'] = __DATAURL__. $pro['photo_x'];
        $pro['photo_d'] = __DATAURL__. $pro['photo_d'];
        $pro['brand'] = M('brand') -> where('id = '. intval( $pro['brand_id'])) ->
```

```
getField('name');
            $pro['cat_name'] = M('category') -> where('id = '.intval($pro['cid']))
-> getField('name');

        //图片轮播数组
        $img = explode(',',trim($pro['photo_string'],','));
        $b = array();
        if ($pro['photo_string']) {
            foreach ($img as $k => $v) {
                $b[] = __DATAURL__.$v;
            }
        }else{
            $b[] = $pro['photo_d'];
        }
        $pro['img_arr'] = $b;//图片轮播数组

        //处理产品属性
        $catlist = array();
        if($pro['pro_buff']){//如果产品属性有值才进行数据组装
            $pro_buff = explode(',',$pro['pro_buff']);
            $commodityAttr = array();//产品库还剩下的产品规格
            $attrValueList = array();//产品所有的产品规格
            foreach($pro_buff as $key => $val){
                $attr_name = M('attribute') -> where('id = '.intval($val)) -> get-
                        Field('attr_name');
                $guigelist = M('guige') -> where("attr_id = ".intval($val).' AND
                        pid = '.intval($pro['id'])) -> field("id,name") ->
                        select();
                $ggss = array();
                $gg = array();
                foreach ($guigelist as $k => $v) {
                    $gg[$k]['attrKey'] = $attr_name;
                    $gg[$k]['attrValue'] = $v['name'];
                    $ggss[] = $v['name'];
                }
                $commodityAttr[$key]['attrValueList'] = $gg;
                $attrValueList[$key]['attrKey'] = $attr_name;
                $attrValueList[$key]['attrValueList'] = $ggss;
            }
        }

        $content = str_replace('/minipetmrschool/Data/',__DATAURL__, $pro
['content']);
```

```
        $pro['content'] = html_entity_decode($content, ENT_QUOTES , 'utf-8');

        echo json_encode(array('status' => 1,'pro' => $pro,'commodityAttr' =>
$commodityAttr,'attrValueList' => $attrValueList));
        exit();

    }


    /**
     * 获取商品详情接口
     */
    public function details(){
        header('Content-type:text/html; Charset=utf8');
        $pro_id = intval($_REQUEST['pro_id']);
        $pro = M('product')->where('id='.intval($pro_id).' AND del=0 AND is_
down=0')->find();
        if(!$pro){
            echo json_encode(array('status' => 0,'err' => '商品不存在或已下架!
'));
            exit();
        }
        //$content = preg_replace("/width:.+?[\d]+px;/","",$pro['content']);
        $content = htmlspecialchars_decode($pro['content']);
        echo json_encode(array('status'=> 1,'content' => $content));
        exit();
    }


    /**
     * 下单信息预处理
     * 处理产品信息,用户地址
     * uid:uid,pid:pro_id,aid:addr_id,sid:shop_id,buff:buff,num:num,price_yh:
price_yh,p_price:p_price,price:z_price,type:pay_type,yunfei:yun_id,cart_id:cart_id,re-
mark:ly
     */
    public function make_order(){
        header('Content-type:text/html; Charset=utf8');
        //产品
        $pro_id = I('request.pro_id');
        //$uid=I('request.uid');
        $uid=1;
        //获得产品信息
```

```
            $pro = M('product')->field('id,shop_id,photo_x,name,price,price_yh')->
                    where('id='.intval($pro_id).' AND del=0 AND is_down=0')->find();
            $pro['photo_x'] = __DATAURL__.$pro['photo_x'];
            if(!$pro){
                echo json_encode(array('status'=> 0,'err'=> '商品不存在或已下架！'));
                exit();
            }
            //获取地址
            $address = "";
            $addr = M("address")->where("uid= $uid")->select();
            if($addr){
                foreach($addr as $k=> $v){
                    if($v['is_default']==1){
                        $address = $address[$k];
                    }
                }
                if(!$address){
                    $address = $addr[0];
                }
            }

            echo json_encode(array('status'=> 1,'pro'=> $pro,'address'=> $address));
            exit();
        }


        /**
         * 获取商品详情接口,并组装产品属性
         */
        public function get_buff(){
            $pro = M('product')->where('id='.intval($_POST['pro_id']).' AND del=0 AND
is_down=0')->find();
            if(!$pro){
                echo json_encode(array('status'=> 0,'err'=> '商品不存在或已下架！'.__
LINE__));
                exit();
            }
            //处理产品属性
            $catlist = array();
            if($pro['pro_buff']){//如果产品属性有值才进行数据组装
```

```
    $pro_buff = explode(',', $pro['pro_buff']);
    $buff = array();
    foreach($pro_buff as $key => $val){
        $attr_name = M('attribute')->where('id = '.intval($val))->get-
                    Field('attr_name');
        $guigelist = M('guige')->where("attr_id = ".intval($val).' AND pid
                   = '.intval($pro['id']))->field("id,name")->select
                  ();
        $gg = array(); $ggss = array();
        foreach ($guigelist as $k => $v) {
            $gg['attrKey'] = $attr_name;
            $gg['attr_id'] = $val;
            $gg['attrValue'] = $v['name'];
            $gg['selectedValue'] = $v['id'];
            $ggss[] = $gg;
        }
        $buff['attrValueList'] = $ggss;
        $catlist[] = $buff;
    }
    echo json_encode(array('status'=> 1,'buff'=> $catlist));
    exit();
  }else{
    echo json_encode(array('status'=> 0));
    exit();
  }
}


/**
 * 获取商品列表接口
 */
 public function lists(){
   $json = "";
   $id = intval($_POST['cat_id']);//获得分类 id 这里的 id 是 pro 表里的 cid
   $brand_id = intval($_POST['brand_id']);
   // $id = 44;
   $type = I('post.type');//排序类型

   $page = intval($_POST['page']) ? intval($_POST['page']) : 0;
   $keyword = I('post.keyword');
   //排序
```

```
 $ order = "addtime desc";//默认按添加时间排序
 if( $ type == 'ids'){
      $ order = "id desc";
 }elseif( $ type == 'sale'){
      $ order = "shiyong desc";
 }elseif( $ type == 'price'){
      $ order = "price_yh desc";
 }elseif( $ type == 'hot'){
      $ order = "renqi desc";
 }
 //条件
 $ where = "1 = 1 AND pro_type = 1 AND del = 0 AND is_down = 0";
 if(intval( $ id)){
      $ where. = " AND cid = ". intval( $ id);
 }
 if (intval( $ brand_id)) {
      $ where. = " AND brand_id = ". intval( $ brand_id);
 }
 if( $ keyword) {
     $ where. = ' AND name LIKE " %'. $ keyword.' % "';
}
if (isset( $ _REQUEST['ptype']) && $ _REQUEST['ptype'] == 'new') {
     $ where . = ' AND is_show = 1';
}
if (isset( $ _REQUEST['ptype']) && $ _REQUEST['ptype'] == 'hot') {
     $ where . = ' AND is_hot = 1';
}
if (isset( $ _REQUEST['ptype']) && $ _REQUEST['ptype'] == 'zk') {
     $ where . = ' AND is_sale = 1';
}


 $ product = M('product') ->where( $ where) ->order( $ order) ->limit( $ page.',
           20') ->select();
 //echo M('product') ->_sql();exit;
 $ json  =  array(); $ json_arr  =  array();
 foreach ( $ product as  $ k  =>  $ v) {
      $ json['id'] = $ v['id'];
      $ json['name'] = $ v['name'];
      $ json['photo_x'] = __DATAURL__. $ v['photo_x'];
      $ json['price'] = $ v['price'];
      $ json['price_yh'] = $ v['price_yh'];
      $ json['shiyong'] = $ v['shiyong'];
      $ json['intro'] = $ v['intro'];
```

```
            $ json_arr[] = $ json;
        }
        $ cat_name = M('category') ->where("id = ". intval( $ id)) ->getField('name');
        $ cat_pic = M('category') ->where("id = ". intval( $ id)) ->getField('bz_2');
        echo json_encode(array('status' => 1,'pro' => $ json_arr,'cat_name' => $ cat_
name,'cat_pic' => $ cat_pic));
        exit();
    }


    /**
     * 商品列表页面 获取更多接口
     */
    public function get_more(){
        $ json = "";
        $ id = intval( $ _POST['cat_id']);//获得分类 id,这里的 id 是 pro 表里的 cid
        // $ id = 44;
        $ type = I('post.type');//排序类型

        $ page = intval( $ _POST['page']);
        if (! $ page) {
            $ page = 1;
        }
        $ limit = intval( $ page * 8) - 8;

        $ keyword = I('post.keyword');
        //排序
        $ order = "addtime desc";//默认按添加时间排序
        if( $ type == 'ids'){
            $ order = "id desc";
        }elseif( $ type == 'sale'){
            $ order = "shiyong desc";
        }elseif( $ type == 'price'){
            $ order = "price_yh desc";
        }elseif( $ type == 'hot'){
            $ order = "renqi desc";
        }
        //条件
        $ where = "1 = 1 AND pro_type = 1 AND del = 0 AND is_down = 0";
        if(intval( $ id)){
            $ where. = " AND cid = ". intval( $ id);
        }
```

```
    if( $ keyword) {
        $ where. = ' AND name LIKE " %'. $ keyword.' % "';
    }
    if (isset( $ _REQUEST['ptype']) && $ _REQUEST['ptype'] == 'new') {
        $ where . = ' AND is_show = 1';
    }
    if (isset( $ _REQUEST['ptype']) && $ _REQUEST['ptype'] == 'hot') {
        $ where . = ' AND is_hot = 1';
    }
    if (isset( $ _REQUEST['ptype']) && $ _REQUEST['ptype'] == 'zk') {
        $ where . = ' AND is_sale = 1';
    }


    $ product = M('product') ->where( $ where) ->order( $ order) ->limit( $ limit.
            ',8') ->select();
    //echo M('product') ->_sql();exit;
    $ json = array(); $ json_arr = array();
    foreach ( $ product as $ k => $ v) {
        $ json['id'] = $ v['id'];
        $ json['name'] = $ v['name'];
        $ json['photo_x'] = __DATAURL__. $ v['photo_x'];
        $ json['price'] = $ v['price'];
        $ json['price_yh'] = $ v['price_yh'];
        $ json['shiyong'] = $ v['shiyong'];
        $ json['intro'] = $ v['intro'];
        $ json_arr[] = $ json;
    }
    $ cat_name = M('category') ->where("id = ". intval( $ id)) ->getField('name');
    echo json_encode(array('pro' => $ json_arr,'cat_name' => $ cat_name));
    exit();
}


/**
 * 获取商品属性价格接口
 */
public function jiage(){
    $ buff = trim( $ _POST['buff'],',');
    $ buff_arr = trim( $ _POST['buff_arr'],',');
    $ pid = intval( $ _POST['pid']);
    $ pro_info = M('product') ->where('id = '. intval( $ pid)) ->find();
```

```
        if ( $ buff_arr && $ pro_info) {
            $ arr = explode(',', $ buff_arr);
            $ str = 0;
            foreach ( $ arr as $ k => $ v) {
                $ price[] = M('guige') -> where('id = '. intval( $ v)) -> getField
                            ('price');
                $ stock[] = M('guige') -> where('id = '. intval( $ v)) -> getField
                            ('stock');
            }

            rsort( $ price);
            sort( $ stock);
            // $ price = implode(',', $ price);
            echo json_encode(array('status' => 1,'price' => $ price[0],'stock' =>
$ stock[0]));
            exit();
        }

        echo json_encode(array('status'=> 0));
        exit();
    }

}
```

9.2.6 购物车接口代码清单

购物车接口控制器文件为 ShoppingController. class. php，主要用于处理购物车商品添加、删除、列表信息的获取等，代码如下：

```
<? php
namespace Api\Controller;
use Think\Controller;

/**
 * 购物车接口
 *
 * 处理购物车信息 API
 * Class ShoppingController
 */
class ShoppingController extends PublicController {
```

```
/**
 * 会员获取购物车列表接口
 */
public function index(){
    $qz = C('DB_PREFIX');
    $shopping = M("shopping_char");
    $shangchang = M("shangchang");
    $product = M("product");
    $user_id = intval($_REQUEST['user_id']);
    if (!$user_id) {
        echo json_encode(array('status'=> 0));
        exit();
    }

    $cart = $shopping->where('uid='.intval($user_id))->field('id,uid,pid,
            price,num')->select();
    foreach ($cart as $k => $v) {
        $pro_info = $product->where('id='.intval($v['pid']))->field('name,
                    photo_x')->find();
        $cart[$k]['pro_name'] = $pro_info['name'];
        $cart[$k]['photo_x'] = __DATAURL__.$pro_info['photo_x'];
    }

    echo json_encode(array('status'=> 1,'cart'=> $cart));
    exit();
}


/**
 * 购物车商品删除
 */
public function delete(){
    $shopping = M("shopping_char");
    $cart_id = intval($_REQUEST['cart_id']);
    $check_id = $shopping->where('id='.intval($cart_id))->getField('id');
    if (!$check_id) {
        echo json_encode(array('status'=> 1));
        exit();
    }

    $res = $shopping->where('id ='.intval($cart_id))->delete(); // 删除
```

```
        if( $ res){
            echo json_encode(array('status'=> 1));
            exit();
        }else{
            echo json_encode(array('status'=> 0));
            exit();
        }
    }


    /**
     * 会员修改购物车数量接口
     */
    public function up_cart(){
        $ shopping = M("shopping_char");
        $ uid = intval( $ _REQUEST['user_id']);
        $ cart_id = intval( $ _REQUEST['cart_id']);
        $ num = intval( $ _REQUEST['num']);

        if (! $ uid || ! $ cart_id || ! $ num) {
            echo json_encode(array('status'=> 0,'err'=> '网络异常.'.__LINE__));
            exit();
        }

        $ check = $ shopping ->where('id= '. intval( $ cart_id)) ->find();
        if (! $ check) {
            echo json_encode(array('status'=> 0,'err'=> '购物车信息错误! '));
            exit();
        }

        //检测库存
        $ pro_num = M('product') ->where('id = '. intval( $ check['pid'])) ->getField
                ('num');
        if( $ num > intval( $ pro_num)){
            echo json_encode(array('status'=> 0,'err'=> '库存不足! '));
            exit();
        }

        $ data = array();
        $ data['num'] = $ num;

        $ res = $ shopping ->where('id = '. intval( $ cart_id). ' AND uid = '. intval
```

```
            ($uid))->save($data);
        if ($res) {
            echo json_encode(array('status'=> 1,'succ'=> '操作成功!'));
            exit();
        }else{
            echo json_encode(array('status'=> 0,'err'=> '操作失败.'));
            exit();
        }

    }


    /**
     * 多个购物车商品删除
     */
    public function qdelete(){
        $uid = intval($_REQUEST['uid']);
        if (!$uid) {
            echo json_encode(array('status'=> 0,'err'=> '网络异常,请稍后再试.'));
            exit();
        }
        $shopping = M("shopping_char");
        $cart_id = trim($_REQUEST['cart_id'],',');
        if (!$cart_id) {
            echo json_encode(array('status'=> 0,'err'=> '网络错误,请稍后再试.'));
            exit();
        }

        $res = $shopping->where('id in ('.$cart_id.') AND uid = '.intval($uid))-
            >delete(); // 删除
        if($res){
            echo json_encode(array('status'=> 1));
            exit();
        }else{
            echo json_encode(array('status'=> 0,'err'=> '操作失败.'));
            exit();
        }
    }


    /**
     * 添加购物车
```

```
*/
public function add(){
    $uid = intval($_REQUEST['uid']);
    if (!$uid) {
        echo json_encode(array('status'=> 0,'err'=> '登录状态异常.'));
        exit();
    }

    $pid = intval($_REQUEST['pid']);
    $num = intval($_REQUEST['num']);
    if (!intval($pid) || !intval($num)) {
        echo json_encode(array('status'=> 0,'err'=> '参数错误.'));
        exit();
    }

    //加入购物车
    $check = $this->check_cart(intval($pid));
    if ($check['status'] == 0) {
        echo json_encode(array('status'=> 0,'err'=> $check['err']));
        exit;
    }

    $check_info = M('product')->where('id='.intval($pid).' AND del=0 AND is_
                down=0')->find();

    //判断库存
    if (intval($check_info['num']) <= $num) {
        echo json_encode(array('status'=> 0,'err'=> '库存不足！'));
        exit;
    }

    $shpp = M("shopping_char");


    //判断购物车内是否已经存在该商品
    $data = array();
    $cart_info = $shpp->where('pid='.intval($pid).' AND uid='.intval
                ($uid))->field('id,num')->find();
    if ($cart_info) {
        $data['num'] = intval($cart_info['num']) + intval($num);
        //判断库存
```

```
        if (intval( $ check_info['num']) < = $ data['num']) {
            echo json_encode(array('status'=> 0,'err'=> '库存不足！'));
            exit;
        }
        $ res =  $ shpp ->where('id='. intval( $ cart_info['id'])) ->save( $ data);
    }else{
        $ data['pid'] = intval( $ pid);
        $ data['num'] = intval( $ num);
        $ data['addtime'] = time();
        $ data['uid'] = intval( $ uid);
        $ data['shop_id'] = intval( $ check_info['shop_id']);
        $ ptype = 1;
        if (intval( $ check_info['pro_type'])) {
            $ ptype =  intval( $ check_info['pro_type']);
        }
        $ data['type'] = $ ptype;
        $ data['price'] =  $ check_info['price_yh'];

        $ res = $ shpp ->add( $ data);
    }

    if( $ res){
        echo json_encode(array('status'=> 1,'cart_id'=> $ res));
        //该商品已成功加入您的购物车
        //记录日志:要写入文件的文件名(可以是任意文件名),如果文件不存在,将会
          创建一个.log.txt位置在项目的根目录下
        $ file   = 'log.txt';
        $ content = "内容:成功加入购物车\n";
        $ content . = "购物车 内容".json_encode( $ data)."\n";
        // 这个函数支持版本(PHP 5)
        if( $ f   = file_put_contents( $ file, $ content,FILE_APPEND)){
            //echo "成功。< br / >";
        }

        exit;
    }else{
        echo json_encode(array('status'=> 0,'err'=> '加入失败.'));
        //记录日志:要写入文件的文件名(可以是任意文件名),如果文件不存在,将会
          创建一个.log.txt位置在项目的根目录下
        $ file   = 'log.txt';
        $ content = "内容:加入购物车失败\n";
        $ content . = "购物车 内容".json_encode( $ res)."\n";
```

```
            // 这个函数支持版本(PHP 5)
            if( $f   = file_put_contents( $file, $content,FILE_APPEND)){
                //echo "成功。
                < br / > ";
            }

            exit;
        }
    }


    /**
     * 会员立即购买下单接口
     */
    public function check_shop(){
        $cart_id = trim( $_REQUEST['cart_id'],',');
        $id = explode(',', $cart_id);
        if (! $cart_id) {
            echo json_encode(array('status' => 0));
            exit();
        }

        foreach ( $id as $k => $v){
            $shoop[ $k] = M("shopping_char") ->where('id = '. intval( $v)) ->field
                        ('shop_id,pid') ->find();
        }

        foreach( $shoop as $key =>  $value){
            $result[ $key] = M("product") ->where('id = '. intval( $value['pid'])) -
                            >field('id,price,price_yh') ->select();
            $price[] = i_array_column( $result[ $key], 'price_yh');
        }
        //dump( $price);exit
        foreach( $price as $keys =>  $va){
            $str . = implode(",", $va).",";
        }
        $str = trim( $str, ",");
        $parr = explode(",", $str);
        if(array_sum( $parr) && in_array("0", $parr)){
            echo json_encode(array('status' => 0));
            exit();
```

```
    }

    $names = i_array_column($shoop, 'shop_id');

    $arr = array_unique($names);
    $val = sizeof($arr);
    if($val == '1'){
        echo json_encode(array('status' => 1));
        exit();
    }else{
        echo json_encode(array('status' => 2));
        exit();
    }
}


/**
 * 检查产品是否存在或删除
 * @param $pid
 * @return array
 */
public function check_cart($pid){
    //检查产品是否存在或删除
    $check_info = M('product') ->where('id = '.intval($pid).' AND del = 0 AND is_
                  down = 0') ->find();
    if (!$check_info) {
        return array('status' => 0,'err' => '商品不存在或已下架.');
    }

    return array('status' => 1);
}


/**
 * 去除 HTML 标签
 * @param $array
 * @return mixed
 */
public function html_entity($array){
    foreach ($array as $key => $value) {
        $array[$key]['content'] = strip_tags(html_entity_decode($value
                                  ['content']));
```

```
        }
        return $array;
    }

}
```

9.2.7 订单接口代码清单

订单接口控制器文件为 OrderController.class.php，主要用于订单信息获取、编辑等，代码如下：

```
<?php
namespace Api\Controller;
use Think\Controller;

/**
 * 订单接口控制器
 *
 * 处理订单信息 API
 * Class OrderController
 */
class OrderController extends PublicController {

    /**
     * 用户获取订单信息接口
     */
    public function index(){
        $uid = intval($_REQUEST['uid']);
        if (!$uid) {
            echo json_encode(array('status'=> 0,'err'=> '登录状态异常'));
            exit();
        }

        //分页
        $pages = intval($_REQUEST['page']);
        if (!$pages) {
            $pages = 0;
        }

        $orders = M("order");
        $orderp = M("order_product");
```

```
$ shangchang = M('shangchang');

//按条件查询
$ condition = array();
$ condition['del'] = 0;
$ condition['back'] = '0';
$ condition['uid'] = intval( $ uid);
$ condition['status'] = 10;
$ order_type = trim( $ _REQUEST['order_type']);
if ( $ order_type) {
    switch ( $ order_type) {
        case 'pay':
            $ condition['status'] = 10;
            break;
        case 'deliver':
            $ condition['status'] = 20;
            break;
        case 'receive':
            $ condition['status'] = 30;
            break;
        case 'evaluate':
            $ condition['status'] = 40;
            break;
        case 'finish':
            $ condition['status'] = array('IN',array(40,50));
            break;
        default:
            $ condition['status'] = 10;
            break;
    }
}

//获取总页数
$ count =  $ orders ->where( $ condition) ->count();
$ eachpage = 7;
//订单状态
$ order_status = array('0' => '已取消','10' => '待付款','20' => '待发货',
                '30' => '待收货','40' => '待评价','50' => '交易完成','51' =
                > '交易关闭');

$ order =  $ orders ->where( $ condition) ->order('id desc') ->field('id,order_
        sn,pay_sn,status,price,type,product_num') ->limit( $ pages.',7') -
```

```
                >select();
        foreach ( $ order as $ n => $ v){
            $ order[ $ n]['desc'] = $ order_status[ $ v['status']];
            $ prolist = $ orderp ->where('order_id = '. intval( $ v['id'])) ->find();
            $ order[ $ n]['photo_x'] = __DATAURL__. $ prolist['photo_x'];
            $ order[ $ n]['pid'] = $ prolist['pid'];
            $ order[ $ n]['name'] = $ prolist['name'];
            $ order[ $ n]['price_yh'] = $ prolist['price'];
            $ order[ $ n]['pro_count'] = $ orderp ->where('order_id = '. intval( $ v
['id'])) ->getField('COUNT(id)');
        }

        echo json_encode(array('status'=> 1,'ord' => $ order,'eachpage' => $ eachpage));
        exit();

    }


    /**
     * 用户获取订单信息,获取更多接口
     */
    public function get_more(){
        $ uid = intval( $ _REQUEST['uid']);
        if (! $ uid) {
            echo json_encode(array('status'=> 0,'err'=> '登录状态异常'));
            exit();
        }

        //分页
        $ pages = intval( $ _REQUEST['page']);
        if (! $ pages) {
            $ pages = 2;
        }
        $ limit = $ pages * 7 - 7;

        $ orders = M("order");
        $ orderp = M("order_product");
        $ shangchang = M('shangchang');

        //按条件查询
        $ condition = array();
```

```
$condition['del'] = 0;
$condition['back'] = '0';
$condition['uid'] = intval($uid);
$condition['status'] = 10;
$order_type = trim($_REQUEST['order_type']);
if ($order_type) {
    switch ($order_type) {
        case 'pay':
            $condition['status'] = 10;
            break;
        case 'deliver':
            $condition['status'] = 20;
            break;
        case 'receive':
            $condition['status'] = 30;
            break;
        case 'evaluate':
            $condition['status'] = 40;
            break;
        case 'finish':
            $condition['status'] = array('IN',array(40,50));
            break;
        default:
            $condition['status'] = 10;
            break;
    }
}

//获取总页数
$count = $orders->where($condition)->count();
$eachpage = 7;

$order_status = array('0'=>'已取消','10'=>'待付款','20'=>'待发货',
            '30'=>'待收货','40'=>'待评价','50'=>'交易完成','51'=
            >'交易关闭');

$order = $orders->where($condition)->order('id desc')->field('id,order_
        sn,pay_sn,status,price,type,product_num')->limit($limit.',7')-
        >select();
foreach ($order as $n=>$v){
    $order[$n]['desc'] = $order_status[$v['status']];
```

```
                $prolist = $orderp->where('order_id='.intval($v['id']))->find();
                $order[$n]['photo_x'] = __DATAURL__.$prolist['photo_x'];
                $order[$n]['pid'] = $prolist['pid'];
                $order[$n]['name'] = $prolist['name'];
                $order[$n]['price_yh'] = $prolist['price'];
                $order[$n]['pro_count'] = $orderp->where('order_id = '.intval($v
                                        ['id']))->getField('COUNT(id)');
            }

            echo json_encode(array('status'=>1,'ord'=>$order));
            exit();

        }

        /**
         * 用户退款退货接口
         */
        public function order_refund(){
            $uid = intval($_REQUEST['uid']);
            if (!$uid) {
                echo json_encode(array('status'=>0,'err'=>'登录状态异常'));
                exit();
            }

            //分页
            $pages = intval($_REQUEST['page']);
            if (!$pages) {
                $pages = 0;
            }

            $orders = M("order");
            $orderp = M("order_product");
            $shangchang = M('shangchang');

            $condition = array();
            $condition['back'] = array('gt','0');
            //获取总页数
            $count = $orders->where($condition)->count();
            $the_page = ceil($count/6);
```

```
        $refund_status = array('1'=>'退款申请中','2'=>'已退款','3'=>'处理中
                    ','4'=>'已拒绝');

        $order = $orders->where($condition)->order('back_addtime desc')->field
('id,price,order_sn,product_num,back,back_addtime')->limit($pages.',6')->select();
        foreach ($order as $n=>$v){
            $order[$n]['desc'] = $refund_status[$v['back']];
            $prolist = $orderp->where('order_id='.intval($v['id']))->find();
            $order[$n]['photo_x'] = __DATAURL__.$prolist['photo_x'];
            $order[$n]['pid'] = $prolist['pid'];
            $order[$n]['name'] = $prolist['name'];
            $order[$n]['price_yh'] = $prolist['price'];
            $order[$n]['back_addtime'] = date("Y-m-d H:i",$v['back_addtime']);
            $order[$n]['pro_count'] = $orderp->where('order_id='.intval($v
                                  ['id']))->getField('COUNT(id)');
        }

        echo json_encode(array('status'=>1,'ord'=>$order));
        exit();
    }


    /**
     * 获取用户退款退货信息接口
     */
    public function get_refund_more(){
        $uid = intval($_REQUEST['uid']);
        if (!$uid) {
            echo json_encode(array('status'=>0,'err'=>'登录状态异常'));
            exit();
        }

        //分页
        $pages = intval($_REQUEST['page']);
        if (!$pages) {
            $pages = 2;
        }
        $limit = $pages*6-6;

        $orders = M("order");
        $orderp = M("order_product");
```

```
        $ shangchang = M('shangchang');

        $ condition = array();
        $ condition['back'] = array('gt','0');
        //获取总页数
        $ count = $ orders ->where( $ condition) ->count();
        $ the_page = ceil( $ count/6);

        $ refund_status = array('1' => '退款申请中','2' => '已退款','3' => '处理中
                        ','4' => '已拒绝');

        $ order = $ orders ->where( $ condition) ->order('back_addtime desc') ->field
('id,price,order_sn,product_num,back,back_addtime') ->limit( $ limit.',6') ->select();
        foreach ( $ order as $ n => $ v) {
            $ order[ $ n]['desc'] = $ refund_status[ $ v['back']];
            $ prolist = $ orderp ->where('order_id = '. intval( $ v['id'])) ->find();
            $ order[ $ n]['photo_x'] = __DATAURL__. $ prolist['photo_x'];
            $ order[ $ n]['pid'] = $ prolist['pid'];
            $ order[ $ n]['name'] = $ prolist['name'];
            $ order[ $ n]['price_yh'] = $ prolist['price'];
            $ order[ $ n]['back_addtime'] = date("Y - m - d H:i", $ v['back_addtime']);
            $ order[ $ n]['pro_count'] = $ orderp ->where('order_id = '. intval( $ v
                                        ['id'])) ->getField('COUNT(id)');
        }

        echo json_encode(array('status' => 1,'ord' => $ order));
        exit();
    }


    /**
     * 用户订单编辑接口
     */
    public function orders_edit(){

        $ orders = M("order");
        $ order_id = intval( $ _REQUEST['id']);
        $ type = $ _REQUEST['type'];

        $ check_id = $ orders ->where('id = '. intval( $ order_id). ' AND del = 0') ->get-
                Field('id');
```

```
        if (! $check_id || ! $type) {
            echo json_encode(array('status'=> 0,'err'=> '订单信息错误.'.__LINE__));
            exit();
        }

        $data = array();
        if ($type=='cancel') {
            $data['status'] = 0;
        }elseif ($type=='receive') {
            $data['status'] = 40;
        }elseif ($type=='refund') {
            $data['back'] = 1;
            $data['back_remark'] = $_REQUEST['back_remark'];
        }

        if ($data) {
            $result = $orders->where('id='.intval($order_id))->save($data);
            if($result !== false){
                echo json_encode(array('status'=> 1));
                exit();
            }else{
                echo json_encode(array('status'=> 0,'err'=> '操作失败.'.__LINE__));
            exit();
            }
        }else{
            echo json_encode(array('status'=> 0,'err'=> '订单信息错误.'.__LINE__));
            exit();
        }
    }


    /**
     * 用户订单详情接口
     */
    public function order_details(){

        $order_id = intval($_REQUEST['order_id']);
        //订单详情
        $orders=M("order");
        $product_dp=M("product_dp");
        $orderp=M("order_product");
        $id=intval($_REQUEST['id']);
```

```
$qz = C('DB_PREFIX');     //前缀

$order_info = $orders->where('id = '.intval($order_id.' AND del = 0')) ->
              field('id,order_sn,shop_id,status,addtime,price,type,post,
               tel,receiver,address_xq,remark') ->find();
if (!$order_info) {
    echo json_encode(array('status' => 0,'err' => '订单信息错误.'));
    exit();
}

//订单状态
$order_status = array('0' => '已取消','10' => '待付款','20' => '待发货','30'
=> '待收货','40' => '已收货','50' => '交易完成');
//支付类型
$pay_type = array('cash' => '现金支付','alipay' => '支付宝','weixin' => '微信支付
');

$order_info['shop_name'] = M('shangchang') ->where('id = '.intval($order_in-
                            fo['shop_id'])) ->getField('name');
$order_info['order_status'] = $order_status[$order_info['status']];
$order_info['pay_type'] = $pay_type[$order_info['type']];
$order_info['addtime'] = date('Y-m-d H:i:s',$order_info['addtime']);
$order_info['yunfei'] = 0;
if ($order_info['post']) {
    $order_info['yunfei'] = M('post') ->where('id = '.intval($order_info
                              ['post'])) ->getField('price');
}

//获取产品
$pro = $orderp ->where('order_id = '.intval($order_info['id'])) ->select();
foreach ($pro as $k => $v) {
    $pro[$k]['photo_x'] = __DATAURL__.$v['photo_x'];
}

echo json_encode(array('status' => 1,'pro' => $pro,'ord' => $order_info));
exit();
}
```

```
}
```

9.2.8 下单结算接口代码清单

下单结算接口控制器文件为 PaymentController. class. php，主要用于处理下单结算等，代码如下：

```
<? php
namespace Api\Controller;
use Think\Controller;

/**
 * 下单结算接口
 *
 * 处理下单、购物车结算等 API
 * Class PaymentController
 */
class PaymentController extends PublicController {

    /**
     * 会员立即购买获取数据接口
     */
    public function buy_now(){
        $uid = intval($_REQUEST['uid']);
        if (!$uid){
            echo json_encode(array('status'=> 0,'err'=> '系统错误.'));
            exit();
        }
        //单件商品结算
        //地址管理
        $address = M("address");
        $city = M("china_city");
        $add = $address->where('uid='.intval($uid))->select();
        $citys = $city->where('tid=0')->field('id,name')->select();
        $shopping = M('shopping_char');
        $product = M("product");
        //运费
        $post = M('post');

        //立即购买数量
        $num = intval($_REQUEST['num']);
```

```
if (! $num) {
    $num = 1;
}

//购物车 id
$cart_id = intval($_REQUEST['cart_id']);
//检测购物车是否有对应数据
$check_cart = $shopping->where('id = '.intval($cart_id).' AND num > = '.
            intval($num))->getField('pid');
if (! $check_cart) {
    echo json_encode(array('status'=> 0,'err'=> '购物车信息错误.'));
    exit();
}
//判断基本库存
$pro_num = $product->where('id = '.intval($check_cart))->getField
        ('num');
if ($num > intval($pro_num)) {
    echo json_encode(array('status'=> 0,'err'=> '库存不足.'));
    exit();
}

$qz = C('DB_PREFIX');//前缀

$pro = $shopping->where(''.$qz.'shopping_char.uid = '.intval($uid).' and
        '.$qz.'shopping_char.id = '.intval($cart_id))->join('LEFT JOIN __
        PRODUCT__ ON __PRODUCT__.id = __SHOPPING_CHAR__.pid')->join('LEFT
        JOIN __SHANGCHANG__ ON __SHANGCHANG__.id = __SHOPPING_CHAR__.shop_
        id')->field(''.$qz.'product.num as pnum,'.$qz.'shopping_char.id,'.
        $qz.'shopping_char.pid,'.$qz.'shangchang.name as sname,'.$qz.
        'product.name,'.$qz.'product.shop_id,'.$qz.'product.photo_x,'.$qz.
        'product.price_yh,'.$qz.'shopping_char.num,'.$qz.'shopping_char.
        buff,'.$qz.'shopping_char.price,'.$qz.'shangchang.alipay,'.$qz.
        'shangchang.alipay_pid,'.$qz.'shangchang.alipay_key')->find();
//获取运费
$yunfei = $post->where('pid = '.intval($pro['shop_id']))->find();

if($pro['buff']! = ''){
    $pro['zprice'] = $pro['price'] * $num;
}else{
    $pro['price'] = $pro['price_yh'];
    $pro['zprice'] = $pro['price'] * $num;
```

```
        }

        //如果需要运费
        if ( $ yunfei) {
            if ( $ yunfei['price_max'] > 0 && $ yunfei['price_max'] < = $ pro['zprice']) {
                $ yunfei['price'] = 0;
            }
        }

        $ buff_text = '';
        if( $ pro['buff']){
            //获取属性名称
            $ buff = explode(',', $ pro['buff']);
            if(is_array( $ buff)){
                foreach( $ buff as $ keys => $ val){
                    $ ggid = M("guige") ->where('id = '. intval( $ val)) ->getField('name');
                    // $ buff_text . = select('name','aaa_cpy_category','id = '. $ val
['id']).':'.select('name','aaa_cpy_category','id = '. $ val['val']).' ';
                    $ buff_text . = ' '. $ ggid.' ';
                }
            }
        }
        $ pro['buff'] = $ buff_text;
        $ pro['photo_x'] = 'http://'. $ _SERVER['SERVER_NAME']. __UPLOAD__. '/'. $ pro
                    ['photo_x'];

        echo json_encode(array('status' => 1, 'citys' => $ citys, 'yun' => $ yunfei,
'adds' => $ add, 'pro' => $ pro, 'num' => $ num, 'buff' => $ buff_text));
        exit();
        // $ this ->assign('citys', $ citys);
    }

    /**
     * 会员立即购买下单接口
     * @throws \Exception
     */
    public function pay_now(){
        $ product = M("product");
        //运费
        $ post = M('post');
        $ order = M("order");
```

```
$order_pro = M("order_product");

$uid = intval($_REQUEST['uid']);
if (!$uid) {
    echo json_encode(array('status' => 0,'err' => '登录状态异常.'));
    exit();
}

//下单
    try {
        $data = array();
        $data['shop_id'] = intval($_POST['sid']);

        $data['uid'] = intval($uid);
        $data['addtime'] = time();
        $data['del'] = 0;
        $data['type'] = trim($_POST['paytype']);
        //订单状态：10,未付款;20,代发货;30,确认收货(待收货);40,交易关
          闭;50,交易完成
        $data['status'] = 10;//未付款

        //dump($_POST);exit;
        $_POST['yunfei'] ? $yunPrice = $post->where('id ='.intval($_
                                          POST['yunfei']))->find() : NULL;
        //dump($yunPrice);exit;
        if(!empty($yunPrice)){
            $data['post'] = $yunPrice['id'];
            $data['price'] = $_POST['price'] + $yunPrice['price'];
        }else{
            $data['post'] = 0;
            $data['price'] = $_POST['price'];
        }

        $adds_id = intval($_POST['aid']);
        if (!$adds_id) {
            echo json_encode(array('status' => 0,'err' => '请选择收货地址.
                                              '.__LINE__));

            exit();
        }

        $adds_info = M('address')->where('id ='.intval($adds_id))->find();
        $data['receiver'] = $adds_info['name'];
```

```
$data['tel'] = $adds_info['tel'];
$data['address_xq'] = $adds_info['address_xq'];
$data['code'] = $adds_info['code'];
$data['product_num'] = intval($_POST['num']);
$data['remark'] = $_POST['remark'];
/*******解决同一订单重复支付问题 **********/
$data['order_sn'] = $this->build_order_no();//生成唯一订单号

if (!$data['product_num'] || !$data['price']) {
    throw new \Exception("System Error !");
}

/*****************************************************/
//dump($data);exit;
$result = $order->add($data);
if($result){
    $date = array();
    $date['pid'] = intval($_POST['pid']);//商品 id
    $date['order_id'] = $result;//订单 id
    $date['name'] = $product->where('id='.intval($date['pid']))
                ->getField('name');//商品名字
    $date['price'] = $product->where('id='.intval($date['pid']))
                ->getField('price_yh');
    $date['pro_buff'] = $_POST['buff'];
    $date['photo_x'] = $product->where('id='.intval($date
                ['pid']))->getField('photo_x');
    $date['pro_buff'] = $_POST['buff'];
    $date['addtime'] = time();
    $date['num'] = intval($_POST['num']);
   //$date['pro_guige'] = $_REQUEST['guige'];
    $res = $order_pro->add($date);
    if(!$res){
        throw new \Exception("下单 失败!".__LINE__);
    }

    //检查产品是否存在,并修改库存
    $check_pro = $product->where('id='.intval($date['pid']).'
            AND del=0 AND is_down=0')->field('num,shiyong')
            ->find();
    if (!$check_pro) {
        throw new \Exception("商品不存在或已下架!");
```

```
                }
                $ up = array();
                $ up['num'] = intval ( $ check _ pro ['num']) - intval ( $ date
                        ['num']);
                $ up['shiyong'] = intval ( $ check _ pro ['shiyong']) + intval
                        ( $ date['num']);
                $ product ->where('id = '. intval( $ date['pid'])) ->save( $ up);

                $ url = $ _SERVER['HTTP_REFERER'];

            }else{
                throw new \Exception("下单 失败!");
            }
        } catch (Exception $ e) {
            echo json_encode(array('status' => 0,'err' => $ e ->getMessage()));
            exit();
        }
        //把需要的数据返回
        $ arr = array();
        $ arr['order_id'] = $ result;
        $ arr['order_sn'] = $ data['order_sn'];
        $ arr['pay_type'] = $ _POST['paytype'];
        echo json_encode(array('status' => 1,'arr' => $ arr));
        exit();
}


/ **
 * 购物车结算 获取数据
 */
public function buy_cart(){
    $ uid = intval( $ _REQUEST['uid']);
    if (! $ uid) {
        echo json_encode(array('status' => 0,'err' => '登录状态异常.'));
        exit();
    }

    $ address = M("address");
    //运费
    $ post = M('post');
    $ qz = C('DB_PREFIX');
    $ add = $ address ->where('uid = '. intval( $ uid)) ->order('is_default desc,id
```

```
        desc')->limit(1)->find();
    $product = M("product");
    $shopping = M('shopping_char');
    $cart_id = trim($_REQUEST['cart_id'],',');
    $id = explode(',', $cart_id);
    if (!$cart_id) {
        echo json_encode(array('status'=> 0,'err'=> '网络异常.'.__LINE__));
        exit();
    }

    $pro = array();
    $pro1 = array();
    foreach($id as $k => $v){
        //检测购物车是否有对应数据
        $check_cart = $shopping->where('id='.intval($v))->getField('id');
        if (!$check_cart) {
            echo json_encode(array('status'=> 0,'err'=> '非法操作.'.__LINE__));
            exit();
        }

        $pro[$k] = $shopping->where(''.$qz.'shopping_char.uid='.intval
                ($uid).' and '.$qz.'shopping_char.id='.$v)->join('LEFT
                JOIN __PRODUCT__ ON __PRODUCT__.id=__SHOPPING_CHAR__.
                pid')->join('LEFT JOIN __SHANGCHANG__ ON __SHANGCHANG__.id
                =__SHOPPING_CHAR__.shop_id')->field(''.$qz.'product.num
                as pnum,'.$qz.'shopping_char.id,'.$qz.'shopping_char.pid,
                '.$qz.'shangchang.name as sname,'.$qz.'product.name,'.
                $qz.'product.shop_id,'.$qz.'product.photo_x,'.$qz.
                'product.price_yh,'.$qz.'shopping_char.num,'.$qz.
                'shopping_char.buff,'.$qz.'shopping_char.price')->find();
        //获取运费
        $yunfei = $post->where('pid='.intval($pro[$k]['shop_id']))->find();
        //dump($yunfei);
        if($pro[$k]['buff']!=''){
            $pro[$k]['zprice'] = $pro[$k]['price'] * $pro[$k]['num'];
        }else{
            $pro[$k]['price'] = $pro[$k]['price_yh'];
            $pro[$k]['zprice'] = $pro[$k]['price'] * $pro[$k]['num'];
        }
        $pro[$k]['photo_x'] = __DATAURL__.$pro[$k]['photo_x'];
```

```
        //获取可用优惠券
        $vou = $this->get_voucher($uid,intval($pro[$k]['pid']),$id);
    }

    //计算总价
    foreach($id as $ks => $vs){
        $pro1[$ks] = $shopping->where(''.$qz.'shopping_char.uid = '.intval
                    ($uid).' and '.$qz.'shopping_char.id = '.$vs)->join
                    ('LEFT JOIN __PRODUCT__ ON __PRODUCT__.id = __SHOPPING_
                    CHAR__.pid')->join('LEFT JOIN __SHANGCHANG__ ON __
                    SHANGCHANG__.id = __SHOPPING_CHAR__.shop_id')->field(''.
                    $qz.'product.num as pnum,'.$qz.'shopping_char.id,'.
                    $qz.'shangchang.name as sname,'.$qz.'product.name,'.
                    $qz.'product.photo_x,'.$qz.'product.price_yh,'.$qz.
                    'shopping_char.num,'.$qz.'shopping_char.buff,'.$qz.
                    'shopping_char.price')->find();
        if($pro1[$ks]['buff']){
            $pro1[$ks]['zprice'] = $pro1[$ks]['price'] * $pro1[$ks]['num'];
        }else{
            $pro1[$ks]['price'] = $pro1[$ks]['price_yh'];
            $pro1[$ks]['zprice'] = $pro1[$ks]['price'] * $pro1[$ks]['num'];
        }
        $price+ = $pro1[$ks]['zprice'];
    }

    //如果需要运费
    if ($yunfei) {
        if ($yunfei['price_max'] > 0 && $yunfei['price_max'] <= $price) {
            $yunfei['price'] = 0;
        }
    }

    if (!$add) {
        $addemt = 1;
    }else{
        $addemt = 0;
    }

    echo json_encode(array('status' => 1,'vou' => $vou,'price' => floatval
($price),'pro' => $pro,'adds' => $add,'addemt' => $addemt,'yun' => $yunfei));
    exit();
```

```
}

/**
 * 购物车结算 下订单
 * @throws \Exception
 */
public function payment(){
    $product = M("product");
    //运费
    $post = M('post');
    $order = M("order");
    $order_pro = M("order_product");
    $shopping = M('shopping_char');

    $uid = intval($_REQUEST['uid']);

    //记录日志:要写入文件的文件名(可以是任意文件名),如果文件不存在,将会创建
      一个.log.txt位置在项目的根目录下
    $file  = 'log.txt';
    $content = "内容:\n";
    $content .= "rand_row 内容".json_encode($_REQUEST)."\n";
    // 这个函数支持版本(PHP 5)
    if($f  = file_put_contents($file, $content,FILE_APPEND)){
        //echo "成功
          <br/>";
    }

    if (!$uid) {
        echo json_encode(array('status'=> 0,'err'=> '登录状态异常.'));
        exit();
    }

    $cart_id = trim($_REQUEST['cart_id'],',');
    if (!$cart_id) {
        echo json_encode(array('status'=> 0,'err'=> '数据异常.'));
        exit();
    }

    //生成订单
      try {
          $qz = C('DB_PREFIX');//前缀
```

```
$cart_id = explode(',', $cart_id);
$shop = array();
foreach($cart_id as $ke => $vl){
    $shop[$ke] = $shopping->where(''.$qz.'shopping_char.uid = '.intval($uid).' and '.$qz.'shopping_char.id = '.$vl)->join('LEFT JOIN __PRODUCT__ ON __PRODUCT__.id = __SHOPPING_CHAR__.pid')->field(''.$qz.'shopping_char.pid,'.$qz.'shopping_char.num,'.$qz.'shopping_char.shop_id,'.$qz.'shopping_char.buff,'.$qz.'shopping_char.price,'.$qz.'product.price_yh')->find();
    $num += $shop[$ke]['num'];
    if($shop[$ke]['buff'] != ''){
        $ozprice += $shop[$ke]['price'] * $shop[$ke]['num'];
    }else{
        $shop[$ke]['price'] = $shop[$ke]['price_yh'];
        $ozprice += $shop[$ke]['price'] * $shop[$ke]['num'];
    }
}


$yunPrice = array();
if ($_POST['yunfei']) {
    $yunPrice = $post->where('id = '.intval($_POST['yunfei']))->find();
}

$data['shop_id'] = $shop[$ke]['shop_id'];

  //记录日志:要写入文件的文件名(可以是任意文件名),如果文件不存在,将会创建一个.log.txt 位置在项目的根目录下
  $file  = 'log.txt';
  $content = "关键内容:\n";
  $content .= "data shop_id 内容".$data['shop_id']."\n";
  $content .= "查询出来的 内容".json_encode($shop)."\n";
  // 这个函数支持版本(PHP 5)
  if($f  = file_put_contents($file, $content,FILE_APPEND)){
      //echo "成功
        <br />";
  }
```

```
            $data['uid'] = intval($uid);

            if(!empty($yunPrice)){
                $data['post'] = $yunPrice['id'];
                $data['price'] = floatval($ozprice) + $yunPrice['price'];
            }else{
                $data['post'] = 0;
                $data['price'] = floatval($ozprice);
            }

            $data['amount'] = $data['price'];
            $vid = intval($_POST['vid']);
            if ($vid) {
                $vouinfo = M('user_voucher')->where('status = 1 AND uid = '.intval
($uid).' AND vid = '.intval($vid))->find();
                $chk = M('order')->where('uid = '.intval($uid).' AND vid = '.intval
                    ($vid).' AND status > 0')->find();
                if (!$vouinfo || $chk) {
                    //throw new \Exception("此优惠券不可用,请选择其他.".__LINE__)
                    echo json_encode(array('status' => 0,'err' => '此优惠券不可用,
                        请选择其他.'));
                    exit();
                }
                if ($vouinfo['end_time'] < time()) {
                    //throw new \Exception("优惠券已过期了.".__LINE__)
                    echo json_encode(array('status' => 0,'err' => "优惠券已过期
                        了.".__LINE__));
                    exit();
                }
                if ($vouinfo['start_time'] > time()) {
                    //throw new \Exception("优惠券还未生效.".__LINE__)
                    echo json_encode(array('status' => 0,'err' => "优惠券还未生
                        效.".__LINE__));
                    exit();
                }
                $data['vid'] = intval($vid);
                $data['amount'] = floatval($data['price']) - floatval($vouinfo
                                  ['amount']);
            }

            $data['addtime'] = time();
```

```
$data['del'] = 0;
$data['type'] = $_POST['type'];
$data['status'] = 10;

$adds_id = intval($_POST['aid']);
if (!$adds_id) {
    throw new \Exception("请选择收货地址.".__LINE__);
}
$adds_info = M('address')->where('id='.intval($adds_id))->find();
$data['receiver'] = $adds_info['name'];
$data['tel'] = $adds_info['tel'];
$data['address_xq'] = $adds_info['address_xq'];
$data['code'] = $adds_info['code'];
$data['product_num'] = $num;
$data['remark'] = $_REQUEST['remark'];
$data['order_sn'] = $this->build_order_no();//生成唯一订单号

$result = $order->add($data)
if($result){
    //$prid = explode(",", $_POST['ids'])
    foreach($cart_id as $key => $var){
        $shops[$key] = $shopping->where(''.$qz.'shopping_char.uid =
                        '.intval($uid).' and '.$qz.'shopping_char.id
                        ='.intval($var))->join('LEFT JOIN __PRODUCT_
                        _ON __PRODUCT__.id=__SHOPPING_CHAR__.pid')-
                       >field(''.$qz.'shopping_char.pid,'.$qz.
                        'shopping_char.num,'.$qz.'shopping_char.shop_
                        id,'.$qz.'shopping_char.buff,'.$qz.'shopping
                        _char.price,'.$qz.'product.name,'.$qz.
                        'product.photo_x,'.$qz.'product.price_yh,'.
                        $qz.'product.num as pnum')->find();
        if($shops[$key]['buff'] == '' || !$shops[$key]['buff']){
            $shops[$key]['price'] = $shops[$key]['price_yh'];
        }

        $buff_text = '';
        if($shops[$key]['buff']){
            //验证属性
            $buff = explode(',',$shops[$key]['buff']);
            if(is_array($buff)){
                foreach($buff as $keys => $val){
```

```
                    $ ggid = M("guige") ->where('id = '. intval( $ val))
                              ->getField('name');
                    $ buff_text . =  $ ggid.' ';
                };
            }
        }

        $ date = array();
        $ date['pid'] = $ shops[ $ key]['pid'];
        $ date['name'] = $ shops[ $ key]['name'];
        $ date['order_id'] = $ result;
        $ date['price'] = $ shops[ $ key]['price'];
        $ date['photo_x'] = $ shops[ $ key]['photo_x'];
        $ date['pro_buff'] = trim( $ buff_text,' ');
        $ date['addtime'] = time();
        $ date['num'] = $ shops[ $ key]['num'];
        $ date['pro_guige'] = '';
        $ res = $ order_pro ->add( $ date);
        if (! $ res) {
            throw new \Exception("下单 失败!".__LINE__);
        }
        //检查产品是否存在,并修改库存
        $ check_pro =  $ product ->where('id = '. intval( $ date['pid']).'
                 AND del = 0 AND is_down = 0') ->field('num,shiyong')
                 ->find();
        $ up = array();
        $ up['num'] = intval ( $ check _ pro ['num']) - intval ( $ date
                   ['num']);
        $ up['shiyong'] = intval ( $ check _ pro ['shiyong']) + intval
                   ( $ date['num']);
        $ product ->where('id = '. intval( $ date['pid'])) ->save( $ up);
        //echo   $ product ->getLastSql();
        //删除购物车数据
        $ shopping -> where ('uid = '. intval ( $ uid). ' AND id = '. intval
                ( $ var)) ->delete();

    }
}else{
    //记录日志:要写入文件的文件名(可以是任意文件名),如果文件不存在,
      将会创建一个.log.txt 位置在项目的根目录下
    $ file  = 'log.txt';
```

```
        $content = "出口 1:\n";
        // 这个函数支持版本(PHP 5)
        if($f    = file_put_contents($file, $content,FILE_APPEND)){
            //echo "成功。<br />";
        }


        throw new \Exception("下单 失败!");
    }
} catch (Exception $e) {
    echo json_encode(array('status'=> 0,'err'=> $e->getMessage()));

    //记录日志:要写入文件的文件名(可以是任意文件名),如果文件不存在,将
      会创建一个.log.txt 位置在项目的根目录下
    $file    = 'log.txt';
    $content = "出口 2:\n";
    $content .= "err 内容".$e->getMessage()."\n";
    // 这个函数支持版本(PHP 5)
    if($f    = file_put_contents($file, $content,FILE_APPEND)){
        //echo "成功。<br />";
    }

    exit();
}

    //把需要的数据返回
    $arr = array();
    $arr['order_id'] = $result;
    $arr['order_sn'] = $data['order_sn'];
    $arr['pay_type'] = $_POST['type'];
    echo json_encode(array('status'=> 1,'arr'=> $arr));
    exit();
}


/**
 * 获取可用优惠券
 * @param $uid
 * @param $pid
 * @param $cart_id
 * @return array
 */
```

```
public function get_voucher( $uid, $pid, $cart_id){
    $qz = C('DB_PREFIX');
    //计算总价
    $prices = 0;
    foreach( $cart_id as $ks => $vs){
        $pros = M('shopping_char') ->where(''. $qz.'shopping_char.uid = '. intval
                ( $uid).' AND '. $qz.'shopping_char.id = '. $vs) ->join('LEFT JOIN
                __PRODUCT__ ON __PRODUCT__. id = __SHOPPING_CHAR__. pid') ->join
                ('LEFT JOIN __SHANGCHANG__ ON __SHANGCHANG__. id = __SHOPPING_
                CHAR__. shop_id') ->field(''. $qz.'shopping_char.num,'. $qz.
                'shopping_char.price,'. $qz.'shopping_char.type') ->find();
        $zprice = $pros['price'] * $pros['num'];
        $prices += $zprice;
    }

    $condition = array();
    $condition['uid'] = intval( $uid);
    $condition['status'] = array('eq',1);
    $condition['start_time'] = array('lt',time());
    $condition['end_time'] = array('gt',time());
    $condition['full_money'] = array('elt',floatval( $prices));

    $vou = M('user_voucher') ->where( $condition) ->order('addtime desc') ->se-
    lect();
    $vouarr = array();
    foreach ( $vou as $k => $v) {
        $chk_order = M('order') ->where('uid = '. intval( $uid).' AND vid = '. in-
                    tval( $v['vid']).' AND status > 0') ->find();
        $vou_info = M('voucher') ->where('id = '. intval( $v['vid'])) ->find();
        $proid = explode(',', trim( $vou_info['proid'],','));
        if (( $vou_info['proid'] == 'all' || $vou_info['proid'] == '' || in_array
                    ( $pid, $proid)) && ! $chk_order) {
            $arr = array();
            $arr['vid'] = intval( $v['vid']);
            $arr['full_money'] = floatval( $v['full_money']);
            $arr['amount'] = floatval( $v['amount']);
            $vouarr[] = $arr;
        }
    }

    return $vouarr;
```

```
    }

    /**
     * 生成唯一订单号
     * @return string  返回16位的唯一订单号
     */
    public function build_order_no(){
        return date('Ymd').substr(implode(NULL, array_map('ord', str_split(substr
(uniqid(), 7, 13), 1))), 0, 8);
    }
}
```

9.2.9 微信支付接口代码清单

微信支付接口控制器文件为 WxpayController.class.php，主要用于调起微信支付和支付回调等，代码如下：

```
<?php
namespace Api\Controller;
use Think\Controller;

/**
 * 微信支付接口
 *
 * 调起微信支付功能
 * Class WxpayController
 */
class WxpayController extends Controller{

    public function _initialize(){
        //php 判断 http 还是 https
        $this->http_type = ((isset($_SERVER['HTTPS']) && $_SERVER['HTTPS'] ==
                            'on') || (isset($_SERVER['HTTP_X_FORWARDED_PROTO']) &&
                            $_SERVER['HTTP_X_FORWARDED_PROTO'] == 'https')) ?
                            'https://' : 'http://';
        vendor('WeiXinpay.wxpay');
    }

    /**
     * 微信支付 接口
     */
```

```
public function wxpay(){

    //记录日志:要写入文件的文件名(可以是任意文件名),如果文件不存在,将会创建
      一个.log.txt位置在项目的根目录下
    $file   = 'log.txt';
    $content = "微信支付请求参数:\n";
    $content .= "请求参数内容".json_encode($_REQUEST)."\n";
    // 这个函数支持版本(PHP 5)
    if($f   = file_put_contents($file, $content,FILE_APPEND)){
        //echo "成功
          <br/>";
    }

    $pay_sn = trim($_REQUEST['order_sn']);
    if (!$pay_sn) {
        echo json_encode(array('status'=> 0,'err'=> '支付信息错误!'));
        exit();
    }

    $order_info = M('order')->where('order_sn="'.$pay_sn.'"')->find();
    if (!$order_info) {
        echo json_encode(array('status'=> 0,'err'=> '没有找到支付订单!'));
        exit();
    }

    if (intval($order_info['status'])!=10) {
        echo json_encode(array('status'=> 0,'err'=> '订单状态异常!'));
        exit();
    }

    //①获取用户 openid
    $tools = new \JsApiPay();
    $openId = M('user')->where('id='.intval($order_info['uid']))->getField
            ('openid');
    if (!$openId) {
        echo json_encode(array('status'=> 0,'err'=> '用户状态异常!'));
        exit();
    }

    //②统一下单
    $input = new \WxPayUnifiedOrder();
    $input->SetBody("信真小铺商品购买_".trim($order_info['order_sn']));
```

```
        $input->SetAttach("信真小铺商品购买_".trim($order_info['order_sn']));
        $input->SetOut_trade_no($pay_sn);
        $input->SetTotal_fee(floatval($order_info['amount']) * 100);
        $input->SetTime_start(date("YmdHis"));
        $input->SetTime_expire(date("YmdHis", time() + 3600));
        $input->SetGoods_tag("信真小铺商品购买_".trim($order_info['order_sn']));
        $input->SetNotify_url('https://mini.laohuzx.com/index.php/Api/Wxpay/notify');
        $input->SetTrade_type("JSAPI");
        $input->SetOpenid($openId);
        $order = \WxPayApi::unifiedOrder($input);

        $arr = array();
        $arr['appId'] = $order['appid'];
        $arr['nonceStr'] = $order['nonce_str'];
        $arr['package'] = "prepay_id=".$order['prepay_id'];
        $arr['signType'] = "MD5";
        $arr['timeStamp'] = (string)time();
        $str = $this->ToUrlParams($arr);
        $jmstr = $str."&key=".\WxPayConfig::KEY;
        $arr['paySign'] = strtoupper(MD5($jmstr));
        echo json_encode(array('status'=> 1,'arr'=> $arr));
        exit();

    }

    /**
     * 支付回调
     */
    public function notify(){

        $res_xml = file_get_contents("php://input");
        libxml_disable_entity_loader(true);
        $ret = json_decode(json_encode(simplexml_load_string($res_xml,
                'simpleXMLElement',LIBXML_NOCDATA)),true);

        $path = "./Data/log/";
        if (! is_dir($path)){
            mkdir($path,0777);  // 创建文件夹 test,并给 777 的权限(所有权限)
        }
        $content = date("Y-m-d H:i:s").'=> '.json_encode($ret);  // 写入的内容
        $file = $path."weixin_".date("Ymd").".log";    // 写入的文件
```

```
        file_put_contents( $ file, $ content,FILE_APPEND);
                                            // 以追加的方式写入文件最简便快捷

        $ data = array();
        $ data['order_sn'] = $ ret['out_trade_no'];
        $ data['pay_type'] = 'weixin';
        $ data['trade_no'] = $ ret['transaction_id'];
        $ data['total_fee'] = $ ret['total_fee'];
        $ result = $ this->orderhandle( $ data);
        if (is_array( $ result)) {
            $ xml = "< xml >< return_code >< ! [CDATA[SUCCESS]]> </return_code
                    >< return_msg >< ! [CDATA[OK]]> </return_msg >";
            $ xml. = "</xml >";
            echo $ xml;
        }else{
            $ contents = 'error => '.json_encode( $ result);  // 写入的内容
            $ files = $ path."error_".date("Ymd").".log";   // 写入的文件
            file_put_contents( $ files, $ contents,FILE_APPEND);
                                            // 以追加的方式写入文件最简便快捷
            echo 'fail';
        }
    }

    /**
     * 订单处理
     * @param $ data
     * @return array|string
     */
    public function orderhandle( $ data){
        $ order_sn = trim( $ data['order_sn']);
        $ pay_type = trim( $ data['pay_type']);
        $ trade_no = trim( $ data['trade_no']);
        $ total_fee = floatval( $ data['total_fee']);
        $ check_info = M('order')->where('order_sn="'. $ order_sn.'"')->find();
        if (! $ check_info) {
            return "订单信息错误...";
        }

        if ( $ check_info['status'] < 10 || $ check_info['back'] > '0') {
            return "订单异常...";
        }
```

```
        if ($check_info['status'] > 10) {
            return array('status' => 1,'data' => $data);
        }

        $up = array();
        $up['type'] = $pay_type;
        $up['price_h'] = sprintf("%.2f",floatval($total_fee/100));
        $up['status'] = 20;
        $up['trade_no'] = $trade_no;
        $res = M('order')->where('order_sn = "'. $order_sn.'"')->save($up);
        if ($res) {
            //处理优惠券
            if (intval($check_info['vid'])) {
                $vou_info = M('user_voucher')->where('uid = '. intval($check_info
                            ['uid']).' AND vid = '. intval($check_info['vid']))->
                            find();
                if (intval($vou_info['status']) == 1) {
                    M('user_voucher')->where('id = '. intval($vou_info['id']))->save
(array('status' => 2));
                }
            }
            return array('status' => 1,'data' => $data);
        }else{
            return '订单处理失败...';
        }
    }

    /**
     * 构建字符串
     * @param $urlObj
     * @return string
     */
    private function ToUrlParams($urlObj)
    {
        $buff = "";
        foreach ($urlObj as $k => $v)
        {
            if($k != "sign"){
                $buff .= $k . "=" . $v . "&";
            }
        }
```

```
        $ buff = trim( $ buff, "&");
        return $ buff;
    }
}
? >
```

9.2.10 地址接口代码清单

收货地址接口控制器文件为 AddressController. class. php，主要用于处理用户收货地址添加、删除和获取等，代码如下：

```
< ? php
namespace Api\Controller;
use Think\Controller;

/ **
  * 用户地址接口控制器
  *
  * 处理用户地址数据相关 API
  * Class AddressController
  * /
class AddressController extends PublicController {

    / **
     * 获取用户地址数据接口
     * /
    public function index(){
        $ user_id = intval( $ _REQUEST['user_id']);
        if (! $ user_id){
            echo json_encode(array('status' => 0,'err' => '网络异常.'.__LINE__));
            exit();
        }

        //所有地址
        $ addressModel = M('address');
        $ adds_list = $ addressModel ->where('uid = '. intval( $ user_id)) ->order('is_
                default desc,id desc') ->select();

        //所有省份
        // $ china_city = M("china_city");
        // $ sheng = $ china_city ->where('tid = 0') ->field('id,name') ->select();
```

```
        echo json_encode(array('status' => 1,'adds' => $adds_list));
        exit();
    }


    /**
     * 会员添加地址接口
     */
    public function add_adds(){
        $user_id = intval($_REQUEST['user_id']);
        if (!$user_id){
            echo json_encode(array('status' => 0,'err' => '网络异常.'.__LINE__));
            exit();
        }

        //接收ajax传过来的数据
        //data:{user_id:uid,receiver:rec,tel:tel,sheng:sheng,city:city,quyu:quyu,
          adds:address,code:code}
        $data = array();
        $data['name'] = trim($_POST['receiver']);
        $data['tel'] = trim($_POST['tel']);
        $data['sheng'] = intval($_POST['sheng']);
        $data['city'] = intval($_POST['city']);
        $data['quyu'] = intval($_POST['quyu']);
        $data['address'] = $_POST['adds'];
        $data['code'] = $_POST['code'];
        $data['uid'] = intval($user_id);
        if (!$data['name'] || !$data['tel'] || !$data['address']) {
            echo json_encode(array('status' => 0,'err' => '请先完善信息后再提交.
'));

            exit();
        }
        if (!$data['sheng'] || !$data['city'] || !$data['quyu']) {
            echo json_encode(array('status' => 0,'err' => '请选择省市区.'));
            exit();
        }
        $check_id = M('address')->where($data)->getField('id');
        if ($check_id) {
            echo json_encode(array('status' => 0,'err' => '该地址已经添加了.'));
            exit();
        }
        $province = M('china_city')->where('id ='.intval($data['sheng']))->get-
```

```
                Field('name');
    $city_name = M('china_city')->where('id='.intval($data['city']))->get-
                Field('name');
    $quyu_name = M('china_city')->where('id='.intval($data['quyu']))->get-
                Field('name');
    $data['address_xq'] = $province.' '.$city_name.' '.$quyu_name.' '.$data
                        ['address'];
    $res = M('address')->add($data);
    if ($res) {
        $arr = array();
        $arr['addr_id'] = $res;
        $arr['rec'] = $data['name'];
        $arr['tel'] = $data['tel'];
        $arr['addr_xq'] = $data['address_xq'];
        echo json_encode(array('status'=>1,'add_arr'=>$arr));
        exit();
    }else{
        echo json_encode(array('status'=>0,'err'=>'操作失败.'));
        exit();
    }
}


/**
 * 会员获取单个地址接口
 */
public function details(){
    $addr_id = intval($_REQUEST['addr_id']);
    if (!$addr_id) {
        echo json_encode(array('status'=>0));
        exit();
    }

    $address = M('address')->where('id='.intval($addr_id))->find();
    if (!$address) {
        echo json_encode(array('status'=>0));
        exit();
    }
    $arr=array();
    $arr['status']=1;
    $arr['addr_id']=$address['id'];
```

```
        $arr['name'] = $address['name'];
        $arr['tel'] = $address['tel'];
        $arr['addr_xq'] = $address['address_xq'];
        echo json_encode($arr);
        exit();
    }


    /**
     * 会员删除地址接口
     */
    public function del_adds(){
        $user_id = intval($_REQUEST['user_id']);
        if (!$user_id){
            echo json_encode(array('status'=> 0,'err'=> '网络异常.'.__LINE__));
            exit();
        }

        $id_arr = trim($_POST['id_arr'],',');
        if ($id_arr) {
            $res = M('address')->where('uid='.intval($user_id).' AND id IN ('.$id
                _arr.')')->delete();
            if ($res) {
                echo json_encode(array('status'=> 1));
                exit();
            }else{
                echo json_encode(array('status'=> 0,'err'=> '操作失败.'));
                exit();
            }
        }else{
            echo json_encode(array('status'=> 0,'err'=> '没有找到要删除的数据.
'));
            exit();
        }
    }


    /**
     * 获取省份数据接口
     */
    public function get_province(){
        //所有省份
```

```
        $china_city = M("china_city");
        $list =  $china_city->where('tid = 0')->field('id,name')->select();

        echo json_encode(array('status' => 1,'list' => $list));
        exit();
    }


    /**
     * 获取城市数据接口
     */
    public function get_city(){
        $sheng = intval($_REQUEST['sheng']);
        if (!$sheng){
            echo json_encode(array('status' => 0,'err' => '请选择省份.'.__LINE__));
            exit();
        }

        //所有省份
        $china_city = M("china_city");
        $list =  $china_city->where('tid = 0')->field('id,name')->select();
        $city =  $china_city->where('tid = '.intval($list[$sheng - 1]['id']))->
                field('id,name')->select();

        echo json_encode(array('status' => 1,'city_list' => $city,'sheng' => intval
($list[$sheng - 1]['id'])));
        exit();
    }


    /**
     * 获取区域数据接口
     */
    public function get_area(){
        $city = intval($_REQUEST['city']);
        if (!$city){
            echo json_encode(array('status' => 0,'err' => '请选择城市.'.__LINE__));
            exit();
        }

        //所有省份
```

```
    $china_city = M("china_city");
    $list  =  $china_city->where('tid = '.intval($_REQUEST['sheng']))->field
            ('id,name')->select();
    $area  =  $china_city->where('tid = '.intval($list[$city - 1]['id']))->
            field('id,name')->select();

    echo json_encode(array('status' => 1,'area_list' => $area,'city' => intval
        ($list[$city - 1]['id'])));
    exit();
}


/**
 * 获取邮政编号接口
 */
public function get_code(){
    $quyu = intval($_REQUEST['quyu']);

    //所有省份
    $china_city = M("china_city");
    $list  =  $china_city->where('tid = '.intval($_REQUEST['city']))->field
            ('id,name')->select();
    $code  =  $china_city->where('id = '.intval($list[$quyu - 1]['id']))->get-
            Field('code');
    echo json_encode(array('status' => 1,'code' => $code,'area' => intval($list
        [$quyu - 1]['id'])));
    exit();
}


/**
 * 设置默认地址
 */
public function set_default(){
    $uid = intval($_REQUEST['uid']);
    if (!$uid){
        echo json_encode(array('status'=> 0,'err'=> '登录状态异常.'));
        exit();
    }

    $addr_id  =  intval($_REQUEST['addr_id']);
```

```
        if (! $addr_id) {
            echo json_encode(array('status'=> 0,'err'=> '地址信息错误.'));
            exit();
        }
        //修改默认状态
        $check = M('address') ->where('uid='. intval( $uid).' AND is_default=1') ->
            find();
        if ( $check) {
            $up1 = M('address') ->where('uid = '. intval( $uid)) ->save(array('is_
                default'=> 0));
            if (! $up1) {
                echo json_encode(array('status'=> 0,'err'=> '设置失败.'.__LINE__));
                exit();
            }
        }

        $up2 = M('address') ->where('id = '. intval( $addr_id).' AND uid = '. intval
            ( $uid)) ->save(array('is_default'=> 1));
        if ( $up2) {
            echo json_encode(array('status'=> 1));
            exit();
        }else{
            echo json_encode(array('status'=> 0,'err'=> '设置失败.'.__LINE__));
            exit();
        }
    }

}
```

9.2.11 优惠券接口代码清单

优惠券接口控制器文件为 VoucherController. class. php，主要用于展现优惠券信息和处理用户领取优惠券，代码如下：

```
<? php
namespace Api\Controller;
use Think\Controller;

/**
 * 优惠券接口
 *
 * 处理优惠券 API
```

```
 * Class VoucherController
 */
class VoucherController extends PublicController {

    /**
     * 所有单页数据接口
     */
    public function index(){
        $condition = array();
        $condition['del'] = 0;
        $condition['start_time'] = array('lt',time());
        $condition['end_time'] = array('gt',time());

        $vou = M('voucher')->where($condition)->order('addtime desc')->select();
        foreach ($vou as $k => $v) {
            $vou[$k]['start_time'] = date("Y.m.d",intval($v['start_time']));
            $vou[$k]['end_time'] = date("Y.m.d",intval($v['end_time']));
            $vou[$k]['amount'] = floatval($v['amount']);
            $vou[$k]['full_money'] = floatval($v['full_money']);
            if ($v['proid']=='all' || empty($v['proid'])) {
                $vou[$k]['desc'] = '店内通用';
            }else{
                $vou[$k]['desc'] = '限定商品';
            }
        }
        echo json_encode(array('status'=>1,'vou'=>$vou));
        exit();
    }


    /**
     * 用户领取优惠券
     */
    public function get_voucher(){
        $vid = intval($_REQUEST['vid']);
        $uid = intval($_REQUEST['uid']);
        $check_user = M('user')->where('id='.intval($uid).' AND del=0')->find();
        if (!$check_user) {
            echo json_encode(array('status'=>0,'err'=>'登录状态异常！err_code:'.
                __LINE__));
            exit();
        }
```

```
$check_vou = M('voucher')->where('id='.intval($vid).' AND del=0')->find
        ();
if (!$check_vou) {
    echo json_encode(array('status'=> 0,'err'=> '优惠券信息错误! err_
        code:'.__LINE__));
    exit();
}

//判断是否已领取过
$check = M('user_voucher')->where('uid='.intval($uid).' AND vid='.intval
        ($vid))->getField('id');
if ($check) {
    echo json_encode(array('status'=> 0,'err'=> '您已经领取过了!'));
    exit();
}

if (intval($check_vou['point'])!= 0 && intval($check_vou['point']) > in-
                                    tval($check_user['jifen'])) {
    echo json_encode(array('status'=> 0,'err'=> '积分余额不足!'));
    exit();
}

if ($check_vou['start_time'] > time()) {
    echo json_encode(array('status'=> 0,'err'=> '优惠券还未生效!'));
    exit();
}

if ($check_vou['end_time'] < time()) {
    echo json_encode(array('status'=> 0,'err'=> '优惠券已失效!'));
    exit();
}

if (intval($check_vou['count']) <= intval($check_vou['receive_num'])) {
    echo json_encode(array('status'=> 0,'err'=> '优惠券已被领取完了!'));
    exit();
}

$data = array();
$data['uid'] = $uid;
$data['vid'] = $vid;
$data['shop_id'] = intval($check_vou['shop_id']);
```

```
        $data['full_money'] = floatval($check_vou['full_money']);
        $data['amount'] = floatval($check_vou['amount']);
        $data['start_time'] = $check_vou['start_time'];
        $data['end_time'] = $check_vou['end_time'];
        $data['addtime'] = time();
        $res = M('user_voucher')->add($data);
        if ($res) {
            //修改会员积分
            if (intval($check_vou['point'])!=0) {
                $arr = array();
                $arr['jifen'] = intval($check_user['jifen']) - intval($check_vou
                               ['point']);
                $up = M('user')->where('id='.intval($uid))->save($arr);
            }

            //修改领取数量
            $arrs = array();
            $arrs['receive_num'] = intval($check_vou['receive_num']) + 1;
            $ups = M('voucher')->where('id='.intval($vid))->save($arrs);

            echo json_encode(array('status'=>1));
            exit();
        }else{
            echo json_encode(array('status'=>0,'err'=>'领取失败！'));
            exit();
        }
    }
}
```

9.2.12 用户接口代码清单

用户接口控制器文件为 UserController.class.php，主要用于处理我的个人中心页用户信息展现、优惠券信息、意见反馈，代码如下：

```
<?php
namespace Api\Controller;
use Think\Controller;

/**
 * 用户接口
 *
 * 处理我的个人中心页面接口
```

```
 * Class UserController
 */
class UserController extends PublicController {

    /**
     * 校验
     */
    Public function verify(){
        $image = new \Org\Util\Image;
        $image->buildImageVerify();
    }


    /**
     * 获取用户订单数量
     */
    public function getorder(){
        $uid = intval($_REQUEST['userId']);
        if (!$uid) {
            echo json_encode(array('status'=> 0,'err'=> '非法操作.'));
            exit();
        }

        $order = array();
        $order['pay_num'] = intval(M('order')->where('uid = '.intval($uid).' AND
                            status = 10 AND del = 0')->getField('COUNT(id)'));
        $order['rec_num'] = intval(M('order')->where('uid = '.intval($uid).' AND
                            status = 30 AND del = 0 AND back = "0"')->getField('COUNT
                            (id)'));
        $order['finish_num'] = intval(M('order')->where('uid = '.intval($uid).' AND
                               status > 30 AND del = 0 AND back = "0"')->getField
                               ('COUNT(id)'));
        $order['refund_num'] = intval(M('order')->where('uid = '.intval($uid).' AND
                               back > "0"')->getField('COUNT(id)'));
        echo json_encode(array('status'=> 1,'orderInfo'=> $order));
        exit();
    }

    /**
     * 找回密码和修改
     */
```

```
public function findfwd_edit(){
    $name     = $_POST['name'];
    $tel      = $_POST['tel'];
    $newpwd   = $_POST['newpwd'];
    $newpwds  = $_POST['newpwds'];
    if(empty($name)){
        $this->error('请输入用户名',U('User/findfwd',array('key'=>$_RE-
                QUEST['key'])));
    }
    if(empty($tel)){
        $this->error('请输入手机号',U('User/findfwd',array('key'=>$_RE-
                QUEST['key'])));
    }
    if($newpwd!=$newpwds){
        $this->error('两次密码输入不同',U('User/findfwd',array('key'=>$_
                REQUEST['key'])));
    }else{
        $name = $_REQUEST['name'];          //账号
        $tel = $_REQUEST['tel'];            //接受短信用户
        $yzm = $_REQUEST['yzm'];            //验证码
        $data['pwd'] = md5(md5($_REQUEST['newpwd'])); //新密码
        $sms_o = file_get_contents('Public/Rand/'.$tel.'.txt');
        if($sms_o!=$yzm){
            $this->error('验证码错误!',U('User/findfwd',array('key'=>$_
                    REQUEST['key'])));
        }else{
            $result=M("user")->where('name = "'.$name.'"')->save($data);
            if($result !== false){
                $this->success('修改成功!',U('User/logo',array('key'=>$_
                        REQUEST['key'])));
            }else{
                $this->error('修改失败!',U('User/logo',array('key'=>$_RE-
                        QUEST['key'])));
            }
        }
    }
}


/**
 * 获取用户信息
```

```
    */
public function userinfo(){
    /* if (! $_SESSION['ID']) {
        echo json_encode(array('status'=> 4));
        exit();
    } */
    $uid = intval( $_REQUEST['uid']);
    if (! $uid) {
        echo json_encode(array('status'=> 0,'err'=> '非法操作.'));
        exit();
    }

    $user = M("user") ->where('id='. intval( $uid)) ->field('id,name,uname,pho-
            to,tel') ->find();
    if ( $user['photo']) {
        if ( $user['source'] == '') {
            $user['photo'] = __DATAURL__. $user['photo'];
        }
    }else{
        $user['photo'] = __PUBLICURL__.'home/images/moren.png';

    $user['tel'] = substr_replace( $user['tel'],'****',3,4);
    echo json_encode(array('status'=> 1,'userinfo'=> $user));
    exit();

    }
}


/**
  * 修改用户信息
  */
public function user_edit(){
        $time = mktime();
        $arr = $_POST['photo'];
        if( $_POST['photo']! = ''){
            $data['photo'] = $arr;
        }

        $user_id = intval( $_REQUEST['user_id']);
        $old_pwd = $_REQUEST['old_pwd'];
        $pwd = $_REQUEST['new_pwd'];
```

```
$old_tel = $_REQUEST['old_tel'];
$uname = $_REQUEST['uname'];
$tel = $_REQUEST['new_tel'];

$user_info = M('user')->where('id ='.intval($user_id).' AND del = 0')->find();
if (!$user_info) {
    echo json_encode(array('status'=> 0,'err'=> '会员信息错误.'));
    exit();
}

//用户密码检测
$data = array();
if ($pwd) {
    $data['pwd'] = md5(md5($pwd));
    if ($user_info['pwd'] && md5(md5($old_pwd))!== $user_info['pwd']) {
        echo json_encode(array('status'=> 0,'err'=> '旧密码不正确.'));
        exit();
    }
}

//用户手机号检测
if ($tel) {
    if ($user_info['tel'] && $old_tel!== $user_info['tel']) {
        echo json_encode(array('status'=> 0,'err'=> '原手机号不正确.'));
        exit();
    }
    $check_tel = M('user')->where('tel ='.trim($tel).' AND del = 0')->count();
    if ($check_tel) {
        echo json_encode(array('status'=> 0,'err'=> '新手机号已存在.'));
        exit();
    }
    $data['tel'] = trim($tel);
}

if ($uname && $uname!== $user_info['uname']) {
    $data['uname'] = trim($uname);
```

```
        }

        if (! $data) {
            echo json_encode(array('status'=> 0,'err'=> '您没有输入要修改的信
                息.'.__LINE__));
            exit();
        }
        //dump( $data);exit;
         $result = M("user") ->where('id = '. intval( $user_id)) ->save( $data);
        //echo M("aaa_pts_user") ->_sql();exit;
        if( $result){
            echo json_encode(array('status'=> 1));
            exit();
        }else{
            echo json_encode(array('status'=> 0,'err'=> '操作失败.'));
            exit();
        }
}


/**
 * 用户反馈接口
 */
public function feedback(){
     $uid = intval( $_REQUEST['uid']);
    if (! $uid) {
        echo json_encode(array('status'=> 0,'err'=> '登录状态异常.'));
        exit();
    }

     $con = $_POST['con'];
    if (! $con) {
        echo json_encode(array('status'=> 0,'err'=> '请输入反馈内容.'));
        exit();
    }
     $data = array();
     $data['uid'] = $uid;
     $data['message'] = $con;
     $data['addtime'] = time();
     $res = M('fankui') ->add( $data);
    if ( $res) {
        echo json_encode(array('status'=> 1));
```

```
        exit();
    }else{
        echo json_encode(array('status'=> 0,'保存失败！'));
        exit();
    }

}


/* *
 * h5 头像上传
 */
public function uploadify(){
    $ imgtype = array(
      'gif'=> 'gif',
      'png'=> 'png',
      'jpg'=> 'jpg',
      'jpeg'=> 'jpeg'
    ); //图片类型在传输过程中对应的头信息
    $ message = $ _POST['message']; //接收以 base64 编码的图片数据
    $ filename = $ _POST['filename']; //自定义文件名称
    $ ftype = $ _POST['filetype']; //接收文件类型
    //首先将头信息去掉,然后解码剩余的 base64 编码的数据
    $ message = base64_decode(substr( $ message,strlen('data:image/'. $ imgtype
             [strtolower( $ ftype)].';base64,')));
    $ filename2 = $ filename.".". $ ftype;
    $ furl = "./Data/UploadFiles/user_img/".date("Ymd");
    if (! is_dir( $ furl)) {
        @mkdir( $ furl, 0777);
    }
    $ furl = $ furl.'/';

    //开始写文件
    $ file = fopen( $ furl. $ filename2,"w");
    if(fwrite( $ file, $ message) == = false){
      echo json_encode(array('status'=> 0,'err'=> 'failed'));
      exit;
    }

    //图片 URL 地址
    $ pic_url = $ furl. $ filename2;
    // $ pic_url = "./Data/UploadFiles/user_img/20170115/0.jpeg";
```

```
    $ image = new \Think\Image();
    $ image ->open( $ pic_url);
    // 生成一个居中裁剪为 150 * 150 的缩略图并保存为 thumb.jpg
    $ image ->thumb(100, 100,\Think\Image::IMAGE_THUMB_SCALE) ->save( $ pic_url);
    /* echo $ pic_url;
    exit(); */

    $ uid = intval( $ _REQUEST['uid']);
    if (! $ uid) {
        echo json_encode(array('status' => 0,'err' => '登录状态异常! error'));
        exit();
    }
    //获取原来的头像链接
    $ oldpic = M('user') ->where('id = '. intval( $ uid)) ->getField('photo');
    $ oldpic2 = './Data/'. $ oldpic;

    $ data = array();
    $ data['photo'] = "UploadFiles/user_img/".date("Ymd").'/'. $ filename2;
    $ up = M('user') ->where('id = '. intval( $ uid)) ->save( $ data);
    if ( $ up) {
        //如果原头像存在就删除
        if ( $ oldpic && file_exists( $ oldpic2)) {
            @unlink( $ oldpic2);
        }
        echo json_encode(array('status' => 1,'urls' => 'Data/'. $ data['photo']));
        exit();
    }else{
        echo json_encode(array('status' => 0,'err' => '头像保存失败.'));
        exit();
    }

}


/**
 * 用户修改密码接口,忘记密码
 */
public function forget_pwd(){
    $ user_name = trim( $ _REQUEST['username']);
```

```
        $tel = trim($_REQUEST['tel']);
        if (!$user_name || !$tel) {
            echo json_encode(array('status'=> 0,'err'=> '请输入账号或手机号.'));
            exit();
        }

        $where=array();
        $where['name'] = $user_name;
        $where['tel'] =  $tel;
        $check = M('user')->where($where)->count();
        if ($check) {
            echo json_encode(array('status'=> 1));
            exit();
        }else{
            echo json_encode(array('status'=> 0,'err'=> '账号不存在.'));
            exit();
        }
    }


    /**
     * 用户修改密码接口,输入新密码
     */
    public function up_pwd(){
        $psw = trim($_POST['psw']);
        if (!$psw) {
            echo json_encode(array('status'=> 0,'err'=> '请输入新密码.'));
            exit();
        }
        $user_name = trim($_POST['user']);
        $tel = trim($_POST['tel']);
        if (!$user_name || !$tel) {
            echo json_encode(array('status'=> 0,'err'=> '系统错误,请稍后再试.'));
            exit();
        }

        $where=array();
        $where['name'] = $user_name;
```

```
        $ where['tel'] =  $ tel;
        $ pwd = md5(md5( $ psw));
        $ up = M('user') ->where( $ where) ->save(array('pwd' => $ pwd));
        if ( $ up) {
            echo json_encode(array('status' => 1));
            exit();
        }else{
            echo json_encode(array('status' => 0,'err' => '账号不存在.'));
            exit();
        }
    }


    /**
     * 获取用户优惠券
     */
    public function voucher(){
        $ uid = intval( $ _REQUEST['uid']);
        if (! $ uid) {
            echo json_encode(array('status' => 0,'err' => '登录状态异常！'.__LINE__));
            exit();
        }

        //获取未使用或者已失效的优惠券
        $ nouse = array(); $ nouses = array(); $ offdate = array(); $ offdates = ar-
                        ray();
        $ vou_list = M('user_voucher') ->where('uid = '. intval( $ uid).' AND status! =
                2') ->select();
        foreach ( $ vou_list as $ k =>  $ v) {
            $ vou_info = M('voucher') ->where('id = '. intval( $ v['vid'])) ->find();
            if (intval( $ vou_info['del']) == 1 || $ vou_info['end_time'] < time()) {
                $ offdate['vid'] = intval( $ vou_info['id']);
                $ offdate['full_money'] = floatval( $ vou_info['full_money']);
                $ offdate['amount'] = floatval( $ vou_info['amount']);
                $ offdate['start_time'] = date('Y. m. d', intval( $ vou_info['start_
                        time']));
                $ offdate['end_time'] = date('Y. m. d', intval( $ vou_info['end_
```

```
                                        time']));
            $offdates[] = $offdate;
        }elseif ($vou_info['end_time'] > time()) {
            $nouse['vid'] = intval($vou_info['id']);
            $nouse['shop_id'] = intval($vou_info['shop_id']);
            $nouse['title'] = $vou_info['title'];
            $nouse['full_money'] = floatval($vou_info['full_money']);
            $nouse['amount'] = floatval($vou_info['amount']);
            if ($vou_info['proid'] == 'all' || empty($vou_info['proid'])) {
                $nouse['desc'] = '店内通用';
            }else{
                $nouse['desc'] = '限定商品';
            }
            $nouse['start_time'] = date('Y.m.d', intval($vou_info['start_
                                        time']));
            $nouse['end_time'] = date('Y.m.d',intval($vou_info['end_time']));
            if ($vou_info['proid']) {
                $proid = explode(',', $vou_info['proid']);
                $nouse['proid'] = intval($proid[0]);
            }
            $nouses[] = $nouse;
        }
    }

    //获取已使用的优惠券
    $used = array(); $useds = array();
    $vouusedlist = M('user_voucher')->where('uid='.intval($uid).' AND status
                    =2')->select();
    foreach ($vouusedlist as $k => $v) {
        $vou_info = M('voucher')->where('id='.intval($v['vid']))->find();
        $used['vid'] = intval($vou_info['id']);
        $used['full_money'] = floatval($vou_info['full_money']);
        $used['amount'] = floatval($vou_info['amount']);
        $used['start_time'] = date('Y.m.d',intval($vou_info['start_time']));
        $used['end_time'] = date('Y.m.d',intval($vou_info['end_time']));
        $useds[] = $used;
    }
```

```
        echo json_encode(array('status' => 1,'offdates' => $offdates,'nouses' =>
            $nouses,'useds' => $useds));
        exit();
    }

}
```

9.3 项目小结

通过整个小程序商城项目的学习,相信读者对于小程序框架基础、框架组件和框架 API 有了更接近于实战层面的理解。其实一般的小程序项目可能还没有这个商城项目大,大多数小程序项目功能都比较聚焦或单一,所以大家在自己学习或开发小程序项目的时候尽量把小程序功能做的更加符合“小而美”的定位,不必一味地对功能求大求全,适当地对小程序功能做一些“减法”,专注解决一个实际需求的小程序也可能会成为下一个流量爆款小程序。

最后希望大家能学会小程序开发,抓住当下小程序的市场红利,用它来解决实际的市场需求问题,早日打造出属于自己的爆款小程序应用。

参考文献

[1] 雷磊. 微信小程序开发入门与实践[M]. 北京 ：清华大学出版社，2017.

[2] 刘刚. 微信小程序开发图解案例教程[M]. 北京 ：人民邮电出版社，2017.

[3] 熊普江，谢宇华. 小程序，巧应用：微信小程序开发实战[M]. 北京：机械工业出版社，2017.

[4] QuestMobile 移动大数据研究院. 2018 年春季微信小程序用户规模 TOP100 榜[OL]，2018. http://www.questmobile.com.cn/research/report-new/5 .

[5] 微信开发团队. 微信小程序接入指南 [OL]，2018. https://developers.weixin.qq.com/miniprogram/introduction/index.html .

[6] 微信开发团队. 微信小程序简易教程 [OL]，2018. https://developers.weixin.qq.com/miniprogram/dev/index.html .

[7] 微信开发团队. 微信小程序框架 [OL]，2018. https://developers.weixin.qq.com/miniprogram/dev/framework/MINA.html .

[8] 微信开发团队. 微信小程序组件 [OL]，2018. https://developers.weixin.qq.com/miniprogram/dev/component/ .

[9] 微信开发团队. 微信小程序 API [OL]，2018. https://developers.weixin.qq.com/miniprogram/dev/api/ .

[10] 微信开发团队. 微信支付开发文档[OL]，2018. https://pay.weixin.qq.com/wiki/doc/api/wxa/wxa_api.php? chapter=9_1.

[11] 阮一峰. 理解 RESTful 架构 [OL]，2011. http://www.ruanyifeng.com/blog/2011/09/restful.html .

[12] 曾健生. App 后台开发运维和架构实践[M]. 北京：电子工业出版社，2016.